Advances in Inorganic and Bioinorganic Mechanisms

Volume 4

Advances in Inorganic and Bioinorganic Mechanisms

Volume 4

edited by
A. G. Sykes

Department of Inorganic Chemistry
The University
Newcastle upon Tyne, England

1986

ACADEMIC PRESS

Harcourt Brace Jovanovich, Publishers

London   Orlando   San Diego   New York
Austin   Montreal   Sydney   Tokyo   Toronto

ACADEMIC PRESS INC. (LONDON) LTD.
24–28 Oval Road
LONDON NW1 7DX

United States Edition published by
ACADEMIC PRESS, INC.
Orlando, Florida 32887

ISBN 0–12–023804–7

PRINTED IN THE UNITED STATES OF AMERICA

86 87 88 89   9 8 7 6 5 4 3 2 1

Contents

Contributors

JOHN H. DAWSON
Department of Chemistry
University of South Carolina
Columbia
South Carolina 29208
USA

KIM SMITH EBLE[1]
Department of Chemistry
University of South Carolina
Columbia
South Carolina 29208
USA

STEPHEN F. LINCOLN
Department of Physical and Inorganic Chemistry
The University of Adelaide
Adelaide
South Australia 5001
Australia

K. L. NASH
U.S. Geological Survey
Denver Federal Center
Denver
Colorado 80225
USA

HIROSHI OGINO
Department of Chemistry
Faculty of Science
Tohoku University
Sendai 980
Japan

[1] Present address: Bowman Gray School of Medicine, Wake Forest University, Winston-Salem, North Carolina 27103, USA.

MAKOTO SHIMURA
Department of Chemistry
Faculty of Science
Tohoku University
Sendai 980
Japan

L. H. SKIBSTED
Chemistry Department
Royal Veterinary and Agricultural University
DK-1871 Copenhagen V
Denmark

J. C. SULLIVAN
Chemistry Division
Argonne National Laboratory
Argonne
Illinois 60439
USA

RICHARD C. THOMPSON
Department of Chemistry
University of Missouri
Columbia
Missouri 65211
USA

ADVANCES IN INORGANIC AND BIOINORGANIC MECHANISMS, VOL. 4

Cytochrome P-450: Heme Iron Coordination Structure and Mechanisms of Action

John H. Dawson and Kim Smith Eble*

*Department of Chemistry
University of South Carolina
Columbia, South Carolina, USA*

---

* Present address: Bowman Gray School of Medicine, Wake Forest University, Winston-Salem, North Carolina 27103, USA.

I. INTRODUCTION

Oxidation–reduction reactions are of vital importance to all biological systems. Redox reactions are involved not only in the essential processes of energy storage and interconversion, but also in the biosynthesis of organic compounds that are necessary for the complete functioning of the cell. Perhaps the most intriguing of the enzymes which catalyze redox reactions are those that directly interact with molecular oxygen (dioxygen).[1] Well over 200 enzymes are known to utilize dioxygen as a substrate.[2] These enzymes have been broken down into two categories based on whether oxygen atoms from dioxygen are directly incorporated into the substrate (the oxygenases) or whether the substrate is oxidized without oxygen incorporation (the oxidases).[3] The oxygenases have been further subdivided into the dioxygenases, which incorporate both oxygen atoms of dioxygen into the organic substrate, and the monooxygenases, which only incorporate one oxygen atom (the other oxygen atom of dioxygen is reduced to water).[3]

This article will examine the most well understood of the monooxygenase enzymes: cytochrome P-450. Cytochrome P-450 is an enzyme that has something for everyone: organic chemists are fascinated by its ability to activate dioxygen, inorganic chemists by its unusual heme iron coordination structure, physical chemists by its unique spectroscopic properties, pharmacologists and medicinal chemists by its important involvement in drug metabolism, environmental chemists by its function in the breakdown of xenobiotics, molecular biologists by its inducibility and existence in multiple forms, and, of course, biochemists by all of the above as well as its role in chemical carcinogenesis, membrane detoxification, and the biosynthesis of steroid hormones, bile acids and prostaglandins. Consequently, cytochrome P-450 has been the subject of one book,[4] numerous conference proceedings,[5–9] as well as an extensive collection of review articles.[10–23] In this review, following a brief general introduction, we will focus on two topics: the coordination structure of the active site heme iron and selected aspects of the enzymatic mechanism of action. For the latter topic, particular emphasis will be placed on the transfer of electrons to the heme iron, the breakdown of the oxygenated intermediate, and the use of alternative substrates.

Cytochrome P-450 is a collective name given to a group of heme-containing enzymes with similar spectral properties, first observed in 1958 in rat liver microsomes.[24,25] The historical development of the field has been nicely reviewed by Mannering.[17] The name P-450 arises from the major absorption band of the ferrous heme–carbon monoxide enzyme complex, which, relative to other heme proteins, occurs at the unusually long wavelength of approximately 450 nm. Unlike most other cytochromes, P-450 does not function merely as an electron carrier, but is also an enzyme capable of catalyzing oxygenation reactions. A general representation of the reaction catalyzed by cytochrome P-450 is shown in Eq. (1).

$$RH + NAD(P)H + H^+ + O_2 \rightarrow ROH + NAD(P)^+ + H_2O \tag{1}$$

The hydroxylation reaction shown here is actually an oversimplification since P-450 is also known to catalyze epoxidation reactions, N-, S-, and O-dealkylations, N-oxidations, sulfoxidations, dehalogenations, and more (Fig. 1).[23]

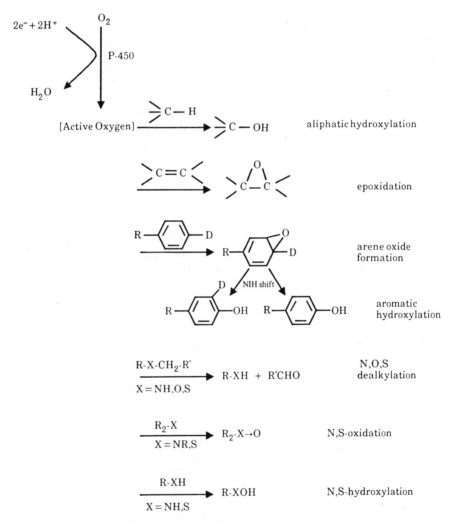

Fig. 1. Major types of oxygen incorporation reactions catalyzed by cytochrome P-450.

Cytochrome P-450 enzymes having widely varied substrate specificities have been isolated from many sources including mammalian tissues (liver, adrenal cortex, kidney, lung, skin, etc.), plants, insects, yeast, and bacteria. Multiple forms (isozymes) of P-450 have been shown to exist in several of the sources just listed; these isozymes have different, although usually overlapping, substrate specificities.[23] Studies involving the membrane-bound, insoluble mammalian P-450 enzymes have demonstrated their involvement in membrane detoxification and the solubilization of membrane-trapped nonpolar molecules, conversion of cholesterol to steroidal hormones, and drug hydroxylations. In addition to these beneficial roles, perhaps the best known group of reactions catalyzed by P-450 is the activation of chemical carcinogens. As shown in Fig. 2, benzo[a]pyrene and other polycyclic aromatic hydrocarbons are converted in a series of reactions catalyzed by liver microsomal P-450 into carcinogenic dihydrodiol epoxide metabolites that can bind to DNA guanine residues.[26,27] The long-range goal of selective inhibition of this detrimental P-450 activity is the rationale behind many of the structural and mechanistic studies of cytochrome P-450.

Investigation of the physical and chemical nature of enzymes is largely dependent on the availability of purified, homogeneous protein preparations. Initial attempts to purify cytochrome P-450 from liver microsomes were unsuccessful because this membrane-bound P-450 was denatured by common purification techniques such as detergent solubilization or the use of proteases or high salt concentrations.[24,25,28] Discovery of a soluble P-450 involved in camphor degradation in the bacterium *Pseudomonas putida*[29] and the subsequent purification of this enzyme to homogeneity[30] greatly facilitated examination of the P-450 enzyme system. Since the catalytic cycle appears to be independent of the P-450 source,[14] many investigators have utilized this

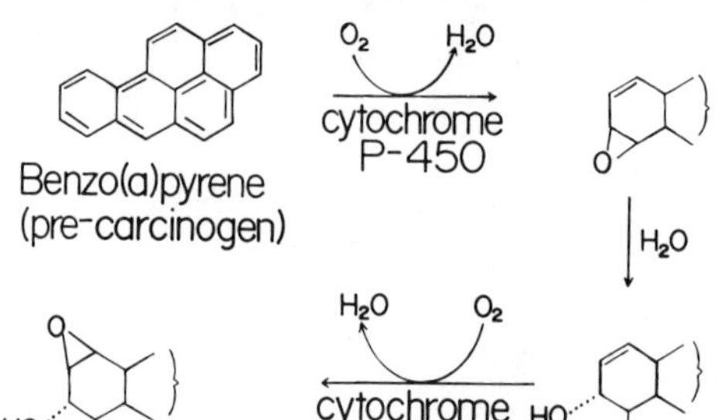

Fig. 2. P-450-catalyzed activation of chemical carcinogens. The conversion of benzo[a]-pyrene to the dihydrodiol epoxide is a three-step reaction involving liver microsomal P-450 and epoxide hydrase. The reactive intermediate thus formed can bind to DNA guanine residues.

more stable and more easily purified bacterial P-450, which has come to be known as P-450-CAM.

Cytochrome P-450 belongs to a subset of the monooxygenases known as hydroxylases, since one oxygen atom from molecular oxygen is incorporated into the substrate in the form of a hydroxyl group. The P-450-catalyzed hydroxylation of camphor in *P. putida* is the first step in camphor deg-

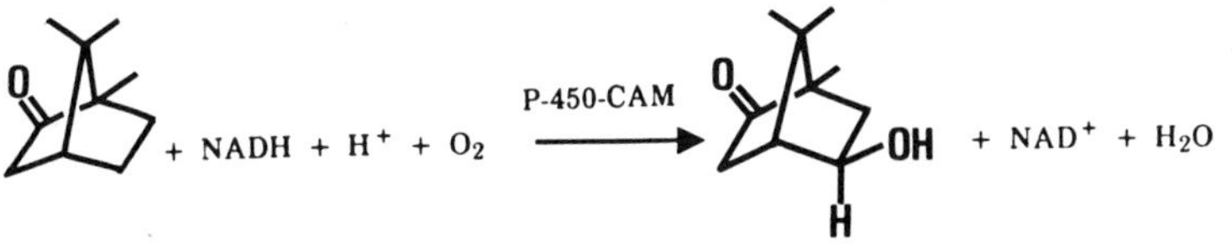

$$\text{camphor} + NADH + H^+ + O_2 \xrightarrow{\text{P-450-CAM}} \text{5-hydroxycamphor} + NAD^+ + H_2O \tag{2}$$

radation as a means for providing energy to the cell. The specific reaction catalyzed by P-450-CAM is shown in Eq. (2).

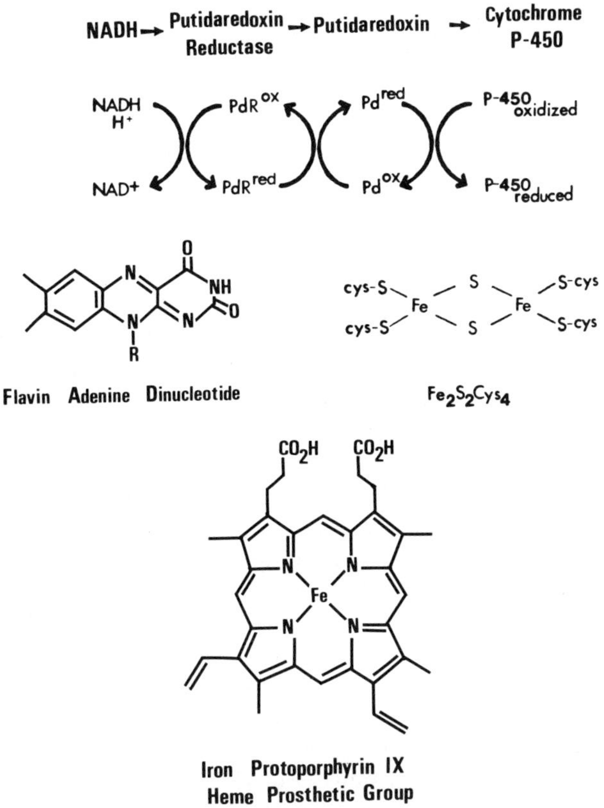

Flavin Adenine Dinucleotide

$Fe_2S_2Cys_4$

Iron Protoporphyrin IX
Heme Prosthetic Group

Fig. 3. The electron transport pathway from NADH to cytochrome P-450 and the structures of the prosthetic groups. Putidaredoxin reductase (PdR) contains flavin adenine dinucleotide (FAD), putidaredoxin (Pd) contains a [2Fe–2S] iron–sulfur center, and P-450 contains an iron protoporphyrin IX unit.

The substrate camphor molecule is stereospecifically and regiospecifically hydroxylated at the exo-5 position, with reduced nicotinamide adenine dinucleotide (NADH) serving as the source of reducing equivalents. As with other P-450 enzymes, P-450-CAM cannot accept electrons directly from NADH. Two other proteins are involved in the transfer of reducing equivalents from NADH to P-450-CAM. As shown in Fig. 3, two electrons are initially transferred from NADH to a flavoprotein, putidaredoxin reductase, which contains a flavin adenine dinucleotide (FAD) prosthetic group. From putidaredoxin reductase, the reducing equivalents pass in sequential one-electron steps to putidaredoxin, an iron–sulfur protein of the [2Fe–2S] class, and then to the iron protoporphyrin IX (heme)-containing cytochrome P-450-CAM. All three of these proteins have been purified to homogeneity[31–33] and reconstitution of the purified components results in a functional hydroxylase system. Analogous, short electron transport systems exist to deliver electrons from NAD(P)H to mammalian P-450 except that in some cases, such as the liver microsomal P-450, no iron–sulfur protein is required.

II. THE HEME IRON COORDINATION STRUCTURE OF CYTOCHROME P-450

A. Overview

Since a thorough understanding of the basic structural features of an enzyme active site is a prerequisite to any attempts to accurately predict the mechanism of action of the enzyme, there has been an extensive effort made to determine the heme iron coordination structure of cytochrome P-450. In fact, it is fair to say that more is known about the active site structure of this enzyme than any other enzyme for which there has not been a crystal structure reported. The reasons for this are twofold. (1) The spectroscopic properties of cytochrome P-450 are quite unusual among heme proteins. Because of this, (2) inorganic chemists interested in building structural models for the active site of heme proteins have focused considerable attention on P-450 and have succeeded in constructing models whose spectroscopic properties closely match those of the native enzyme. In this two-prong protein/model system approach, no single spectroscopic method has provided sufficient information to delineate the active site structure of P-450. Instead, the combined results from investigations with a large number of spectroscopic techniques have led to the picture of the active site that will be presented here. Although there are still some points of uncertainty to be worked out, as far as the basic metal coordination unit of cytochrome P-450 is concerned, the crystal structure of P-450-CAM should not contain any surprises.[34]

It is generally agreed that Mason was the first to suggest that some of the unusual properties of P-450 were indicative of a sulfur-containing amino acid axial ligand.[35] As shown in Fig. 4, the active site structure of P-450 has been established to consist of a cysteinate ligated heme iron which goes through a reaction cycle consisting of four states. The evidence for the sequence of intermediates in this cycle has been amply reviewed elsewhere.[4–23] Here, we will focus on the evidence for the structures of these intermediates. The cycle begins with the native substrate-free, low-spin ferric enzyme (state **1** in Fig. 4) which is converted to a high-spin form (**2**) upon substrate binding. Electron transfer leads to the high-spin ferrous state (**3**) which can bind oxygen to yield oxy-P-450 (**4**). Although it is not part of the reaction cycle, the ferrous-CO complex (**5**) is also shown because it is this form of the enzyme that has its absorption maximum near 450 nm from which the name P-450 was derived.[28]

Additional support for the suggestion of Mason that a sulfhydryl group is coordinated to the heme iron of P-450[35,36] was provided by the similarity between the EPR properties of thiolate-bound myoglobin [(His)imidazole–ferric heme–SR] and low-spin P-450 [(Cys)RS–ferric heme–L].[37–39] Despite these indirect clues as to the identity of the key axial ligand to the heme iron of P-450, in the opening paragraph of the 1975 publication of the

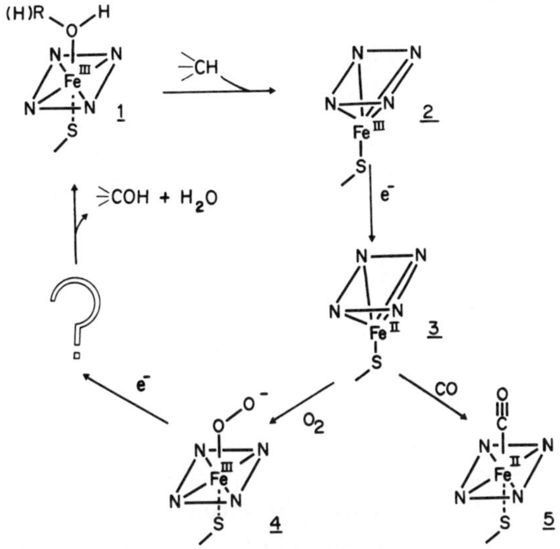

Fig. 4. Catalytic cycle of P-450, and the postulated structures of the iron site for the isolable intermediates in the P-450 catalytic cycle. Oxy-P-450 (state **4**) is shown as a complex of ferric iron and superoxide anion but could also be described as an adduct of neutral dioxygen and ferrous iron. The porphyrin ring is abbreviated as a parallelogram with nitrogens at the corners.

first crystal structure of a heme iron–thiolate complex, Holm stated, "The possibility... of axial sulfur ligation... has proved difficult to assess in the absence of fully characterized sulfur-bound porphyrins."[40] That work by Holm and co-workers,[40–42] similar investigations by Collman, Dolphin, and their co-workers[43–48] and by others,[49–53] as well as subsequent model studies most importantly by Weiss and co-workers[54–56] have provided homogeneous, often crystalline, active site analogs for each of the reaction states of P-450. Together with important parallel studies of the spectroscopic properties of the protein, their investigations have led to our current picture of the active of P-450 that is to be reviewed in this section.

Rather than discuss the evidence for each structure in Fig. 4 independently, this section has been organized by spectroscopic method with all the key results published with each particular technique grouped together. Following a discussion of the evidence for the active site structure of P-450 derived from the most commonly employed form of spectroscopy, namely electronic absorption, the remaining material will be presented in order of decreasing energy of the form of electromagnetic radiation utilized, from Mössbauer (γ rays) to NMR (radio waves) spectroscopy. Finally, evidence derived from magnetic susceptibility, from chemical modification studies, from examination of cobalt-substituted P-450, from theoretical calculations, and from proton balance ligand and redox titration experiments will be discussed. A thorough, but not exhaustive, coverage of the literature has been attempted.

B. Electronic absorption spectroscopy

The electronic absorption spectral properties of states **1–5** of cytochrome P-450 (Table 1) were what first attracted interest to this protein.[24,25] A number of these properties are unusual relative to hemoglobin/myoglobin[63] and to other heme proteins such as cytochrome c[64] and b_5.[65] Most striking, of course, is the red-shifted Soret peak for the ferrous-CO state of P-450, which is at 450 nm rather than near 420 nm. The spectrum of the ferrous-CO protein is also atypical in the presence of an intense near-UV (δ) band at 366 nm[44,66,67] and in the coalescence of the normally distinct α and β bands into a single peak. For the ferric low-spin enzyme, the slightly red-shifted Soret peak and the relative intensity of the α and β bands ($\alpha/\beta > 1$) represent spectral differences that distinguish this P-450 state from other low-spin ferric heme proteins. For high-spin ferric and ferrous P-450, the Soret peaks are substantially blue-shifted in comparison to other high-spin heme proteins.[63] Finally, the spectrum of oxy-P-450 is puzzling because its Soret peak[58,61,62] is at the same wavelength as oxymyoglobin and yet, in contrast to oxymyoglobin, it has a prominent δ band and coalesced α and β transitions.

TABLE 1

Electronic absorption spectral properties of cytochrome P-450-CAM

Structure[a]	Oxidation state	Spin state	δ	Soret	β	α		Ref.[c]
1	Ferric	Low	356 (32)	417 (115)	536 (10.6)	569 (11.1)	—	[57–59]
2	Ferric	High	—	391 (102)	510 (13.0)	540 (11.2)	646 (5.4)	[31, 59, 68]
3	Ferrous	High	—	408 (87)	542 (16)		—	[31, 59, 68]
5	Ferrous-CO	Low	366 (52)	446 (120)	551 (14)		—	[60, 58, 59]
4	Ferrous-O$_2$	Low	353 (46)	419 (82)	554 (16)		—	[61, 58, 62]

Peak positions[b] [nm (ε_{mM}, mM^{-1}, cm^{-1})]

[a] See Fig. 4.

[b] Dash indicates the absence of a band in this region.

[c] The first reference listed is the one from which the data in the table are taken. The other references refer to other studies on that particular P-450 state.

1. Ferric P-450

The report by Koch *et al.*[40] of a crystalline, five-coordinate, high-spin ferric porphyrin–thiolate complex allowed for a direct comparison of the electronic absorption properties of the model complex and high-spin ferric P-450 (Fig. 5). The similarity between the two spectra is striking. In contrast, the spectra of five-coordinate, high-spin ferric porphyrin model complexes with alkoxide and carboxylate ligands[40,41] and of high-spin ferric myoglobin[63] are rather different from those shown in Fig. 5 and allow for the conclusion that P-450 state **2** has an axial thiolate ligand. Comparison of the other spectroscopic properties of the high-spin ferric porphyrin–thiolate model complex and P-450 state **2** will be deferred until later. The only drawback to the Holm model complex was the necessity to use an aromatic thiolate rather than an alkyl thiolate since, with protoheme as the porphyrin, alkyl thiolates led to metal reduction with concomitant disulfide formation.[40] Ogoshi *et al.*[49] circumvented this difficulty through use of an iron porphyrin complex that apparently was more difficult to reduce, and thus were able to prepare alkylthiolate complexes of ferric octaethylporphyrin. Sakurai *et al.*[69] have reported on the properties of alkylthiolate complexes of ferric protoporphyrin and observed an optical absorption spectrum similar to high-spin ferric P-450 when examined within 1 min of complex formation or at 77K.

Another approach that has been used to demonstrate the presence of an axial thiolate ligand to ferric P-450 has been to study six-coordinate ligand complexes of the protein itself where the added ligand is also a thiolate. Ruf and Wende[51] have reported that bisthiolate–ferric heme model complexes have

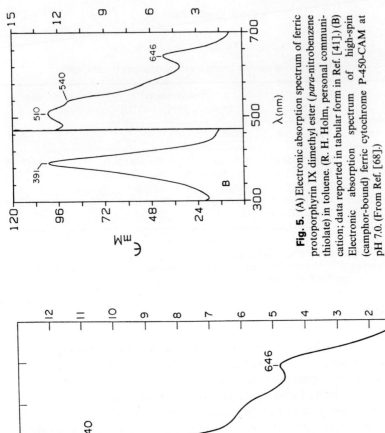

Fig. 5. (A) Electronic absorption spectrum of ferric protoporphyrin IX dimethyl ester (*para*-nitrobenzene thiolate) in toluene. (R. H. Holm, personal communication; data reported in tabular form in Ref. [41].) (B) Electronic absorption spectrum of high-spin (camphor-bound) ferric cytochrome P-450-CAM at pH 7.0. (From Ref. [68].)

very unusual absorption spectra in which there are two Soret bands (~ 380, ~ 470 nm) with equal integrated intensity. The crystal structure of such a complex has recently been published.[70] With biomimetic nitrogen, oxygen, and sulfur donor sixth ligands and a thiolate fifth ligand, the bisthiolate case is the only one giving rise to this unique split Soret (hyperporphyrin) absorption spectrum. Over the last 10 years extensive P-450 ligand binding studies have been reported by Ullrich,[52,71–74] Dawson,[57,75,76] and others.[77,78] In these investigations, only thiolate adducts of ferric P-450 have been found to reproduce the hyperporphyrin spectrum observed in bisthiolate model complexes by Ruf and Wende[51] and more recently by others.[79,80] The dramatic spectral change observed upon thiol addition to ferric P-450 is displayed in Fig. 6 for the addition of an acidic thiol, *para*-chlorobenzenethiol, to either P-450 state **1** or **2**. The generation of such a hyper spectrum provides strong support for the presence of an endogeneous thiolate ligand trans to the added thiolate.

A similar approach has been used to identify the sixth ligand to P-450 state **1**. The UV-visible absorption spectrum of low-spin ferric P-450 is shown in Fig. 7A. In the model system approach, Holm,[41] Ullrich,[52,73,74] and, more recently, Sakurai[81,82] and their co-workers have studied numerous low-spin six-coordinate complexes of the composition [RS–ferric heme–X] where X is nearly every conceivable biomimetic ligand. First of all, it was found that the thiolate ligand was necessary to obtain absorption spectra that at all resembled that of low-spin ferric P-450, further substantiating the presence of the endogeneous thiolate ligand. Beyond this, it was found, most convincingly by Ruf et al.,[52] that the best fit of the absorption spectrum of P-450 state **1** was obtained was X equal to an oxygen donor ligand.[52,73,74,81,82] Similar, equally extensive studies of the absorption spectra of ligand complexes of ferric P-450 having the composition [(Cys)RS–ferric heme–X] by Dawson and co-workers,[57,76,83] White and Coon,[78] and Yoshida et al.[84] have led to a similar conclusion; i.e., the sixth ligand to ferric P-450 is an oxygen donor, most likely from an alcohol containing amino acid or possibly from water. The UV-visible absorption spectrum of the 1-pentanol complex of ferric P-450 is displayed in Fig. 7B and is easily seen to be quite similar to the spectrum of P-450 state **1** shown in Fig. 7A. Competition experiments were done in the Dawson laboratory to be certain that the oxygen donor ligands were indeed coordinating to the P-450 heme iron.[57,83] Alternatively, White and Coon[78] as well as Yoshida et al.[84] used P-450 isozymes that are five-coordinate, high-spin as isolated; in such cases ligand binding and concomitant conversion of the heme iron to the low-spin state were easily verified by the large spectral changes that were observed. As will be discussed later, in some cases results from other laboratories using different techniques have led to further support for an oxygen donor as the sixth ligand to ferric P-450, while in other cases contradictory conclusions have been reached.

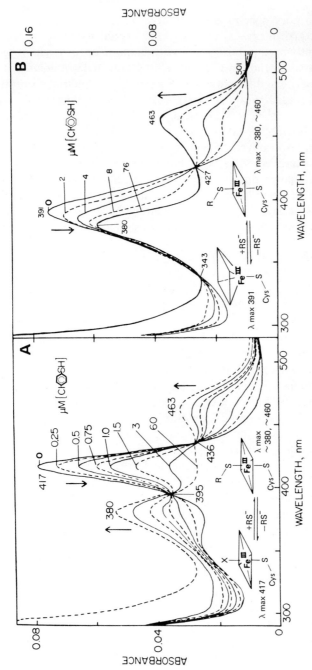

Fig. 6. Spectrophotometric titrations of camphor-free (A) and camphor-bound (B) ferric cytochrome P-450-CAM with *para*-chlorothiophenol. Titrations were performed in 0.1 M potassium phosphate buffer, pH 6.0 (A) or pH 7.0 (B), at 4°C in the absence or presence of 80 μM camphor, respectively. Total concentrations of the sulfur donor ligands added stepwise are shown for each spectrum. The arrows indicate the decrease (↓) or increase (↑) in absorbance upon additions of the ligands. The changes in the coordination structures of the heme iron accompanying the spectral changes are depicted at the bottom of each figure together with absorption maxima of the complexes (X denotes the sixth ligand of the native enzyme). (From Ref. [75].)

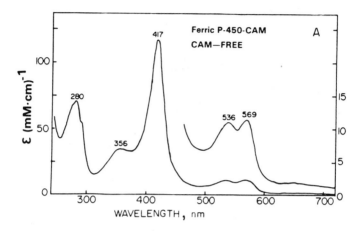

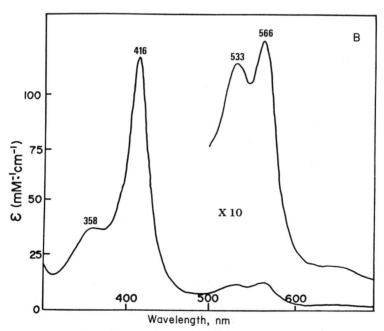

Fig. 7. (A) Electronic absorption spectrum of low-spin (camphor-free) ferric cytochrome P-450-CAM at pH 7.4 (From Ref. [57].) (B) Electronic absorption spectrum of the 1-pentanol complex of ferric cytochrome P-450-CAM at pH 7.4. (From Ref. [57].)

2. Ferrous P-450

Turning to high-spin ferrous P-450, once again model porphyrin complexes have been prepared in an attempt to duplicate the absorption spectral properties of this state of the enzyme. Chang and Dolphin[47,48] were the first to prepare a five-coordinate, high-spin ferrous–thiolate complex with an electronic absorption spectrum $[\lambda_{max} (\varepsilon_{mM}) = 408 (87.5), 555 (15)]^{[48]}$ that closely matches that of P-450 state **3** (Table 1) and clearly contrasts with the spectrum of deoxymyoglobin.[63] Similar results have since been obtained in several other laboratories.[85–87] The crystal structure of such a complex has also been published by Caron *et al.*[54] While these results support the assignment of a thiolate ligand to the heme iron of P-450 state **3**, such a conclusion must be qualified by the knowledge that not all possible biomimetic five-coordinate, high-spin ferrous heme systems have been generated.

3. Ferrous-CO P-450

The unusually red-shifted Soret absorption maximum of ferrous-CO P-450, as noted earlier, was the first physical property of P-450 to gain attention. As might be expected, attempts to provide an explanation for this property both by producing a model complex with duplicate spectral properties and by providing a theoretical basis for the red-shift have been the object of considerable effort. The first model system attempt to explain the red-shifted spectrum was published by Jefcoate and Gaylor, who observed that a peak at about 450 nm could be obtained in the presence of CO and nitrogenous axial ligands at high protoheme concentrations.[88] This led to the suggestion that heme–heme interactions might account for the red shift of the ferrous-CO Soret peak. However, the characterization of a soluble, clearly monomeric P-450 (P-450-CAM)[30,58] ruled out heme-heme interaction as an explanation for the 450-nm peak.

In 1974, a significant breakthrough occurred with the report by Stern and Peisach of a ferrous porphyrin model complex with a peak at 450 nm in the presence of CO.[89] The requirement for thiol and strong base suggested ligation by thiolate. Even though the peak at 450 nm was not the major transition in the Soret region and more than one peak was observed in the α,β region, nonetheless, a peak at 450 nm had been observed and appeared to depend on the presence of a thiolate ligand. This study was quickly followed by reports by Collman and Sorrell[44] and by Chang and Dolphin[47,48] of homogeneous thiolate–ferrous heme–CO model complexes whose absorption spectra closely matched that of ferrous-CO P-450, as can be seen in

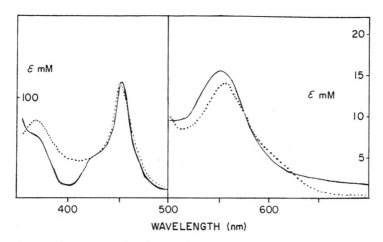

Fig. 8. Electronic absorption spectra of ferrous-CO P-450-LM2 at pH 7.4 (———) and of ferrous protoporphyrin IX dimethyl ester (methane thiolate) (CO) in benzene (·····). (Adapted from Ref. [45].)

Fig. 8.[45] Most importantly, the model work just cited as well as subsequent studies[45,50,53,90,91] have failed to find any ligand complex other than with thiolate trans to CO whose absorption spectrum matches that of the ferrous-CO enzyme so closely, including the intense δ peak near 370 nm and the coalescence of α and β peaks into a single peak near 550 nm. The crystal structure of a thiolate–ferrous heme–CO model complex has subsequently been published by Caron *et al.*[54] A thiolate-ligated model complex with an absorption spectrum very similar to that of NO-bound ferrous P-450[92] has also been reported by Stern and Peisach.[50]

4. Oxy-P-450

The identity of the ligand trans to dioxygen in oxy-P-450 has been the most difficult to establish. This has been primarily due to the lack of an extensive set of ferrous-O_2 model heme complexes with various possible biomimetic ligands trans to dioxygen. In the initial reports of the preparation of oxy-P-450,[58,62] it was observed that the Soret peak in the absorption spectrum was at 418 nm, the same wavelength as for the Soret peak of oxymyoglobin (Fig. 9).[63] This led to speculation that both oxygenated proteins might have histidine imidazole trans to dioxygen.[93] As discussed by Dawson and Cramer,[94] careful examination of the two spectra argues against this similarity (see Section II,F for further discussion of this topic). In particular, the ratio of intensities of the δ band to the Soret band is much greater for oxy-P-450 than for oxymyoglobin, such as would occur if the spectrum of oxy-P-450 is classified as a hyper spectrum[67] as might be expected (see Ref. [60], however) for a thiolate–ferrous heme–O_2 complex. In addition, only one band is observed for oxy-P-450 in the α,β region consistent with the hyper spectrum assignment,[60] while two transitions are seen in this region for oxymyoglobin. Similar arguments have been put forth by Dolphin *et al.*,[95] in part based on trends in the spectral properties of oxy-P-450 observed with different porphyrins other than protoheme incorporated into P-450-CAM.

In the area of model chemistry, the best work to date has, unfortunately from the point of view of UV-visible absorption spectroscopy, involved the use of tetraaryl porphyrins whose absorption spectra typically are shifted from those of protoheme-based models. Nonetheless, Weiss and co-workers[55,56] have published the preparation and X-ray crystal structure of a thiolate–ferrous heme–O_2 complex as a model for oxy-P-450 (see Section II,C). Earlier, Chang and Dolphin[96] had reported the preparation of such a model complex and, based on substantial differences between its absorption spectrum and that of oxy-P-450, had concluded that oxy-P-450 does not have a thiolate ligand. Unfortunately, the Chang and Dolphin complex was most likely a bisthiolate–ferric porphyrin species[51,94,95] and not an oxygen complex. More recent spectroscopic studies of purported thiolate–ferrous heme–O_2 complexes by Budyka *et al.*[85] and Okubo *et al.*[86] have provided support for the presence of a thiolate ligand trans to oxygen in oxy-P-450, and will be discussed more extensively in Section II,F.

C. Mössbauer spectroscopy

Due to the requirement for ^{57}Fe-enriched samples for Mössbauer experiments, such measurements have only been carried out on P-450-CAM. As no new protein data have been reported since 1976[97–99] and since the available protein data were nicely reviewed by Coon and White in 1980,[15] the discussion of the Mössbauer data on P-450 and model complexes presented here will be brief. Emphasis will be placed on a discussion of the Mössbauer characteristics of the models for P-450 states **3** and **4** recently reported by Weiss and co-workers.[55,56,100]

Because the Mössbauer technique is primarily sensitive to the properties of the iron nucleus, the technique has not, for the most part, provided any definitive evidence for the axial ligands to the heme iron of P-450. This is because the Mössbauer parameters of quadrupole splitting, ΔE_q, and isomer shift, δ, are not that distinctive for P-450 relative to other heme proteins.[15,97–99] Perhaps the most diagnostic feature of the Mössbauer properties of P-450 is the temperature dependence of the quadrupole splitting. Here, for P-450 states **3** and **4**, a clear difference is seen relative to deoxy- and oxymyoglobin, respectively. In both cases, the P-450 states display very little temperature dependence while the analogous myoglobin states clearly are temperature dependent.[97–99] The recently prepared crystalline models for P-450 states **3** and **4** have been examined by Mössbauer spectroscopy and also display temperature-independent quadrupole splittings.[56,100] Interestingly, three crystalline samples of the model for oxy-P-450 have been reported which vary in the nature of the crown ether or cryptand used to solubilize the counterion and on the nature of the counterion (K^+ or Na^+). Only one of these three oxy-P-450 models has a quadrupole splitting that is as temperature independent as the protein state itself, despite the presence of the thiolate ligand in all three cases.[55,56,100] Clearly, the temperature independence is not a property imposed upon the system by the thiolate ligand.

D. Extended X-ray absorption fine structure spectroscopy

Extended X-ray absorption fine structure (EXAFS) spectroscopy is a newly developed technique that has proven to be particularly useful in determining the ligand coordination structure of metalloenzymes including the number, identity, and distance of the donor atoms surrounding the central, X-ray-absorbing, metal ion.[101] Extensive testing with model porphyrins of known structure has indicated that, through use of curve-fitting procedures based on transferrable phase and amplitude parameters, absorber–ligand distances can, in favorable circumstances, be determined to an accuracy of ± 0.02 Å or better and coordination number to ± 25–35% (± 1 atom in 3–4). EXAFS is the only technique that can determine metal ligand bond distances in the absence of crystalline samples. Furthermore, although information can only be obtained about atoms within 3.5–4.0 Å of the central metal, the resulting accuracy in the bond distances determined by this technique are considerably better than can be obtained with even the best resolved *protein* crystal structures. Unfortunately, EXAFS is insensitive to small changes in the atomic number of the backscattering donor atom and so cannot be used to distinguish C, N, O, or F atoms (usually only O and N are biologically reasonable) or to distinguish Si, P, S, or Cl, although only S is a biologically reasonable donor atom from the last group. No geometrical information can be directly extracted from the EXAFS data. However, the derived Fe–N (porph) bond distances coupled with prior knowledge of spin state can indirectly provide evidence[102] for the coordination number and therefore the geometry.

Through the collaborative efforts of the Dawson and Hodgson laboratories, EXAFS data have been obtained on low-spin ferric mammalian liver P-450[103] and on P-450-CAM states **1**, **2**, **3**, and **5**.[104–106] The Fe–N (porph) and Fe–S (axial) bond distances and number of donor atoms determined from analysis of the EXAFS data and, for appropriate model complexes, from X-ray crystallography are displayed in Table 2. Curve fitting of the P-450 EXAFS data clearly indicates the presence of a sulfur donor atom in the coordination sphere of the central iron. In fact, in all cases, an acceptable fit could only be obtained by including a sulfur atom. These data represented the first *direct* observation of sulfur ligation to the heme iron of P-450. Although the data do not unambiguously allow for a distinction between thiol or thiolate sulfur ligation, in each case the Fe–S bond distance determined by EXAFS is equal to or shorter than that observed for analogous thiolate-ligated model compounds and so is most consistent with thiolate (cysteinate) ligation, with the caveat that only limited information is currently available about Fe–S (thiol) bond distances. The Fe–N (porph) bond distances

TABLE 2

Structural details for P-450-CAM and relevant porphyrin model compounds[a]

	Fe–N (porphyrin)		Fe (axial)		
	R (Å)	N^b	R (Å)	N^b	Ref.
Low-spin Fe(III)					
P-450 state **1**	2.00	5.0	2.22	0.6	[106]
1 (mammalian P-450)	2.00	4.8	2.19	0.8	[103]
Fe(TPP)(SC$_6$H$_5$)$_2$[c]	2.008	4	2.336	1	[70]
Fe(TPP)(HSC$_6$H$_5$)(SC$_5$H$_5$)[c]			2.27[d]	1	[46]
High-spin Fe(III)					
P-450 state **2**	2.06	5.2	2.23	0.8	[106]
Fe(PPIXDME)(SC$_6$H$_4$NO$_2$)[c]	2.064	4	2.324	1	[41]
High-spin Fe(II)					
P-450 state **3**	2.08	3.0	2.34–2.38	0.6	[106]
Fe(TPP)(SC$_2$H$_5$)[c]	2.096	4	2.360	1	[54]
Low-spin Fe(II)					
P-450 state **5**	1.98	3.3	2.32	1.0	[106]
Fe(TPP)(SC$_2$H$_5$)(CO)[c]	1.993	4	2.352	1	[54]

[a] All parameters are from EXAFS measurements except where noted, and were obtained from curve fitting. Abbreviations: TPP, tetraphenylporphyrin; PPIXDME, protoporphyrin IX dimethyl ester.

[b] The number (N) of ligands at the distance indicated.

[c] Data from crystal structure determination.

[d] The Fe–S (thiolate) bond distance; Fe–S (thiol) distance equals 2.43 Å.

determined for the low-spin ferric and ferrous states (**1** and **5**) and for the high-spin ferric and ferrous forms (**2** and **3**) provide further evidence[102] for six and five coordination, respectively, of the heme iron, as shown in Fig. 4.

E. Circular dichroism spectroscopy

Because circular dichroism (CD) spectroscopy is sensitive to the *combined* effects of heme electronic structure and the coupling of the heme chromophore to its optically active protein environment, it has not generally proven to be helpful in yielding information about the ligand environment of metal complexes beyond the simple determination of electronic transition energies. For example, Andersson *et al.*[107] have recently compared the CD spectra of imidazole-bound ferric P-450-CAM with thiolate-bound ferric myoglobin and found that their CD spectra are rather different despite the fact that both have similar thiolate/imidazole coordinated ferric heme structures. Andersson *et al.*[107] did observe that, among the large number of ferric P-450-CAM ligand complexes examined with CD spectroscopy, those with oxygen donor sixth ligands gave CD spectra that most closely matched that of the native low-spin enzyme. The CD properties of P-450 have recently been reviewed[15,107] and will not be discussed further here because, except for the example just cited, this technique has not provided useful information for structural assignments.

F. Magnetic circular dichroism spectroscopy

Magnetic circular dichroism (MCD) spectra arise from the effect of an external magnetic field on the electronic transitions of a chromophore.[108,109] For this reason, in contrast to CD spectroscopy, MCD spectra are primarily influenced by the electronic structure of a metal complex, are therefore sensitive to changes in physical structure that effect electronic structure, and are relatively insensitive to environmental effects. Because asymmetry is not required for an MCD spectrum to be observed, both model compounds as well as protein states display MCD spectra. MCD spectra can be seen at ambient as well as low temperature, are observed for all oxidation states of iron, and do not require isotopic labeling. Finally, because MCD spectra are difference spectra (the technique measures the difference in absorbance of left and right circularly polarized light), at any wavelength twice as much information is available (sign *and* intensity) as in an absorption spectrum (just intensity). Consequently, MCD spectra are more intricate than absorption spectra and serve as much better fingerprints for comparison of one chromophore with another. It is this latter approach that has found greatest utility in the study of

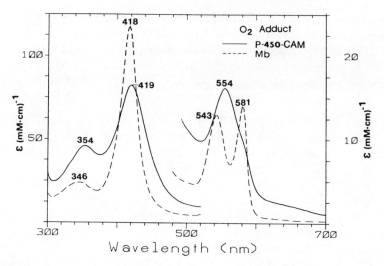

Fig. 9. Electronic absorption spectra at $-30°C$, pH 7.4, of oxymyoglobin (---) and of oxy-P-450-CAM (—). (From Ref. [61]; same conditions as used in Ref. [94].)

cytochrome P-450 through the comparison of the MCD spectra of structurally defined model porphyrins and proteins with the spectra of structurally undefined P-450 states.

The application of MCD spectroscopy to heme iron systems in general and to P-450 in specific has recently been reviewed,[109] and so only those MCD studies of P-450 that have revealed information about the active site structure will be discussed here. As will be seen, the conclusions about the metal coordination structure of P-450 in nearly all cases are based on parallel data from the same studies already presented in Section II,B (electronic absorption spectroscopy). However, due to the increased fingerprinting power of MCD spectroscopy relative to electronic absorption spectroscopy as just discussed, structural assignments based on comparison of MCD spectra are more likely to be accurate than are those based solely on absorption spectroscopy. This is particularly evident in the discussion of the ligand structure of oxy-P-450 (see below) for which a different conclusion was originally reached from examination of the absorption spectra of oxy-P-450 and oxy-myoglobin (Fig. 9) than was later reached from comparing their MCD spectra (see Fig. 14).

1. Low-spin ferric P-450

There are two MCD studies that have directly shed light upon the identity of the fifth ligand to low-spin ferric P-450. Shimizu *et al.*[110,111] have examined the MCD spectra of thiolate adducts of ferric myoglobin. These complexes have a thiolate–ferric heme–imidazole composition and display MCD spectra that are generally similar to low-spin ferric P-450, suggesting similar coordination environments, i.e., the presence of a thiolate ligand in ferric P-450. More recently, Sono *et al.* reported the MCD spectra of thiolate adducts of ferric P-450.[75] As discussed in Section II,B, the spectral properties of these complexes match those of bisthiolate–ferric prophyrin model complexes and provide indirect evidence for the presence of a thiolate fifth ligand in low-spin ferric P-450.

In order to identify the ligand trans to thiolate in low-spin ferric P-450, Dawson *et al.*[57,76,83,112] examined an extensive series of ferric P-450 ligand adducts with the goal of determining which ligand type(s) would produce MCD spectra that most closely matched that of native low-spin ferric P-450. Competition experiments were carried out in a few cases to make certain that the added ligands were indeed coordinating to the heme iron. After examining numerous nitrogen, oxygen, and sulfur donor ligands from essentially all possible biomimetic ligand classes, it was concluded that the oxygen donor adducts have MCD spectra that most closely match that of low-spin ferric P-450 (Fig. 10). An alcohol-containing amino acid and water were suggested as the most likely candidates for the sixth ligand. A similar, although less extensive study by Shimizu *et al.*[113] concluded that a sterically hindered nitrogen donor was the most likely sixth ligand although an oxygen donor was not ruled out. Unfortunately, in the absence of competition experiments, they were not certain that the oxygen donor ligand actually coordinated the heme iron. Dawson *et al.* also considered a sterically hindered nitrogen donor ligand as the possible sixth ligand to ferric P-450 but were able to rule it out, in part, on the basis of EPR studies.[57]

2. High-spin ferric P-450

As discussed in Section II,B, Holm and co-workers prepared a series of five-coordinate high-spin ferric porphyrin complexes as models for P-450 state **2**.[40,41] Dawson *et al.*[42] examined these complexes by MCD spectroscopy and found that only the five-coordinate ferric heme–thiolate complex displayed a strong negative MCD signal in the Soret region (near 400 nm) as is observed for high-spin P-450[114] (Fig. 11). In contrast, the other high-spin models and high-spin ferric myoglobin derivatives exhibited either positive MCD signals in the same wavelength region or very weakly negative intensity. Thus the conclusion was reached that high-spin P-450 has a thiolate axial ligand.

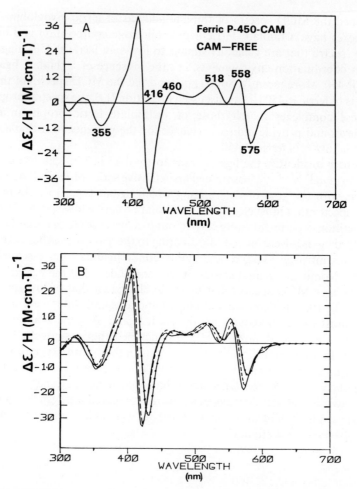

Fig. 10. (A) MCD spectrum of low-spin (camphor-free) ferric cytochrome P-450-CAM at pH 7.4. (Adapted from Ref. [57].) (B) MCD spectra of low-spin ligand complexes of ferric cytochrome P-450-CAM at pH 7.4: (——) 1-pentanol adduct; (– – –) *N*-methylpyrrolidinone adduct; (•–•–•) acetate adduct. (From Ref. (57].)

3. High-spin ferrous P-450

As discussed in Section II,B, the set of five-coordinate high-spin ferrous porphyrin complexes that have been prepared to date is not as extensive as would be desirable in order to consider all possible ligand types as candidates for the axial ligand in ferrous P-450. The MCD spectrum of ferrous P-450 is

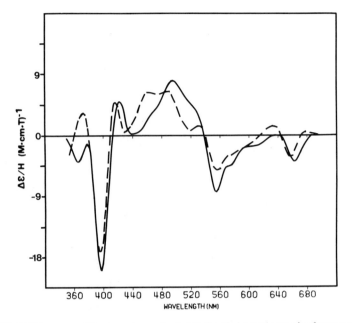

Fig. 11. MCD spectra of ferric protoporphyrin IX dimethyl ester (*para*-nitrobenzene thiolate) in toluene (——) and of high-spin (camphor-bound) ferric cytochrome P-450-CAM at pH 7.0 (– – –). (From Ref. [42]; protein data replotted from Ref. [114].)

displayed in Fig. 12A and is compared to a ferrous heme–thiolate model complex.[116] As can be seen, the two spectra are quite similar. In contrast, the MCD spectrum of high-spin ferrous myoglobin (Fig. 12B)[68,117] is completely different in shape and intensity throughout the spectral range examined. Thus, the most likely candidate for the axial ligand in high-spin ferrous P-450 is a thiolate. However, it would clearly be desirable if more high-spin ferrous porphyrins with biomimetic ligands could be synthesized and examined spectroscopically.

4. Ferrous-CO P-450

Collman and Sorrell[44] and Chang and Dolphin[47,48] were the first to report the synthesis of homogeneous thiolate–ferrous heme–CO complexes whose absorption spectra included a substantially red-shifted Soret peak at ~450 nm. As discussed in Section II,B, a series of such complexes with biomimetic ligands trans to CO was studied and only the thiolate ligand produced an adduct with the correct absorption spectrum. A series of these complexes was also examined by MCD spectroscopy[45]; the MCD spectrum

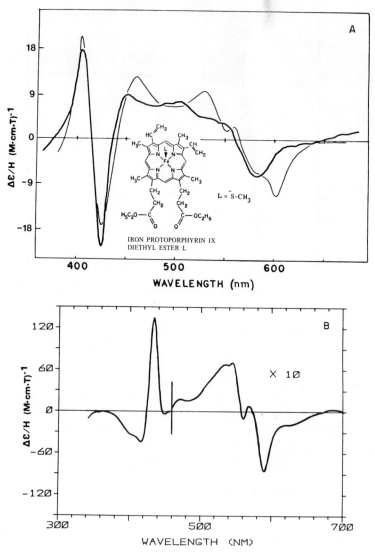

Fig. 12. (A) MCD spectra of ferrous cytochrome P-450-LM2 at pH 7.4 (———) and of ferrous protoporphyrin IX diethyl ester (*n*-propylthiolate) in dimethyl acetamide/H_2O (———) with crown ether to solubilize the thiolate. (From Ref. [115 and 116], respectively.) (B) MCD spectrum of ferrous myoglobin. (From Ref. [68].)

of the thiolate/CO complex is shown in Fig. 13 overplotted with the spectrum of ferrous-CO P-450. As can be seen, the two spectra are virtually superimposable. The spectra obtained with imidazole and thiol trans to CO were quite different.[45] The remarkably close spectral similarity between the thiolate/CO

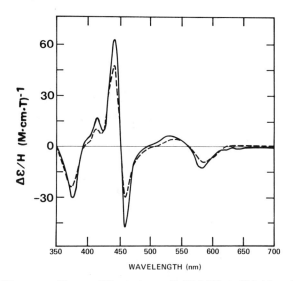

Fig. 13. MCD spectra of ferrous-CO cytochrome P-450-LM2 at pH 7.4 (——) and of ferrous protoporphyrin diethyl ester (methane thiolate) (CO) in benzene (– – –) with crown ether to solubilize the thiolate. (From Ref. [45].)

model and the protein provides very strong support for the assignment of the ligand trans to CO in the protein as a thiolate. An earlier MCD study by Shimizu *et al.*,[110] like the earlier electronic absorption study by Stern and Peisach[89] (see Section II,B), reached the same conclusion from studying a model porphyrin system that had some of the characteristic spectral properties of ferrous-CO P-450 but which was obviously a mixture of species. More recently, Collman and Groh[91] have compared the MCD spectra of a thiolate tail–ferrous heme–CO complex to that of ferrous-CO P-450 and obtained additional evidence for thiolate ligation in P-450 state **5**.

5.　Oxy-P-450

The last stable state in the reaction cycle of P-450 (Fig. 4) is the dioxygen bound state. The absorption spectra of oxy-P-450 and oxymyoglobin were presented in Fig. 9. As discussed in Section II,B, based on the close similarity in the Soret peak positions of these two proteins, it was earlier suggested that both proteins might have histidine imidazole trans to dioxygen.[93] However, as reported by Dawson and Cramer[94] and displayed in Fig. 14, the MCD spectra of these two oxygenated proteins (note the different intensity scales) are so different as to exclude any possibility of them both having histidine ligation. As such, this provides an excellent example of the utility of MCD spectroscopy to distinguish chromophores that may have some common features in their absorption spectra. Of course, this does not establish what the axial ligand to oxy-P-450 actually is; it only establishes that it is *not* histidine.

Unfortunately, as was the case for ferrous P-450, only a few oxygen-bound heme complexes with biomimetic trans ligands have been characterized, and the best characterized examples are with tetraaryl porphyrins which are less suited for optical spectroscopic comparisons. Budyka *et al.*[85] and Okubo *et al.*[86] have reported the MCD spectra of partially characterized oxy-P-450 model protoheme complexes. Both species are described as thiolate–ferrous

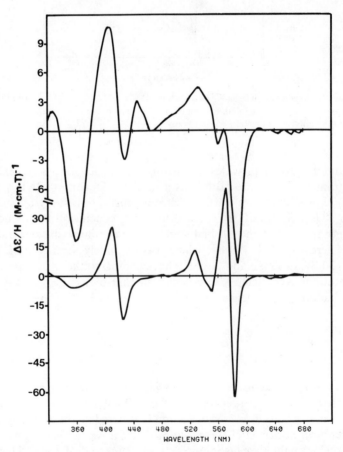

Fig. 14. MCD spectra at −30°C, pH 7.4, of oxygenated ferrous cytochrome P-450-CAM (top) and of oxygenated ferrous myoglobin (bottom). Please note that the scales are different for the two spectra. (From Ref. [94].)

heme–O_2 adducts, and the spectrum reported by Budyka *et al.* bears considerable resemblance to the MCD spectrum of oxy-P-450,[94] suggesting that thiolate is retained as the axial ligand upon dioxygen binding. Clearly more model systems need to be examined in order to firmly establish the identity of the axial ligand in oxy-P-450.

G. Infrared spectroscopy

The use of IR spectroscopy to study P-450 has so far been limited to the measurement of the CO stretching frequency in the ferrous-CO protein and model compounds and of the O–O stretch in models. For several different thiolate–ferrous heme–CO model compounds, the CO stretch has been observed from 1923 to 1956 cm^{-1}.[44,45,48,54,55] Thiol-, imidazole-, and thioether-ligated ferrous-CO models have displayed CO stretches from 1961 to 1980 cm^{-1}.[44,45,91] Studies with several isozymes of the protein have revealed CO stretches ranging from 1940 to 1954 cm^{-1} with the exception of camphor-free ferrous-CO P-450-CAM, which has CO stretches at 1942 *and* 1963 cm^{-1}.[118–120] On the surface, these numbers seem to suggest a correlation of the CO stretch energy and the trans ligand with values of ~ 1945 cm^{-1} typical of thiolate-ligated systems and values above 1960 characteristic of non-thiolate-bound complexes. However, a significant factor contributing to the energy of the CO stretch which is only known in the crystalline model systems from the Weiss laboratory[54,55] is the Fe–C–O bond angle. Shifts in the CO stretch as a function of the Fe–C–O angle have been postulated by Caughey.[118] In fact, the CO stretch in ferrous-CO hemoglobin is at 1951 cm^{-1}, presumably as a result of steric interactions.[21] Thus, it is not possible to reach firm conclusions about the ligand trans to CO from the energy of the CO stretch.[118] As for the O–O stretch in the oxy-P-450 models, it was observed at 1139 cm^{-1} in $^{16}O_2$ and at 1076 in $^{18}O_2$.[55] Unfortunately, the O–O stretch in oxy-P-450 has not been measured. However, in oxyhemoglobin models the O–O stretch is at 1163 cm^{-1}[121]; once again the presence of a thiolate ligand leads to a lowering of the stretching frequency of the trans ligand.

H. Resonance Raman spectroscopy

Given the extensive use that resonance Raman spectroscopy has found in the study of heme proteins,[122] it is somewhat surprising that more studies of cytochrome P-450 and thiolate-containing model compounds have not appeared. It has also been surprising that, in contrast to the studies of P-450 with electronic absorption, MCD, and EPR spectroscopy, in which studies of the protein and model compounds were often done in parallel, until very recently there had been no resonance Raman studies of synthetic thiolate-ligated iron porphyrins. The resonance Raman studies of cytochrome P-450 to date[123–129] have primarily focused on identification of the porphyrin structure-sensitive bands that correlate with oxidation and spin state of the heme iron.[122] Recently, a series of model compounds for cytochrome P-450 has been examined by resonance Raman spectroscopy,[130–133] in some cases confirming conclusions reached from the protein studies, in some cases not.

The initial resonance Raman study of P-450, by Ozaki *et al.*,[123] centered on the unusually low frequency of the so-called "oxidation state marker band" (v_4)[122] for high-spin reduced P-450-CAM. This band, which has subsequently been identified as a C–N breathing mode,[122] was found at 1346 cm^{-1} whereas most other reduced heme proteins have the analogous spectral feature at 1355–1360 cm^{-1} (Table 3).[125] Conversion of intact P-450 into an enzymatically inactive P-420 form shifted this band to

TABLE 3

Resonance Raman oxidation state marker bands
of ferrous and ferrous-CO heme proteins and model compounds[a]

	Ligand	v_4 (cm^{-1})	Ref.
High-spin ferrous			
P-450	Thiolate	1341–1347	[123, 125, 126, 128]
CPO	Thiolate (?)	1348	[125]
Mb, Hb, HRP	Imidazole	1355–1358	[123, 125, 136, 138]
P-420	?	1359, 1360	[125, 126, 128]
Fe(TPP)2-MeIm	Imidazole	1342, 1345	[134, 135]
Fe(PFP)2-MeIm	Imidazole	1344	[134]
Fe(PP)2-MeIm	Imidazole	1357	[136]
Fe(MP)2-MeIm	Imidazole	1359	[137]
Fe(PFP)SR	Thiolate	1341–1343	[133]
Fe(PFP)OR	Phenolate	1343	[133]
Low-spin ferrous-CO			
P-450	Thiolate	1365–1368	[125, 126, 128]
Mb, Hb	Imidazole	1370–1372	[138]
P-420	?	1371	[126]
Fe(TPP)Pyr	Pyridine	1365	[135]
Fe(MP)Pyr	Pyridine	1371	[137]
Fe(PFP)SR	Thiolate	1363, 1364	[133]

[a] Abbreviations: CPO, chloroperoxidase; Mb, myoglobin; Hb, hemoglobin; HRP, horseradish peroxidase; TPP, tetraphenylporphyrin; PFP, picket fence porphyrin[134] (a tetraarylporphyrin); PP, protoporphyrin; MP, mesoporphyrin; SR, $C_6HF_4S^-$; OR, $C_6HF_4O^-$; Pyr, pyridine.

~ 1360 cm^{-1}.[125,126,128] The v_4 band has been found from 1341 to 1347 cm^{-1} in several P-450 enzymes from different sources.[125,126,129] The lower energy of this Raman band has been attributed to the presence of a strongly electron releasing π base such as thiolate which might be expected to cause electron delocalization into the π^* (e_g) orbitals of the porphyrin.[123,125,126,128] Analogous, but smaller, shifts have been seen in the high-spin ferric and ferrous–CO states as well,[125,126,128] although the lowering in the frequency of v_4 for high-spin ferric P-450 is only clearly seen with the excitation into the high-energy side of the Soret electronic absorption band.[125] The energy of v_4 is "normal" for low-spin ferric P-450 and for a low-spin metyrapone-bound derivative of ferrous P-450.[123–128]

Recently, Chottard *et al.*[133] have questioned the correlation between the observation of a lower frequency v_4 band and the presence of a thiolate ligand (Table 3: line 1 vs line 3). They studied a high-spin five-coordinate thiolate-bound ferrous tetraarylporphyrin heme model and observed the oxidation state marker line (v_4) at 1341 cm^{-1}. Although this energy compares well with that found for the protein, it agrees equally well with the energy for the v_4 band in high-spin ferrous tetraarylporphyrin–imidazole complexes (Table 3). On the other hand, the analogous ferrous *protoheme*–imidazole has v_4 at 1357 cm^{-1}, in good agreement with deoxymyoglobin as well as P-420. Unfortunately, the resonance Raman spectrum of a ferrous protoheme–thiolate complex has not been reported. Clearly, for the high-spin tetraarylporphyrin models, there is nothing special about the thiolate ligand; thiolate-bound models have v_4 at the same energy as imidazole- and, for that matter, phenolate-bound adducts (Table 3).

For the ferrous-CO case, although the shift to lower frequencies on going from myoglobin/hemoglobin to P-450 is small, it has been considered as evidence for the presence of an electron-releasing π-base ligand such as thiolate for P-450 and not for P-420.[126,128] Again, ferrous-CO complexes of tetraarylporphyrins with thiolate and nitrogenous trans ligands have v_4 at essentially the same energy despite the different trans ligand. With ferrous-CO mesoporphyrin, the nitrogenous ligand leads to a band at the same energy as ferrous-CO myoglobin/hemoglobin; no ferrous-CO protoheme or mesoheme–thiolate system has yet been studied by resonance Raman spectroscopy. Thus, from the available data on proteins and model systems for the high-spin ferrous and low-spin ferrous-CO systems, it is not clear whether the energy of v_4 is diagnostic for a thiolate ligand as has been proposed. Chottard *et al.*[133] have recently examined a thiolate-ligated model complex for oxy-P-450 and have identified the O—O stretch at 1140 cm^{-1}. Unfortunately, no data have yet been published on this state of the protein.

Another approach that was attempted to test the identity of the axial ligand in P-450 was by Ozaki *et al.* who compared the high frequency resonance

Raman spectrum of ferric myoglobin–thiolate with that of low-spin ferric P-450. Unfortunately, even though the spectra they reported were very similar, so was the spectrum of ferric myoglobin–azide.[126]

As discussed by Champion *et al.*, significant changes occur in the high-frequency spin state-sensitive Raman bands upon substrate binding. This allows the high-spin/low-spin equilibrium between P-450 states 1 and 2 to be readily monitored at ambient temperatures.[125]

More recently, Shimizu *et al.*[128] have carefully examined the structure-sensitive high-frequency bands in the resonance Raman spectrum of P-450-SCC. Through comparison of trends in several of these bands in both five- and six-coordinate high-spin ferric species they reached the provocative conclusion that high-spin ferric P-450-SCC is six coordinate. Ferric P-450-CAM also had the same bands at energies closer to those indicative of six coordination while high-spin liver microsomal ferric P-450 fit in the five-coordinate group. In a similar fashion, Anzenbacher *et al.*[130] have compared high-spin P-450 from rat liver microsomes with Holm's five-coordinate high-spin ferric thiolate model[40,41] by resonance Raman spectroscopy and concluded that ferric P-450-CAM exists in a five-coordinate state. Extensive ligand binding studies with ferric P-450-CAM have so far failed to reveal any ligands that produce a six-coordinate high-spin state.[57,75,83,139] Ferric chloroperoxidase, another heme protein that likely has a thiolate ligand,[103,140-142] is able to bind ligands and form a six-coordinate high-spin state.[68] Nonetheless, given the extensive parallels between the spectral properties of high-spin P-450 (including high-spin P-450-SCC) and *five*-coordinate thiolate-bound ferric heme, in particular with MCD spectroscopy,[42] the existence of a six-coordinate high-spin ferric form of P-450 seems unlikely.

Another recent resonance Raman study by Anzenbacher and co-workers[131] involved the comparison of the high-frequency bands of low-spin ferric P-450 with thiolate–ferric heme–X complexes[41] having X equal to oxygen and nitrogen donors. A somewhat closer match in band positions occurred with the oxygen donor complexes, suggesting that such a ligand is trans to thiolate in the low-spin ferric protein as well.

The resonance Raman studies outlined so far have in common the emphasis on the high-frequency region where oxidation and spin state-sensitive bands are located. Understandably, studies of this type provide only indirect evidence at best for the identity of axial ligands. However, studies in the low-frequency region (150–700 cm^{-1}) can provide much more direct information about ligand types.[143] A preliminary report of a comparison of the low-frequency vibrations of low-spin ferric P-450 and various six-coordinate low-spin ferric heme models has appeared.[132]

The most direct approach to identification of heme ligands with resonance Raman spectroscopy is to look for isotope-sensitive bands in the low-

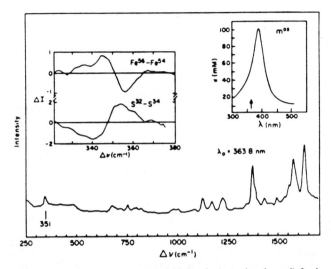

Fig. 15. The resonance Raman spectrum of high-spin (camphor-bound) ferric cytochrome P-450-CAM (lower curve). The absorption spectrum of ferric P-450-CAM is shown in the upper right-hand corner; the vertical line denotes the position of the resonant laser excitation (363.8 nm). The Raman difference spectra in the region of the 351 cm^{-1} Raman mode are displayed in the upper left-hand corner. The vertical scales for the difference spectra are given in units of 1000 counts channel^{-1}. (From Ref. [129].)

frequency region. As demonstrated by Champion *et al.*[129] the Raman band at 351 cm^{-1} in high-spin P-450 shifts upon substitution of either ^{34}S or ^{54}Fe for the naturally occurring isotopes (Fig. 15). Although an earlier attempt to carry out this exact experiment had failed,[125] the more recent study clearly provides compelling evidence for a sulfur donor fifth ligand in high-spin ferric P-450.

I. Electron paramagnetic resonance spectroscopy

Electron paramagnetic resonance (EPR) spectroscopy has played a major role in the study of the active site structure of cytochrome P-450. Because EPR spectra are most readily detected for systems having an odd number of unpaired electrons, this has meant that most EPR studies have been on ferric P-450. However, the ferrous-NO system also has an odd number of unpaired electrons and so its examination by EPR spectroscopy will also be discussed.

As shown in Fig. 4, the reaction cycle of P-450 involves two ferric states: the low-spin resting form (**1**) and the high-spin substrate-bound form (**2**). Although evidence for the interaction of substrate with P-450 was first obtained with electronic absorption spectroscopy,[144,145] the identification of the iron as low-spin in substrate-free P-450 and as high-spin in the substrate-bound enzyme was derived from EPR spectroscopy on crude preparations of P-450 and ultimately with purified samples.[35,36,59,146-153] The spectrum of the high-spin enzyme was particularly difficult to obtain because it essentially disappears above 60K.[152,153] These studies, by themselves, did not contribute to our understanding of the active site structure of ferric P-450 but they did provide the characteristic EPR spectra of these two states of P-450: low-spin P-450 has EPR g values at about 1.4, 2.25, and 1.9 and high-spin P-450 at about 8, 4, and 1.8. The g values for the low-spin enzyme are unusual among heme systems for their narrow range; the rhombic EPR spectrum of the high-spin enzyme, with two g values evenly spaced about $g = 6$, distinguishes it from other high-spin ferric heme systems having only a single g value in that region at $g \approx 6$.

The first evidence directly concerning the active site structure of ferric P-450 derived from EPR studies was the observed similarity between the g values of thiolate complexes of myoglobin and hemoglobin and of P-450.[38,39,154,155] Extensive ligand binding studies of myoglobin and hemoglobin revealed that only thiolate adducts and no other class of ligand adducts gave rise to the characteristic g values of low-spin P-450. While comparison of the electronic absorption spectra of thiolate-bound myoglobin and P-450 state **1** suggested that the coordination spheres of the iron in the two cases are not completely identical,[38] nonetheless the presence of a thiolate ligand seemed critical to producing the narrow-spread g values characteristic of low-spin P-450. Initial model studies also indicated that thiolate ligation leads to narrow-spread g values.[37,39] This requirement for thiolate ligation to generate an EPR spectrum like that of P-450 has subsequently been verified in extensive model studies.[40,41,43,51,52,73,74,79-82,93]

The identity of the sixth ligand to P-450 state **1** has been a matter of greater controversy. Comparison of the g values of thiolate–ferric heme–X complexes having biomimetic X ligands with the g values of ferric P-450 state **1** has led to the following divergent conclusions: (1) insufficient differences among the several X adducts examined were observed to narrow the choice down to one ligand class,[40,41,43] (2) the sixth ligand is histidine,[79,93] and (3) the sixth ligand is an oxygen donor.[52,74,81,82] An extensive examination by Dawson *et al.* of ferric P-450 adducts with biomimetic ligands led to the conclusion that the sixth ligand is an oxygen donor.[57] This conclusion was based in part on a comparison of the EPR properties of the ligand adducts and of low-spin P-450 such as in Fig. 16, where the spectrum of the 1-pentanol adduct is compared to that of the native enzyme. Given the range of g values reported for P-450 enzymes from several different sources and under different conditions[15] (see Ref. [83], however), it may not be possible to use EPR comparisons of this type as the sole method upon which to determine the sixth

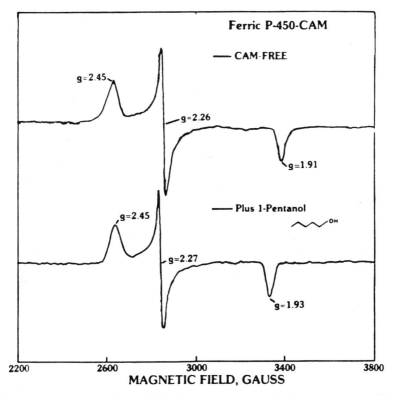

Fig. 16. EPR spectra at pH 7.4, 77K, of low-spin (camphor-free) ferric cytochrome P-450-CAM (top) and of the 1-pentanol adduct of ferric cytochrome P-450-CAM (bottom). (Adapted from Ref. [57].)

ligand. Clearly, a consensus on the identity of the sixth ligand has not been achieved from these EPR investigations, although more studies favor an oxygen donor than favor histidine.

Less controversy exists concerning the identity of the fifth ligand in P-450 state **2**. Model studies by Holm, Collman, Ogoshi, and their co-workers[40,41,43,49] have demonstrated that ferric porphyrin–thiolate complexes have g values that match those of substrate-bound ferric P-450. Although Sato and Kon reported a nonthiolate high-spin model system with rhombic g values like high-spin ferric P-450, subsequent power saturation studies distinguished it from P-450.[157,158] Overall, the EPR studies leave little doubt about the presence of a thiolate ligand to P-450 state **2**.

The use of nitric oxide with its one unpaired electron to transform the usually EPR-nondetectable ferrous heme state into an EPR-detectable form has been utilized with P-450 by Mason, Peterson, and their co-workers.[92,159,160] O'Keeffe *et al.* have concluded that the absence of superhyperfine splitting of the EPR signal of the ferrous-NO complex beyond that expected from the nitrogen of NO itself is consistent with retention of the thiolate ligand trans to NO in the ferrous enzyme.[160]

J. Electron nuclear double resonance spectroscopy

In 1980, LoBrutto *et al.* reported the only publication to date on the electron nuclear double resonance (ENDOR) properties of ferric P-450.[161] They examined both camphor-free and camphor-bound P-450-CAM in protonated and deuterated buffer in order to look for the presence of strongly coupled exchangeable protons.

For low-spin (camphor-free) ferric P-450, ENDOR lines are observed in the proton region maximally at g_z. These signals are seen in protonated buffer and not in deuterated buffer. Analysis of the data led to the conclusion that an exchangeable proton (or possibly two protons) is present at a distance of 2.6–2.9 Å from the iron. This proton could come from water or from amino acid side chains containing ROH, RNH_2, $RCONH_2$, or, possibly, RSH functionalities. Given the extensive evidence available at the time of this study and already discussed in this review for a deprotonated cysteine (cysteinate) fifth ligand to ferric P-450, the authors assumed that the exchangeable proton was present on the sixth ligand. Their data specifically ruled out histidine as the sixth ligand.

The ENDOR data on high-spin ferric P-450 did not reveal any exchangeable protons coupled to the heme iron. Further analysis of the ENDOR data and comparison to five- and six-coordinate high-spin ferric heme systems allowed the authors to rule out histidine and water as ligands to high-spin ferric P-450 and to conclude that the iron is five-coordinate.

In summary, the ENDOR data are consistent with the large body of evidence described in this review suggesting that the high-spin ferric state is five-coordinate. Examination of the low-spin enzyme with this technique led to the determination that the ligand trans to thiolate could be the oxygen donor (H_2O, ROH) proposed in several other studies or that it could be an amine, an amide, or a thiol and that it was definitely *not* a histidine.

K. Electron spin echo spectroscopy

As in the case of ENDOR spectroscopy, there has only been one report of the use of electron spin echo (ESE) spectroscopy to study the ligand environment of ferric P-450.[162] Groh *et al.* have employed ESE spectroscopy to examine substrate binding to P-450-SCC.[163] The ESE study by Peisach *et al.* examined the nuclear modulation pattern in the envelope of electron spin echoes for a number of low-spin ferric heme proteins and model compounds and reached the conclusion that the sixth ligand to the heme iron of P-450 state 1 is an imidazole group from histidine. This conclusion was based on the observation of nuclear modulation of the low-spin heme iron by a nonpyrrole nitrogen atom. However, neither an imidazole–ferric heme iron–thiolate model complex or ferric myoglobin–thiolate gave rise to ESE parameters that matched those of low-spin P-450. Curiously, the authors did not examine imidazole-bound ferric P-450 or alcohol-bound ferric P-450 to obtain data on systems that clearly do and do not, respectively, have a nitrogenous sixth ligand.

Both Dawson *et al.*[57] and White and Coon[78] have suggested that an oxygen donor sixth ligand hydrogen bonded to histidine imidazole might provide a nitrogen to couple to the ferric iron of P-450 and account for the ESE data of Peisach *et al.*[162] This model, first suggested by Ristau *et al.*,[164] would be consistent with evidence from various laboratories for an oxygen donor sixth ligand and the ESE evidence for a weakly coupled nitrogen atom. However, coupling across a hydrogen bond as suggested by that model has not been demonstrated yet with ESE spectroscopy. Clearly, the interpretation of Peisach *et al.* that the sixth ligand is histidine is inconsistent with the evidence already presented and that which is described in the next section for an oxygen donor as in the sixth ligand of low-spin ferric P-450.

L. Nuclear magnetic resonance spectroscopy

Although nuclear magnetic resonance (NMR) spectroscopy has not been extensively employed in the study of cytochrome P-450, there have been three problems of relevance to the active site structure of the enzyme addressed by this technique: (1) the effect of the paramagnetic ferric P-450 iron on the relaxation properties of the bulk water protons and the implications of such data on the identity of the sixth ligand, (2) the ^{1}H NMR properties of high-spin ferrous P-450 and model porphyrin complexes, and (3) the relationship between the ^{13}C NMR resonance of carbon monoxide bound to ferrous heme iron and the ligand trans to CO.

Griffin and Peterson[165] were the first to examine the effect of the paramagnetic iron in ferric P-450 on the relaxation of the bulk water protons. Working with P-450-CAM, they observed that in the low-spin camphor-free ferric case, the paramagnetic iron has a significant effect on the relaxation rate of the solvent protons. From theory, they were able to calculate that a proton located 2.0–2.6 Å from the iron must be in rapid exchange with the solvent.

Therefore, they concluded that water is the sixth ligand to low-spin ferric P-450. Only outer-sphere relaxation effects were observed for the high-spin camphor-bound ferric enzyme consistent with the five-coordinate structure usually assigned to P-450 state **2** (Fig. 4). Similar data were later reported by Philson *et al.*[166] who suggested that an amino acid-derived alcohol, amine, or amide ligand would also be compatible with the data and should be included with water as candidates for the sixth ligand to low-spin ferric P-450. With liver microsomal P-450, on the other hand, Grasdalen *et al.* observed very different results.[167] They found that substrate binding increased proton relaxation rather than diminishing it. However, in both substrate-free and substrate-bound cases the effect appeared to be outer sphere; no evidence for a water ligand or other ligand having a bound proton in rapid exchange with the solvent was found.

Keller *et al.* examined the ^{1}H NMR of P-450 state **3** and obtained the first strong evidence for the high-spin nature of the heme iron in that state.[168] More recently, Parmely and Goff[169] have studied the proton NMR of a ferrous heme–thiolate complex and reported evidence that such a complex is five-coordinate and high-spin. Unfortunately, in part due to the absence of well-defined ferrous heme–thiol complexes and in the absence of very high field NMR data on the ferrous protein, the authors did not speculate on the identity of the fifth ligand to ferrous P-450.

For the ferrous-CO case, Berzinis and Traylor examined the ^{1}H and ^{13}C NMR properties of ligand-bound model heme iron complexes.[170] They observed that the thiolate-containing model complexes had ^{13}C NMR resonances at 200 ppm or less while ferrous-CO hemoglobins had the resonance at 206–208 ppm. With a mercaptan ligand trans to CO, the NMR peak is at 204.7 ppm. Thus it seems likely that a ^{13}C chemical shift for bound CO of less than or equal to approximately 200 would correspond to presence of a thiolate ligand. For P-450-CAM, the resonance is at 200.3 ppm.[171]

M. Other approaches

This review has emphasized the spectroscopic studies of the heme iron of P-450 and of model compounds as a means of determining the active site structure of the enzyme. There have been other methods used to approach this problem and they will be briefly presented here.

Magnetic susceptibility has been used to study P-450 states **2** and **3**. Peterson used this technique to show that camphor-bound ferric P-450-CAM is high-spin,[59] confirming the assignment made from electronic absorption and EPR spectroscopy of the substrate-free and substrate-bound enzyme as discussed previously. More recently, Champion *et al.* examined the magnetic susceptibility of the camphor-bound ferrous enzyme and also found it to be high-spin.[172] The vast majority of high-spin ferric and ferrous porphyrins are five-coordinate,[102] consistent with the structures assigned for P-450 states **2** and **3** in Fig. 4.

The chemical modification approach has been employed by Jänig *et al.* in order to identify the sixth ligand to ferric P-450-LM2.[173,174] Using reagents to nitrate and acetylate tyrosine residues in the protein, they obtained evidence that such an amino acid is the sixth ligand. Dus has also employed chemical modification methods to obtain evidence for the cysteine ligand to P-450 and to probe the substrate binding site.[175]

Theoretical studies of the electronic structure of cytochrome P-450 by Hanson, Jung, Loew, and their co-workers[67,176–181] coupled with single crystal absorption spectra of the protein[67,182] have provided considerable insight into the relationship between spectroscopic properties and physical structure in the P-450 system. Many of the unusual spectroscopic properties of low-spin ferric and ferrous P-450 ligand complexes can be attributed to mixing of a charge-transfer transition from the lone pair $p^\dagger$ orbital of the cysteinate sulfur to the porphyrin π^* orbital with the normal porphyrin π-π^* Soret transition. In the high-spin ferrous case, however, calculations have suggested that the cysteinate ligand is protonated.[176] As discussed earlier in this review, most currently available evidence is consistent with retention of cysteinate as the ligand in high-spin ferrous P-450. The question of the proton balance in the interconversion of P-450 states **3**, **4**, and **5** has been directly addressed by Dolphin *et al.*[95]; no evidence for proton uptake or release was found, suggesting that the cysteinate ligand that has been well established for P-450 state **5** is also the ligand in **3** and **4**.

Finally, the investigation of cobalt porphyrin model complexes and of cobalt-substituted P-450 has provided information concerning the active site structure of P-450. Studies of the cobalt-substituted protein by Wagner *et al.*[183] have revealed the following information: (1) a "hyper" spectrum is obtained with certain cobalt(III) P-450 ligand complexes as has been observed for the native ferric case and attributed to presence of an endogenous thiolate ligand and (2) the EPR spectra of cobalt(II) P-450 with and without O_2 bound are most consistent with the presence of a nonnitrogeneous ligand such as thiolate. These conclusions are substantially strengthened by the parallel results recently reported by Weiss and co-workers with cobalt porphyrin model complexes.[184,185]

N. Summary

Extensive investigations of the spectroscopic properties of cytochrome P-450 and model porphyrin complexes have led to the determination of the active site structures of the intermediates in the P-450 reaction cycle (Fig. 4).

Although the coordination by an axial cysteinate ligand will not be completely established until it is observed by X-ray crystallography, there is near-unanimous agreement that such a ligand must be present. While the studies supporting that assignment are far too numerous to summarize here, results from EXAFS and resonance Raman spectroscopy have provided the most direct evidence for the presence of such a ligand, with MCD and EPR spectral comparisons of the protein and model compounds also strongly contributing to that assignment. The presence of the cysteinate ligand is certain beyond reasonable doubt in P-450 states **1, 2,** and **5**; the evidence in P-450 states **3** and **4**, while less compelling, is nonetheless quite strong. Some controversy still exists with regard to the identity of the sixth ligand in ferric P-450 state **1**; the majority of the data concerning that assignment are consistent with it being an oxygen donor ligand such as water or an alcohol-containing amino acid. This entire problem has provided an excellent example of how much information can be determined about the active site structure of a metalloenzyme in the absence of X-ray crystallography through the concerted efforts of spectroscopists, protein physical biochemists, and inorganic chemists.

III. MECHANISM OF ACTION OF CYTOCHROME P-450: SELECTED ASPECTS

A considerable amount of information is known about the mechanism of action of cytochrome P-450 in part because the discovery of a soluble, bacterial P-450 (P-450-CAM) that could be more easily purified has facilitated such investigations. In most cases, parallel studies between the bacterial and mammalian P-450 enzymes have yielded similar results about the structure and mechanism of P-450. In fact, for mechanistic studies, the two systems constitute an ideal combination: P-450-CAM is very substrate specific and forms a more stable oxygen complex while liver microsomal P-450 hydroxylates virtually any hydrocarbon substrate. Thus, for mechanistic experiments in which the details of substrate binding or the properties of the oxygen complex are desired, P-450-CAM is the system of choice. Furthermore, the substrate specificity narrows the product distribution, simplifying interpretation. Alternatively, for experiments in which a particular carefully designed substrate is to be examined in order to test hypotheses about the reaction mechanism, liver microsomal P-450 is the best choice since product formation from virtually any substrate is assured.

There are six steps in the proposed reaction cycle for P-450 as shown in Fig. 17. These steps are discussed in the following sections:

A. Substrate binding
B. First electron transfer

C. Dioxygen binding
D. Second electron transfer
E. Cleavage of the oxygen–oxygen bond
F. Product formation

Species **1** through **4** have been characterized, whereas the exact nature of **6** and **7** remains to be elucidated. Because the reaction mechanism of P-450 has been recently reviewed,[14,15] only selected aspects of the mechanism will be discussed in detail here, along with new developments in the other areas. Mechanistic work on cytochrome P-450-CAM will be emphasized, although similar studies on liver microsomal P-450 will be mentioned. Topics to be discussed in greater detail include the transfer of electrons to the heme iron, the breakdown of the oxygenated intermediate, and the use of alternative substrates.

A. Substrate binding

The resting form of P-450, as discussed in Section II, consists of a low-spin six-coordinate ferric heme iron (P-450 state **1**) that undergoes a change in coordination structure to a high-spin five-coordinate state (**2**) upon substrate binding. The substrate binding process and accompanying spin state change have been extensively discussed and recently reviewed.[14,186,187] Although significant changes occur in the coordination structure of the heme iron, Lewis and Sligar have recently reported structural studies of P-450 in which they used low-angle X-ray scattering to show that both protein states have the same radius of gyration.[188] Sligar has also shown that ferric P-450-CAM is actually a mixture of low- and high-spin forms in both the absence and presence of the substrate, camphor, with substrate binding dramatically favoring the high-spin form over the low-spin form that predominates in the absence of substrate.[189,190] Similar conclusions have been reached about mammalian P-450 except that substrate binding does not usually lead to more than a 1:1 mixture of high-spin and low-spin forms.[115,164,187] The spin state equilibrium in the presence and absence of substrate has been examined by temperature-jump kinetics for both bacterial and mammalian P-450.[191,192] Finally, the effect of temperature, pH, and pressure has been thoroughly investigated by the Paris low-temperature enzymology group of Lange, Debey, Hui Bon Hoa, and co-workers.[193–198] Ionizable amino acids with pK values of 5.4 and >6.0 were observed to exert an influence on the high-spin/low-spin equilibrium of P-450-CAM. Peterson has clearly demonstrated the significant role played by cations, particularly potassium, in the binding of camphor to ferric P-450-CAM.[59]

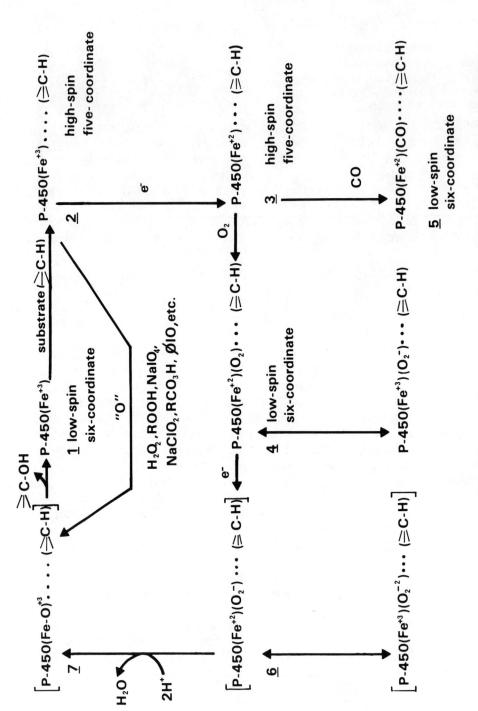

Fig. 17. The reaction cycle of cytochrome P-450. The individual steps of the reaction cycle are discussed in the text. The numbering of the intermediates corresponds to that used in Fig. 4.

B. First electron transfer

The cytochrome P-450-CAM reaction cycle involves two separate steps in which electrons are delivered to P-450 from putidaredoxin (Fig. 17).[59] The binding of camphor to ferric P-450-CAM $(1 \rightarrow 2)$ has been shown to dramatically alter the redox potential of the heme iron of the cytochrome from -303 mV for the substrate-free state to -170 mV with substrate bound; similar, although smaller, changes in reduction potential have been observed with liver microsomal P-450.[189,190,199] The reduction potential of putidaredoxin is -239 mV.[190,200] Formation of a tight complex between putidaredoxin and cytochrome P-450 has been demonstrated by both the Gunsalus and Peterson laboratories.[190,200–203] The reduction potential of camphor-bound P-450-CAM changes only slightly in the presence of putidaredoxin from -170 to -173 mV while the reduction potential of putidaredoxin changes more noticeably from -239 to -196 mV (Fig. 18). The binding of substrate to P-450-CAM also leads to a 170-fold increase in the rate constant

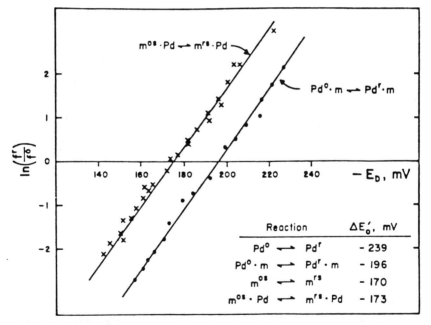

Fig. 18. Oxidation/reduction potentials of putidaredoxin and cytochrome P-450-CAM (cytochrome *m*), free, and bound as a dienzyme complex. The natural log of the fraction of protein reduced over the fraction oxidized is plotted as a function of the total system potential E_D. A substantial shift in the potential of putidaredoxin is observed on binding to the cytochrome. (From Ref. [190].)

for reduction of ferric P-450-CAM by reduced putidaredoxin.[204] The facilitation of electron transfer to ferric P-450 appears to result from the high-spin nature of the substrate-bound enzyme. In mammalian P-450, where substrate binding results in a mixture of high- and low-spin forms, burst kinetics have been observed, consistent with the conclusion that reduction of the high-spin form occurs more rapidly than the low-spin form under conditions in which the high-spin/low-spin equilibrium is relatively slow.[205–207] In that regard, it would be interesting to determine whether camphor derivatives such as 5-*exo*-bromocamphor, which do not shift ferric P-450-CAM completely to the high-spin state upon binding and, when bound, produce a reduction potential (-246 mV) intermediate between that of camphor-free and camphor-bound P-450-CAM,[208] also exhibit burst kinetics as observed for mammalian P-450.

The binding of putidaredoxin to P-450-CAM is strongly dependent upon the oxidation states of the two proteins. Sligar and Gunsalus have determined a dissociation constant of 2.9 μM for the two oxidized proteins and of 0.49 μM for the two reduced proteins.[190,200] Peterson and co-workers determined the affinity of the two oxidized proteins for each other under somewhat different conditions to be 10 μM[203]; the affinity was found to be quite dependent on ionic strength and changed by a factor of six upon changing the potassium phosphate ionic strength from 10 to 100 mM. More importantly, from an analysis of the kinetics of the reduction of camphor-bound ferric P-450-CAM by reduced putidaredoxin they determined the dissociation constant for the complex of *reduced* putidaredoxin and oxidized P-450-CAM to be 0.08 μM.[203] Thus reduced putidaredoxin binds to oxidized P-450-CAM over 100 times more tightly than does oxidized putidaredoxin. Both the Gunsalus and Peterson laboratories have observed saturation kinetics (Fig. 19) in the reduction of P-450-CAM by putidaredoxin[201–203] as expected for a reaction involving tight complex formation followed by intracomplex electron transfer.[203] A Michaelis constant of 0.63 μM was determined by Pederson and Gunsalus (Fig. 19).[201] At subzero temperatures in mixed solvents, Hui Bon Hoa *et al.* found complex formation between P-450·CAM and putidaredoxin to be rate limiting and suggested that the cryogenic solvent system might be responsible for the diminishing of the affinity of the two proteins for each other.[209]

The reduction of P-450-CAM with nonphysiological reductants has also been studied. Hintz and Peterson examined the reduction with sodium dithionite and found that camphor-free P-450 is reduced four times faster than camphor-bound (high-spin) or metyrapone-bound (low-spin) P-450-CAM.[210] Clearly, the rate of reduction by dithionite is not a function of spin state as in the case for enzymatic reduction. Hintz and Peterson presented evidence for an outer-sphere reduction mechanism with the bound camphor

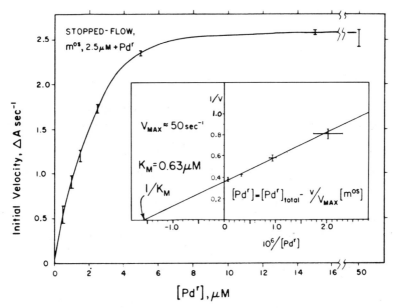

Fig. 19. First electron transfer from reduced putidaredoxin to camphor-bound oxy-P-450-CAM (m^{os}): K_m for reduced putidaredoxin. Saturation and K_m determination for reduced putidaredoxin (Pd^r) in the first electron transfer reaction. Initial velocities at 20°C observed at 390 nm in 50 mM phosphate (pH 7.0), 200 mM K^+, and 0.25 mM camphor. Pd was reduced with dithionite, 100μM. (From Ref. [201].)

and metyrapone restricting access to the heme iron. In another study, Bazin *et al.* used a laser photolysis method to reduce P-450-CAM, presumably by photoionization of an amino acid near the heme iron, and found that only the high-spin form was reduced.[211]

The next three steps of the P-450 cycle are not as well characterized. The binding of molecular oxygen, second electron transfer, and oxygen–oxygen bond cleavage collectively make up the process referred to as oxygen activation. Inherent in the understanding of the mechanism of oxygen activation of P-450 is the explanation of why these heme-containing enzymes activate oxygen and catalyze oxygenation reactions while other heme proteins such as hemoglobin and myoglobin simply bind oxygen reversibly.

C. Dioxygen binding

Reduced substrate-bound P-450 can bind dioxygen to form oxy-P-450 (**4**) and initiate the activation of oxygen. This state of the P-450 reaction cycle was discovered in 1970 for both P-450-CAM and liver microsomal P-450, although only as a transient intermediate in the latter case.[62,212–215] The bimolecular rate constant for the formation of oxy-P-450-CAM is $7.7 \times 10^5 M^{-1} s^{-1}$ at 4°C.[214] The breakdown of oxy-P-450 is also very rapid. Autoxidation to the ferric enzyme occurs with a half-life of only a few minutes at room temperature. Douzou and co-workers have shown that the oxygenated protein can be handled at subzero temperatures in cryogenic solvents where oxy-P-450-CAM has a half-life of several hours.[216,217] It was also found that having camphor present added substantially to the stability of the oxygenated enzyme.[216,217] Similar techniques were used to stabilize the oxygenated form of liver microsomal P-450[218] and P-450-SCC.[219] In other cases, oxygenated P-450 intermediates have been detected with stopped-flow methods.[220]

The breakdown of oxygenated P-450 has been the subject of considerable controversy, with some investigations leading to the conclusion that free superoxide anion is produced concomitantly with the decomposition of oxy-P-450 and other studies, suggesting that superoxide is not produced. Sligar *et al.* detected chemiluminescence during autoxidation of oxy-P-450-CAM and attributed it to decay of excited singlet dioxygen produced from a superoxide radical precursor.[221] However, Bonfils *et al.* were unable to detect any evidence for superoxide during the autoxidation of chromatographically purified oxy-P-450-CAM.[222] The autoxidation of oxygenated liver microsomal P-450 isozyme 4 led directly to hydrogen peroxide without any detectable superoxide even in the presence of detergents that convert the protein from its usual aggregated state to a monomeric form.[220] In contrast, several studies reported by Estabrook, Ullrich, and their co-workers on liver microsomal P-450 provided evidence for the formation of hydrogen peroxide from superoxide via breakdown of oxy-P-450.[223,224] Clearly, the issue of superoxide production from oxy-P-450 has not been completely settled.

D. Second electron transfer

The transfer of a second electron from putidaredoxin to oxy-P-450-CAM has been found to be the rate-limiting step of the overall monoxygenation reaction.[204,225] The reduced oxygen complex that presumably results from this second electron transfer is formally equivalent to a ferrous iron–superoxide complex [$Fe^{2+}-O_2^-$] (6), although the actual electron distribution is unknown since no stable complexes beyond oxy-P-450 have been observed. Iron heme model compounds having the composition $Fe^{2+}-O_2^-$ have been prepared by Valentine, Dolphin, Reed, and their co-workers.[226–228] The most extensively characterized model, [$Fe(octaethylporphyrin)O_2$]$^-$, has been formulated as a high-spin ferric peroxide complex with the peroxide bound in a side-on bidentate fashion with the presence or absence of a trans axial ligand not clearly established.[226] The potential for transfer of an electron to a ferrous-dioxygen complex in order to form the above model complex was -0.04 V.[227] Reports from the Gunsalus laboratory of the reduction potential (E'_0) of oxy-P-450-CAM (**4** → **6**) have indicated a value near 0 V,[58] although determination of a true potential for the protein system is probably not possible because the reaction is not easily reversed.

The role of Pd in the second electron transfer to oxy-P-450 has been examined by stopped-flow kinetics studies of the interaction of reduced Pd with this intermediate.[201,209] As with the first electron transfer, saturation kinetics were observed, consistent with formation of a complex between reduced putidaredoxin and oxy-P-450-CAM (Fig. 20).[201,209,229] The disappearance of oxy-P-450 was seen to coincide with the reoxidation of Pd and the formation of oxidized P-450, as shown in Fig. 21.[201] The rates observed for the appearance of oxidized P-450 under conditions of limited or saturating Pd concentration were 15 and 17.5 s^{-1}, respectively, which agrees with the P-450-CAM turnover number under steady-state conditions.[229]

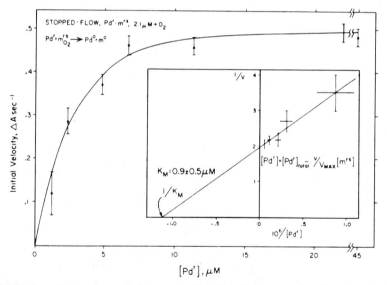

Fig. 20. Second electron transfer from reduced putidaredoxin to camphor-bound oxy-P-450-CAM ($m_{O_2}^{rs}$): K_m for reduced putidaredoxin. Saturation and K_m determination for reduced putidaredoxin (Pd^r) in the second electron transfer reaction. The reaction was limited by one equivalent camphor; P-450 and enzymes were reduced by NADH, 400 μM; other conditions: 50 mM phosphate buffer, pH 7.0, 200 mM K$^+$. Initial velocities were observed at 417 nm. (From Ref. [201].)

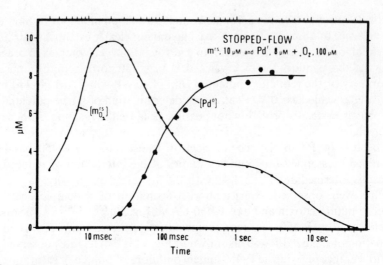

Fig. 21. Dynamics of heme and iron–sulfur center reactions: ligand and redox dynamics accompanying the second electron transfer reaction. The formation and decomposition of camphor-bound oxy-P-450-CAM ($m_{O_2}^{rs}$) were determined from absorbance changes at 395 nm, using $\Delta\varepsilon = -20\,\mathrm{m}M^{-1}\,\mathrm{cm}^{-1}$ for ferrous P-450 → oxy-P-450 and $\Delta\varepsilon = +30\,\mathrm{m}M^{-1}\,\mathrm{cm}^{-1}$ for oxy-P-450 → ferric P-450. Oxidation of reduced putidaredoxin ($Pd^r \rightarrow Pd^o$) was observed at 525 nm using $\Delta\varepsilon = 2.0\,\mathrm{m}M^{-1}$. Other conditions as described in Fig. 20. (From Ref. [201].)

Putidaredoxin has also been suggested to play a role as an effector protein in facilitating product formation in the P-450 system in addition to simply delivering electrons.[200,229,230] Evidence for this role has been obtained from examination of the formation of product from camphor-bound oxy-P-450-CAM. As this state of the P-450 cycle is one electron reduced from the starting point, one would expect to be able to form up to one-half an equivalent of 5-*exo*-hydroxycamphor. However, no product was formed from camphor-bound oxy-P-450-CAM unless oxidized putidaredoxin was present.[200,229,230] Several other "effector" molecules were identified including some dithiols such as dihydrolipoic acid. Curiously, apocytochrome b_5 was the most efficient catalyst for the formation of product from oxy-P-450-CAM. Hintz and Peterson have questioned the evidence for an effector role for putidaredoxin.[203] A photochemical reductant system has also been developed to achieve partial turnovers of the P-450-CAM system.[230]

Recently, phenazine methosulfate (PMS) has been successfully used by Eble and Dawson[231] to mediate *electron transfer* from NADH and achieve oxygen-dependent multiple turnovers of cytochrome P-450-CAM in the absence of putidaredoxin and putidaredoxin reductase. The PMS-mediated turnover of P-450-CAM resulted in the formation of 5-*exo*-hydroxycamphor (the normal, enzymatic product) as the only product. Product formation was inhibited by carbon monoxide and metyrapone, known inhibitors of the fully reconstituted P-450 system, while catalase, superoxide dismutase, and hydroxyl radical scavengers did not have any effect on the rate of product formation in agreement with the lack of effect these probes have on the reconstituted P-450 system. These results, as well as the observation that oxygen bubbling increased product formation, supported the hypothesis that hydroxylation catalyzed by the NADH/PMS/P-450-CAM system occurred via an oxygen-dependent enzymatic pathway. The PMS-mediated hydroxylation is the only oxygen-dependent system to achieve multiple turnovers of P-450-CAM in the absence of the other two proteins. Although the turnover rate is quite a bit lower than the fully reconstituted three protein system, it does still prove that product can be formed repeatedly in the absence of the other two electron transfer proteins.

Addition of one electron to oxy-P-450 provides the last ingredient necessary for product formation. Consequently, no stable intermediates have been characterized beyond oxy-P-450. In a tightly coupled system, all electrons would go on to form the products: alcohol and water. However, in many cases less alcohol is formed than expected based on oxygen and electrons consumed. Mammalian liver P-450 is particularly "leaky," with the extra electrons primarily going to produce hydrogen peroxide.[224,232,233] Whether the hydrogen peroxide comes from superoxide via breakdown of oxy-P-450 or directly from breakdown of the one electron reduced form of oxy-P-450 is not entirely clear.

E. Cleavage of the oxygen–oxygen bond

Hydrogen peroxide, alkyl peroxides, iodosobenzene, and other single oxygen atom donors have been shown to be capable of replacing NADH and molecular oxygen in P-450-catalyzed reactions[11,13–15,234] with similar product distributions. As shown in Fig. 22, this suggests that cleavage of the oxygen–oxygen bond occurs to form an active oxygen complex containing a single oxygen atom (7) similar to Compound I of the peroxidase enzymes. Investigation of the interaction of P-450-CAM and *meta*-chloroperoxybenzoic acid with rapid-scan absorption spectroscopy has provided evidence for the formation of a species with some spectral similarities to the peroxidase Compound I iron-oxo complex (Fig. 23).[236] With P-450-LM, the nature of the complex(es) formed upon interaction of oxygen donors with the ferric enzyme is even less certain.[237] Other low-temperature and rapid-scan experiments designed to examine intermediates formed upon addition of oxygen donors to P-450-CAM have shown the diminished absorption and broadening of the Soret band that are characteristic of

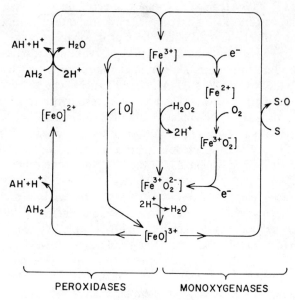

Fig. 22. Iron–oxygen states of P-450 and peroxidases with emphasis on the possibility of a common intermediate $[(FeO)^{3+}]$. (Adapted from Ref. [235].)

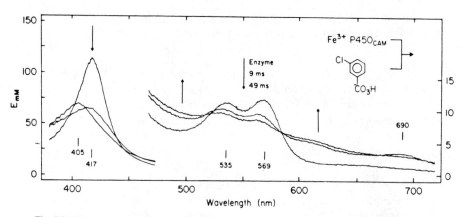

Fig. 23. Electronic absorption spectra of native low-spin ferric P-450-CAM and of the enzyme in the presence of *meta*-chloroperoxybenzoic acid after 9 and 49 ms. Initial conditions after mixing: 0.012 mM (Soret region) or 0.018 mM (visible region) P-450-CAM, 0.26 mM *meta*-chloroperoxybenzoic acid, 50 mM potassium phosphate buffer, pH 7.0, 20.0 $\pm$ 0.2°C. (From Ref. [236].)

peroxidase Compound I formation.[201,238] However, evidence against a Compound I-type intermediate also exists. McCarthy and White compared the oxidative reactivity of several P-450 enzymes with that of peroxidase enzymes and concluded that if both enzyme mechanisms involve an iron-oxo intermediate, then these intermediates are somehow very different, in both their mode of formation and reactivity.[239] These results also suggested that the P-450 active oxygen intermediate results from homolytic cleavage of the oxygen bond of the peroxy complex **6**, while the peroxidase catalytic cycle involves heterolytic oxygen–oxygen bond cleavage. Earlier work by White *et al.*[240] had provided evidence in support of the homolytic cleavage of the oxygen–oxygen bond of iron-bound peroxy acids using peroxyphenylacetic acid. Homolytic cleavage of the O–O bond of this peroxy acid leads to formation of an unstable carboxyl radical which decarboxylates to benzyl alcohol. Unfortunately, a subsequent study by McCarthy and White showed the decarboxylation route to be on a separate pathway not involving any intermediates in the hydroxylation pathway beyond ferric P-450.[241] Studies of the dioxygen-dependent P-450 reaction cycle (Fig. 17) have also failed to provide much information about intermediates beyond oxy-P-450. Coon and co-workers observed two intermediates starting from ferrous P-450 (**3**) in rapid kinetics experiments, but were unable to identify either intermediate.[242]

Because the oxygen–oxygen bond of molecular oxygen must be cleaved at some point during the P-450 reaction cycle, numerous proposals have been put forth to explain how this process occurs. Following Hamilton,[243] Sligar *et al.* have proposed that cleavage of the O–O bond takes place via an enzyme-bound peracid (or peramide) generated by reaction of the reduced P-450 ferrous dioxygen complex (**6**) with an amino acid carboxylate (or amide) as shown in Fig. 24.[235] Evidence in support of this proposal was obtained from single turnover experiments in which the incorporation of an oxygen atom from labeled dioxygen into lipoic acid, an effector molecule previously shown to promote product formation from oxy-P-450-CAM,[229] was found.[235] As shown in Fig. 24, acylation of the bound dioxygen with loss of water (O^{2-}) would result in a bound peracid that could either dissociate and transfer an oxygen atom or cleave heterolytically releasing the carboxylic acid and $(FeO)^{3+}$ followed by oxygen transfer. Either way, one of the labeled oxygen atoms from dioxygen ends up in the released carboxylic acid. Clearly, the general applicability of this intriguing mechanism must still be demonstrated. No particular role for the axial thiolate ligand of P-450 was postulated in this mechanism.

It has often been suggested that the factor(s) responsible for the unusual spectroscopic properties of P-450 also contribute substantially to its unique reactivity. As detailed in Section II, it is clear that a thiolate ligand derived

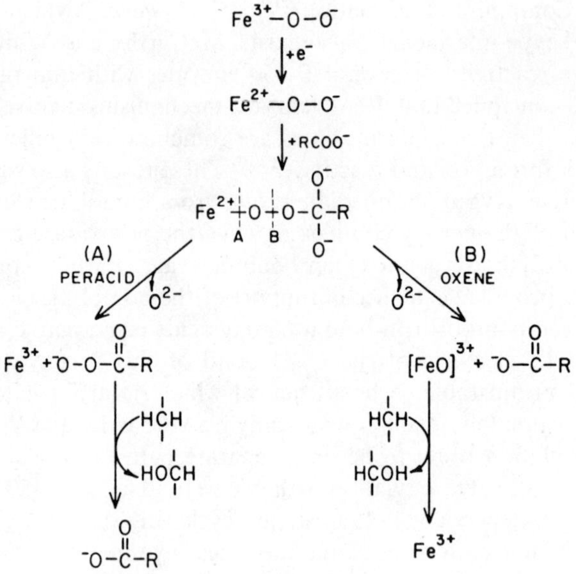

$$Fe^{3+}-O-O^-$$

$$\downarrow +e^-$$

$$Fe^{2+}-O-O^-$$

$$\downarrow +RCOO^-$$

$$Fe^{2+}\overset{|}{+}O\overset{|}{+}O-\overset{O^-}{\underset{|}{\underset{O^-}{C}}}-R$$

A   B

(A) PERACID   (B) OXENE

O^{2-} O^{2-}

$$Fe^{3+}+\overset{O}{\overset{||}{O}}-O-\overset{||}{C}-R \qquad\qquad [FeO]^{3+}+\overset{O}{\overset{||}{O}}-O-\overset{||}{C}-R$$

HCH   HCH
HOCH   HCOH

$$\overset{O}{\underset{-O-\overset{||}{C}-R}{}} \qquad\qquad Fe^{3+}$$

Fig. 24. Acylation of heme-bound dioxygen: mechanisms for peracid generation and for heterolytic O–O bond cleavage. (From Ref. [235].)

from cysteine is responsible for the special spectroscopic properties of P-450. Thus, it is logical to expect that a role for the thiolate ligand in the P-450 reaction cycle would have been suggested. In fact, in 1976 Dawson et al.[42] proposed that the relatively polarizable thiolate ligand might push electron density into the oxygen–oxygen bond, thereby facilitating its cleavage (Scheme 1). Furthermore the thiolate ligand, for the same reasons, might be expected to stabilize the highly electrophilic $(FeO)^{3+}$ species that results after heterolytic O–O bond cleavage. Evidence for the electron-releasing nature of the thiolate ligand of P-450 relative to the imidazole ligand of myoglobin has recently been obtained by comparison of the properties of ligand complexes of the two proteins. Sono and Dawson have reported marked differences in anionic ligand affinities between ferric P-450 and myoglobin that were

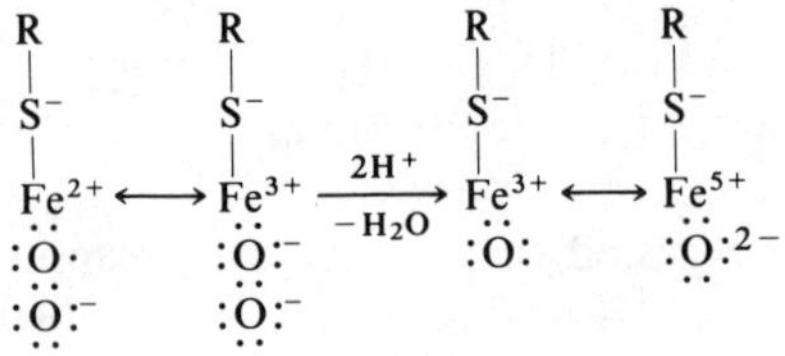

$$\begin{array}{cccc} R & R & R & R \\ | & | & | & | \\ S^- & S^- & S^- & S^- \\ | & | & | & | \\ Fe^{2+} & \longleftrightarrow Fe^{3+} & \xrightarrow[-H_2O]{2H^+} Fe^{3+} & \longleftrightarrow Fe^{5+} \\ \cdot\cdot & \cdot\cdot & \cdot\cdot & \cdot\cdot \\ :O\cdot & :O:^- & :O: & :O:^{2-} \\ \cdot\cdot & \cdot\cdot & & \\ :O:^- & :O:^- & & \end{array}$$

Scheme 1

ascribed to the increased electron density at the heme iron of P-450 as a result of the thiolate ligand.[139] The substantially increased acidity of thiol ligands bound trans to imidazole (myoglobin) in comparison to thiolate (P-450) was also attributed to the electron-releasing nature of the cysteine thiolate ligand of P-450.[244] More recently, White and Coon have carried the concept a step further and suggested complete electron transfer from the thiolate ligand during O–O bond cleavage, producing a transient thiyl radical.[14] Ullrich has also suggested that the major role of the thiolate ligand is to stabilize the proposed $(FeO)^{3+}$ species.[16]

The mechanism of the oxygen–oxygen bond cleavage step in P-450-catalyzed reactions is still unresolved. The importance of characterization of the active oxygen intermediate makes this one of the most interesting and challenging areas of P-450 research.

F.　Product formation

The final step in the P-450 mechanism is product formation and regeneration of the ferric resting form of the enzyme. All of the chemistry discussed to this point has taken place at the heme iron, with the substrate presumably bound nearby to the heme iron. Although the issue is hardly settled, it seems most likely that the catalytic species responsible for hydroxylating alkanes is the $(FeO)^{3+}$ species (7) discussed in the previous subsection. Evidence that this intermediate is the key hydroxylation catalyst has been reviewed.[10–16,245] A recent study that supports the involvement of a single oxygen atom adduct of iron, presumably $(FeO)^{3+}$, as the hydroxylation catalyst has been reported by Sligar *et al.*[245] They observed that hydroxylation of camphor with iodosobenzene as the source of the oxygen atom proceeds via a ping-pong mechanism with iodosobenzene donating a single oxygen atom to the iron and dissociating prior to substrate binding. A ternary iodosobenzene: camphor: P-450 complex was not required to fit the data, in contrast to the normal O_2-dependent pathway. Unfortunately, the generality of this new evidence for a single oxygen atom adduct of P-450 is not clear, since iodosobenzene-driven hydroxylation is the only case of all the exogenous oxidant-dependent or O_2-dependent hydroxylation reactions for which isotope labeling studies have shown that solvent water oxygen atoms are found in the product.[246]

Product formation in P-450 aliphatic hydroxylations is thought to involve the abstraction of a hydrogen atom from the substrate by $(FeO)^{3+}$ and generation of a carbon radical and $(FeOH)^{3+}$. Coupling of the carbon radical and the iron-bound hydroxyl radical then produces the product. Once again, the evidence for this process has been recently reviewed; only the major experiments that initially led to that conclusion and the very recent results of relevance will be presented here.

In 1977, a paper of particular importance to the mechanism of action of cytochrome P-450 was published by Hjelmeland *et al.*[247] In this paper, the kinetic isotope effect for benzylic hydroxylation by liver microsomal P-450 was reported to be much larger, $k_H/k_D = 11$, than previously thought. This value was determined by studying a symmetrical molecule with two equivalent benzylic sites, one deuterated and one not. Thus, the isotope effect was determined in an *intra*molecular fashion that factors out all other steps of the reaction cycle except the step in which hydroxylation occurs. At the time of this report, the most generally accepted mechanism of hydroxylation by P-450 was a concerted insertion mechanism.[248] Previous *inter*molecular isotope effect studies had consistently observed low values,[247] as expected for an insertion mechanism in which the C-H bond is not broken prior to oxygenation. This larger isotope effect was much more indicative of a radical abstraction/recombination mechanism. A large intramolecular isotope effect was subsequently reported by Groves *et al.* in the aliphatic hydroxylation of norbornane.[249,250] In addition, they also found that a significant amount of epimerization took place during hydroxylation. Thus, they observed an *exo*-norborneol product with deuterium retained in the product even though the deuterium had been in the exo position in the substrate. Such a deuterium should be lost in a concerted insertion process or in a mechanism involving complete retention of configuration at carbon. Together, these results[247,249,250] provided rather strong evidence for a radical abstraction/recombination mechanisms with a planar carbon radical intermediate. Extensive, parallel model system studies by Groves and co-workers[251] and others[252] have provided additional support for the radical-based mechanism of action of cytochrome P-450.

Investigations with P-450-CAM have also provided evidence for the radical abstraction/recombination mechanism. Gelb *et al.*[253] synthesized 5-*exo*- and 5-*endo*-deuterocamphor and examined its conversion to product in the NADH/O$_2$ fully reconstituted system as well as in the iodosobenzene, *meta*-chloroperoxybenzoic acid, and hydrogen peroxide exogenous oxidant-dependent pathways. In all cases they observed that, while either exo and endo hydrogen (or deuterium) atom could be abstracted, only the exo alcohol product was formed. A "significant," although smaller than expected, *intra*molecular isotope effect was also observed. These observations are best rationalized by the intermediacy of a planar carbon radical. The difference in specificities of the hydrogen abstraction and alcohol forming steps suggested the possibility that different sites on the enzyme were responsible for the two separate steps. Finally, the observation of essentially identical results in all these experiments (including *inter*molecular isotope effect studies), regardless of whether the fully reconstituted NADH/O$_2$-dependent system or one of the exogenous oxidant systems was being studied, argued for a common pathway for all such processes.

Finally, it is interesting to note that evidence has recently been presented by Ullrich, Coon, and their co-workers that under certain circumstances, liver microsomal P-450 can catalyze the four-electron reduction of molecular oxygen to two molecules of water.[224,233] The amount of this oxidase activity is very dependent on the substrate being used and therefore represents an example of uncoupling of electron transfer from product formation. Presumably, in such cases, the $(FeO)^{3+}$ species is formed and then is reduced by two electrons to form Fe^{3+} and water in a process analogous to the reduction of peroxidase Compound I back to its ferric resting state (Fig. 22). P-450-CAM-catalyzed product formation from a variety of substituted camphors will be discussed in the next subsection. Product formation is followed by dissociation of the product from the enzyme active site, regenerating the resting form of P-450 (**1**).

G. Metabolism of substrate analogs of camphor

A significant amount of information about the mechanism of action of P-450 has resulted from studies of the metabolism of analogs and derivatives of camphor. Since camphor is stereo- and regiospecifically hydroxylated by P-450-CAM at the 5-exo position (Scheme 2), the metabolism of 5-exo-substituted camphors has been examined (Table 4). 5-*exo*-Bromocamphor

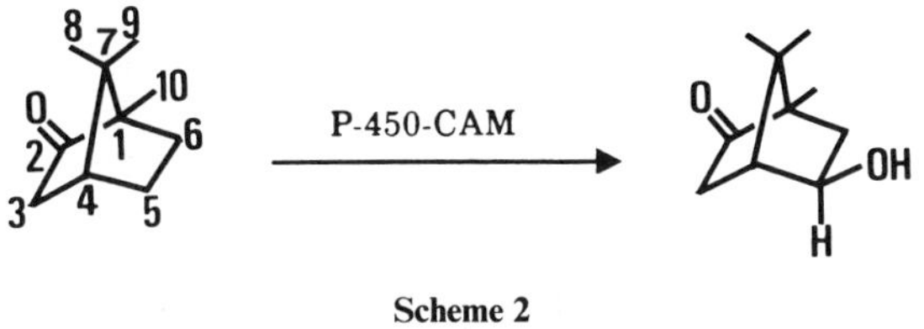

Scheme 2

was metabolized by P-450-CAM to 2,5-diketocamphor, presumably through the intermediacy of a transient 5-bromo-5-hydroxycamphor species followed by loss of HBr.[208] When the 5-exo hydrogen of camphor was replaced by iodine or a hydroxyl group, the product of P-450-CAM metabolism was likewise found to be the 2,5-diketone.[245] These studies suggest that hydrogen abstraction occurs from the 5-endo position when the 5-exo position is substituted. Whether the suggested hydroxylation occurred at the 5-exo or 5-endo position could not be determined since the proposed initial product was not detected. The 2,5-diketone is also formed when 5-*endo*-hydroxy camphor is the substrate.[245]

Evidence to date suggests that oxygen delivery to the substrate occurs from the exo side of the camphor molecule. This is supported by the fact that *exo*-5-hydroxycamphor is the only hydroxylated product observed upon camphor

TABLE 4

TABLE 4

Metabolism of camphor analogs by cytochrome P-450-CAM

Substrate	Products	Reference
		[208]
		[245]
		[245]
		[245]
		[254]
		[255]
		[245]
		[253]
		[253]

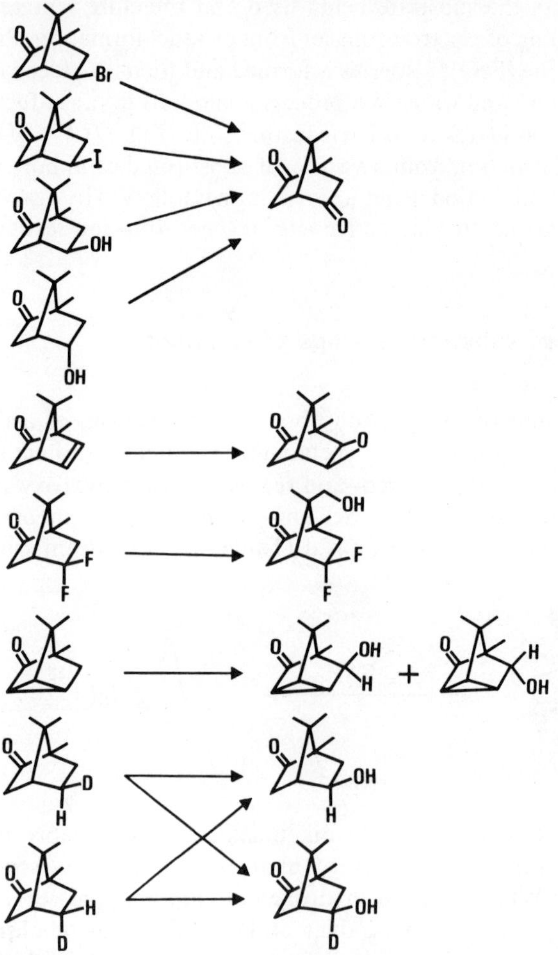

metabolism. In addition, 5,6-*exo*-epoxycamphor is the enzymatic product formed from 5,6-dehydrocamphor.[254] As previously discussed, deuterium labeling of camphor has indicated that, although loss of hydrogen (or deuterium) can occur from either the 5-exo or 5-endo positions of camphor, the exo-hydroxylated product is always formed.[253] When 5,5-difluorocamphor, a substrate analog in which both the 5-exo and 5-endo hydrogens of camphor have been replaced with fluorine, was reacted with P-450-CAM the product observed was 5,5-difluoro-9-hydroxycamphor.[255] The formation of 9-hydroxylated product is the first reported example of methyl hydroxylation catalyzed by P-450-CAM and represents a departure from the strict regio-selectivity normally observed for P-450-CAM-catalyzed reactions. Hydroxy-lation of the 9 position is consistent with oxygen delivery to the substrate from the exo side of the camphor molecule. The product formation mechanism which is most consistent will all these results is shown in Scheme 3.

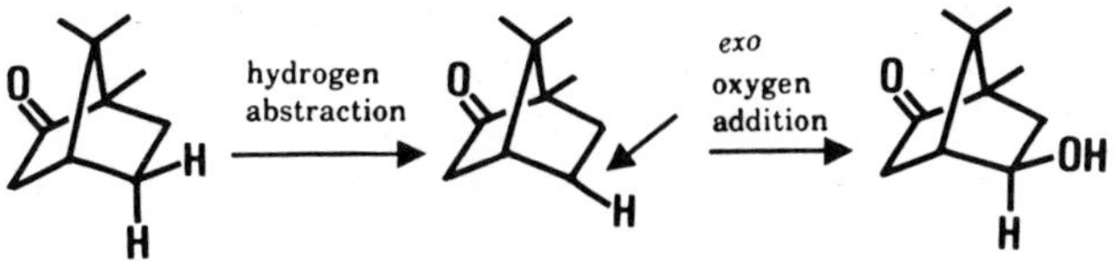

Scheme 3

Hydrogen abstraction can occur from either the 5-exo or 5-endo positions to form a planar carbon radical. As discussed above, exo addition of the "active oxygen intermediate" to the substrate carbon radical would then result in formation of the exo-hydroxylated product. Presumably, when both 5 hydrogens are blocked, a 9 hydrogen is abstracted, followed by oxygen addition from the exo face of the molecule to form the 9-hydroxy product.

Interestingly, metabolism of pericyclocamphanone (Table 4, entry 7) leads to products hydroxylated at the 6 position (6-*exo*- and 6-*endo*-pericyclocamphenol).[245] This hydroxylation is not accompanied by any ring-opened products as might be expected for a mechanism involving a long-lived radical intermediate at a carbon immediately adjacent to a three-membered ring. Thus, oxygen addition to the radical must be fast compared to the known, rapid rates for such cyclopropane rings openings.[245] Because the hydroxy-lation of pericyclocamphanone is the only example with P-450-CAM for which two products are formed during substrate hydroxylation, Sligar *et al.* examined the ratio of the two products as a function of oxidant. They found a 1:1 ratio of 6-*exo*- to 6-*endo*-pericyclocamphenol when the fully reconstituted NADH/O$_2$-dependent system was employed, while the iodosobenzene and *meta*-chloroperoxybenzoic acid-dependent systems yielded ratios of approx-imately 1.5 and 1.8, respectively.[245] This dependence of the ratio of products on the oxidant employed is unexpected, given the similarity in isotope effect parameters observed (see Section III,F) for the different oxidants.[253] The authors suggested two possible explanations: direct participation in the hydroxylation step by the exogenous oxidant or a steric effect by the deoxygenated oxidant.

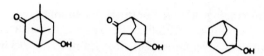

Fig. 25. P-450-CAM hydroxylation products viewed from above the cyclohexane ring. Left: 5-*exo*-hydroxycamphor; middle: 5-hydroxyadamantanone; right: 1-adamantanol. (From Ref. [256].)

The metabolism of compounds structurally similar to camphor by P-450-CAM has also been examined. Norcamphor, a bicyclo[2.2.1] molecule lacking the three methyl groups of camphor, is not hydroxylated by P-450-CAM.[229] Adamantanone and adamantane were found to be hydroxylated by P-450-CAM to form 5-hydroxyadamantanone and 1-adamantanol, products which are isosteric with the product of camphor hydroxylation, 5-*exo*-hydroxycamphor (Fig. 25).[256] These results suggest a rigid enzyme–substrate complex in which hydrogen abstraction can only occur from one substrate site.

H. Summary

The reaction cycle of cytochrome P-450 has been well established to involve four isolable intermediates (Fig. 17, **1 → 4**). While it is clear that a great deal of chemistry occurs beyond the oxygenated intermediate, **4**, in the process of forming the hydroxylated product, there is still much to be learned about the exact details of product formation. The consensus of most recent studies directed at this question is that, following transfer of the second electron, the reduced oxygenated ferrous P-450 intermediate (**6**) loses water to form the active hydroxylation catalyst [(FeO)$^{3+}$,**7**]. This species then abstracts a hydrogen atom from the substrate, yielding a carbon radical and hydroxyl radical bound to iron. Product is formed by recombination of the two radical fragments. Some evidence has been obtained for the intermediacy of a species such as (FeO)$^{3+}$ but it would be more convincing if such a species could be trapped out in some fashion. In addition, the presence of a carbon radical during turnover, while consistent with most currently available evidence, has not been directly observed. Finally, the controversy over whether the fully reconstituted NADH/O$_2$-dependent and the exogenous oxidant-dependent pathways of P-450 catalyzed hydroxylation involve a common intermediate remains unresolved. The question of homolytic vs heterolytic cleavage of the oxygen–oxygen bond also continues to be unsettled. Clearly, despite considerable progress, many exciting challenges still remain to be solved.

Acknowledgments

We thank Drs. Masanori Sono and Laura A. Andersson for helpful discussions and assistance in the preparation of this article, and Professors Richard Holm, James Collman, and Thomas Sorrell for permission to reproduce unpublished results (Figs. 5A and 12A). J.H.D. is a Camille and Henry Dreyfus Teacher-Scholar (1982-1987), an Alfred P. Sloan Foundation Research Fellow (1983–1987), and a recipient of a National Institutes of Health Research Career Development Award (1982–1988); studies of cytochrome P-450 in his laboratory are supported by the NIH (GM-26730) and NSF (PCM 82-16799) grants.

References

[1] Malmström, B. G. *Annu. Rev. Biochem.* **1982**, *51*, 21.

[2] Keevil, T.; Mason, H. S. *In* "Methods in Enzymology"; Fleischer, S.; Packer, L., Eds.; Academic Press, New York, 1978, Vol. 52, p. 3.

[3] Hayaishi, O., Ed.; "Molecular Mechanisms of Oxygen Activation"; Academic Press: New York, 1974.

[4] Sato, R.; Omura, T., Eds.; "Cytochrome P-450"; Academic Press: New York, 1978.

[5] Ullrich, V.; Roots, I.; Hildebrandt, A.; Estabrook, R. W.; Conney, A. H., Eds.; "Microsomes and Drug Oxidations"; Pergamon: Oxford, 1977.

[6] Coon, M. J.; Conney, A. H.; Estabrook, R. W.; Gelboin, H. V.; Gillette, J. R.; O'Brien, P. J., Eds.; "Microsomes, Drug Oxidations, and Chemical Carcinogenesis"; Academic Press: New York, 1980.

[7] Gustafsson, J.-A.; Carlstedt-Duke, J.; Mode, A.; Rafter, J., Eds.; "Biochemistry, Biophysics and Regulation of Cytochrome P-450"; Elsevier: Amsterdam, 1980.

[8] Sato, R.; Kato, R., Eds.; "Microsomes, Drug Oxidations, and Drug Toxicity"; Japan Scientific Societies Press: Tokyo, 1982.

[9] Hietanen, E.; Laitinen, M.; Hänninen, O., Eds.; "Cytochrome P-450, Biochemistry, Biophysics and Environmental Implications"; Elsevier: Amsterdam, 1982.

[10] Peterson, J. A. *In* "Fundamental Research in Homogeneous Catalysis", Vol. 3; Tsutsui, M., Ed.; Plenum: New York, 1979, p. 729.

[11] Gunsalus, I. C.; Sligar, S. G. *Adv. Enzymol.* **1978**, *47*, 1.

[12] Groves, J. T. *Adv. Inorg. Biochem.* **1979**, *1*, 119.

[13] Ullrich, V. *Top. Curr. Chem.* **1979**, *83*, 85.

[14] White, R. E.; Coon, M. J. *Annu. Rev. Biochem.* **1980**, *49*, 315.

[15] Coon, M. J.; White, R. E. *Metal Ions Biol.* **1980**, *2*, 73.

[16] Ullrich, V. *J. Mol. Catal.* **1980**, *7*, 159.

[17] Mannering, G. J. *In* "Concepts in Drug Metabolism", Part B; Jenner, P., Testa, B., Eds.; Dekker: New York, 1981, p. 53.

[18] Alexander, L. S.; Goff, H. M. *J. Chem. Educ.* **1982**, *59*, 179.

[19] Holland, H. L. *Chem. Soc. Rev.* **1982**, *11*, 371.

[20] Estabrook, R. W. *In* "Oxygenases and Oxygen Metabolism"; Nozaki, M.; Yamamoto, S.; Ishimura, Y.; Coon, M. J.; Ernster, L.; Estabrook, R. W., Eds.; Academic Press: New York, 1982, p. 371.

[21] Dolphin, D.; James, B. R. *ACS Symp. Ser.* **1983**, *211*, 99.
[22] Mansuy, D. *In* "The Coordination Chemistry of Metalloenzymes"; Bertini, I.; Drago, R. S.; Luchinat, C., Eds.; Reidel, Dordrecht: 1983, p. 343.
[23] Coon, M. J. *Trans. N.Y. Acad. Sci.* **1983**, *41*, 1.
[24] Klingenberg, M. *Arch. Biochem. Biophys.* **1958**, *75*, 376.
[25] Garfinkel, D. *Arch. Biochem. Biophys.* **1958**, *77*, 493.
[26] Coon, M. J.; Vatsis, K. P. *In* "Polycyclic Hydrocarbons and Cancer", Vol. 1; Gelboin, H. V.; Ts'o, P.O.P., Eds.; Academic Press: New York, 1978, p. 335.
[27] Jerina, D. M.; Yagi, H.; Lehr, R. E.; Thakker, D. R.; Schaefer-Ridder, M.; Karle, J. M.; Levin, W.; Wood, A. W.; Chang, R. L.; Conney, A. H. *In* "Polycyclic Hydrocarbons and Cancer", Vol. 1; Gelboin, H. V.; Ts'o, P.O.P., Eds.; Academic Press: New York, 1978, p. 173.
[28] Omura, T.; Sato, R. *J. Biol. Chem.* **1964**, *239*, 2370.
[29] Hedegaard, J.; Gunsalus, I. C. *J. Biol. Chem.* **1965**, *240*, 4038.
[30] Katagiri, M.; Ganguli, B. N.; Gunsalus, I. C. *J. Biol. Chem.* **1968**, *243*, 3543.
[31] Gunsalus, I. C.; Wagner, G. C. *In* "Methods in Enzymology"; Fleischer, S.; Packer, L., Eds.; Academic Press, New York, 1978, Vol. 52, p. 166.
[32] O'Keeffe, D. H.; Ebel, R. E.; Peterson, J. A. *In* "Methods in Enzymology"; Fleischer, S.; Parker, L., Eds.; Academic Press, New York, 1978, Vol. 52, p. 151.
[33] Roome, P. W.; Philley, J. C.; Peterson, J. A. *J. Biol. Chem.* **1983**, *258*, 2593.
[34] Poulos, T. L.; Finzel, B. C.; Gunsalus, I. C.; Wagner, G. C.; Kraut, J. *J. Biol. Chem.*, in press.
[35] Mason, H. S.; North, J. C.; Vanneste, M. *Fed. Proc. Fed. Am. Soc. Exp. Biol.* **1965**, *24*, 1172.
[36] Murakami, K.; Mason, H. S. *J. Biol. Chem.* **1967**, *242*, 1102.
[37] Bayer, E.; Hill, H. A. O.; Röder, A.; Williams, R. J. P. *J. Chem. Soc. Chem. Commun.* **1969**, 109.
[38] Jefcoate, C. R. E.; Gaylor, J. L. *Biochemistry* **1969**, *8*, 3464.
[39] Röder, A.; Bayer, E. *Eur. J. Biochem.* **1969**, *11*, 89.
[40] Koch, S.; Tang, S. C.; Holm, R. H.; Frankel, R. B.; Ibers, J. A. *J. Am. Chem. Soc.* **1975**, *97*, 917.
[41] Tang, S. C.; Koch, S.; Papaefthymiou, G. C.; Foner, S.; Frankel, R. B.; Ibers, J. A.; Holm, R. H. *J. Am. Chem. Soc.* **1976**, *98*, 2414.
[42] Dawson, J. H.; Holm, R. H.; Trudell, J. R.; Barth, G.; Linder, R. E.; Bunnenberg, E.; Djerassi, C.; Tang, S. C. *J. Am. Chem. Soc.* **1976**, *98*, 3707.
[43] Collman, J. P.; Sorrell, T. N. *J. Am. Chem. Soc.* **1975**, *97*, 913.
[44] Collman, J. P.; Sorrell, T. N. *J. Am. Chem. Soc.* **1975**, *97*, 4133.
[45] Collman, J. P.; Sorrell, T. N.; Dawson, J. H.; Trudell, J. R.; Bunnenberg, E.; Djerassi, C. *Proc. Natl. Acad. Sci. U.S.A.* **1976**, *73*, 6.
[46] Collman, J. P.; Sorrell, T. N.; Hodgson, K. O.; Kulshrestha, A. K.; Strouse, C. E. *J. Am. Chem. Soc.* **1977**, *99*, 5180.
[47] Chang, C. K.; Dolphin, D. *J. Am. Chem. Soc.* **1975**, *97*, 5948.
[48] Chang, C. K.; Dolphin, D. *Proc. Natl. Acad. Sci. U.S.A.* **1976**, *73*, 3338.
[49] Ogoshi, H.; Sugimoto, H.; Yoshida, Z. *Tetrahedron Lett.* **1975**, 2289.
[50] Stern, J. O.; Peisach, J. *FEBS Lett.* **1976**, *62*, 364.
[51] Ruf, H. H.; Wende, P. *J. Am. Chem. Soc.* **1977**, *99*, 5499.
[52] Ruf, H. H.; Wende, P.; Ullrich, V. *J. Inorg. Biochem.* **1979**, *11*, 189.
[53] Traylor, T. G.; Mincey, T. C. *Acta Biol. Med. Ger.* **1979**, *38*, 351. Traylor, T. G.; Mincey, T. C.; Berzinis, A. P. *J. Am. Chem. Soc.* **1981**, *103*, 7084.
[54] Caron, C.; Mitschler, A.; Rivière, G.; Ricard, L.; Schappacher, M.; Weiss, R. *J. Am. Chem. Soc.* **1979**, *101*, 7402.
[55] Schappacher, M.; Ricard, L.; Weiss, R.; Montiel-Montoya, R.; Bill, E.; Gonser, U.; Trautwein, A. *J. Am. Chem. Soc.* **1981**, *103*, 7646.

[56] Ricard, L.; Schappacher, M.; Weiss, R.; Montiel-Montoya, R.; Bill, E.; Gonser, U.; Trautwein, A. *Nouv. J. Chim.* **1983**, *7*, 405.

[57] Dawson, J. H.; Andersson, L. A.; Sono, M. *J. Biol. Chem.* **1982**, *257*, 3606.

[58] Gunsalus, I. C.; Meeks, J. R.; Lipscomb, J. D.; Debrunner, P.; Münck, E. *In* "Molecular Mechanism of Oxygen Activation"; Hayaishi, O., Ed.; Academic Press: New York, 1974, p. 559.

[59] Peterson, J. A. *Arch. Biochem. Biophys.* **1971**, *144*, 678.

[60] Dawson, J. H.; Andersson, L. A.; Sono, M. *J. Biol. Chem.* **1983**, *258*, 13637.

[61] Sono, M.; Eble, K. S.; Dawson, J. H.; Hager, L. P. *J. Biol. Chem.* **1985**, *260*, 15530.

[62] Ishimura, Y.; Ullrich, V.; Peterson, J. A. *Biochem. Biophys. Res. Commun.* **1971**, *42*, 140.

[63] Antonini, E.; Brunori, M. "Hemoglobin and Myoglobin in their Reactions with Ligands"; North-Holland Publ.: Amsterdam, 1971.

[64] Margoliash, E.; Frohwert, E. *Biochem. J.* **1959**, *71*, 570.

[65] Ozols, J.; Strittmatter, P. *J. Biol. Chem.* **1964**, *239*, 1018.

[66] Guengerich, F. P.; Ballou, D. P.; Coon, M. J. *J. Biol. Chem.* **1975**, *250*, 7405.

[67] Hanson, L. K.; Eaton, W. A.; Sligar, S. G.; Gunsalus, I. C.; Gouterman, M.; Connell, C. R. *J. Am. Chem. Soc.* **1976**, *98*, 2672.

[68] Sono, M.; Dawson, J. H. Unpublished results.

[69] Sakurai, H.; Shimomura, S.; Ishizu, K. *Chem. Pharm. Bull.* **1977**, *25*, 199.

[70] Byrn, M. P.; Strouse, C. E. *J. Am. Chem. Soc.* **1981**, *103*, 2633.

[71] Ullrich, V.; Nastainczyk, W.; Ruf, H. H. *Biochem. Soc. Trans.* **1975**, *3*, 803.

[72] Nastainczyk, W.; Ruf, H. H.; Ullrich, V. *Chem. Biol. Interact.* **1976**, *14*, 251.

[73] Ullrich, V.; Ruf, H. H.; Wende, P. *Croat. Chem. Acta* **1977**, *49*, 213.

[74] Ullrich, V.; Sakurai, H.; Ruf, H. H. *Acta Biol. Med. Ger.* **1979**, *38*, 287.

[75] Sono, M.; Andersson, L. A.; Dawson, J. H. *J. Biol. Chem.* **1982**, *257*, 8308.

[76] Dawson, J. H.; Andersson, L. A.; Sono, M. *In* "Cytochrome P-450, Biochemistry, Biophysics and Environmental Implications"; Hietanen, E.; Laitinen, M.; Hänninen, O., Eds.; Elsevier: Amsterdam, 1982, p. 589.

[77] Illing, H. P. A. *Biochem. Soc. Trans.* **1978**, *6*, 89.

[78] White, R. E.; Coon, M. J. *J. Biol. Chem.* **1982**, *257*, 3073.

[79] Sakurai, H.; Shimomura, S.; Sugiura, Y.; Ishizu, K. *Chem. Pharm. Bull.* **1979**, *27*, 3022.

[80] Sakurai, H.; Yoshimura, T. *Inorg. Chim. Acta* **1981**, *56*, L49.

[81] Sakurai, H.; Yoshimura, T. *Inorg. Chim. Acta* **1982**, *66*, L25.

[82] Sakurai, H.; Hatayama, E.; Yoshimura, T.; Maeda, M.; Tamura, H.; Kawasaki, K. *Biochem. Biophys. Res. Commun.*, **1983**, *115*, 590.

[83] Andersson, L. A.; Dawson, J. H. *Xenobiotica* **1984**, *14*, 49.

[84] Yoshida, Y.; Imai, Y.; Hashimoto-Yutsudo, C. *J. Biochem.* **1982**, *91*, 1651.

[85] Budyka, M. F.; Khenkin, A. M.; Shteinman, A. A. *Biochem. Biophys. Res. Commun.* **1981**, *101*, 615.

[86] Okubo, S.; Nozawa, T.; Hatano, M. *Chem. Lett.* **1981**, *1625*.

[87] Battersby, A. R.; Howson, W.; Hamilton, A. D. *J. Chem. Soc. Chem. Commun.* **1982**, 1266.

[88] Jefcoate, C. R. E.; Gaylor, J. L. *J. Am. Chem. Soc.* **1969**, *91*, 4611.

[89] Stern, J. O.; Peisach, J. *J. Biol. Chem.* **1974**, *249*, 7495.

[90] Mincey, T.; Traylor, T. G. *J. Am. Chem. Soc.* **1979**, *101*, 765.

[91] Collman, J. P.; Groh, S. E. *J. Am. Chem. Soc.* **1982**, *104*, 1391.

[92] Ebel, R. E.; O'Keeffe, D. H.; Peterson, J. A. *FEBS Lett.* **1975**, *55*, 198.

[93] Chevion, M.; Peisach, J.; Blumberg, W. E. *J. Biol. Chem.* **1977**, *257*, 3637.

[94] Dawson, J. H.; Cramer, S. P. *FEBS Lett.* **1978**, *88*, 127.

[95] Dolphin, D.; James, B. R.; Welborn, C. *J. Mol. Catal.* **1980**, *7*, 201.

[96] Chang, C. K.; Dolphin, D. *J. Am. Chem. Soc.* **1976**, *98*, 1609.

[97] Sharrock, M.; Münck, E.; Debrunner, P. G.; Marshall, V.; Lipscomb, J. D.; Gunsalus, I. C. *Biochemistry* **1973**, *12*, 258.

[98] Champion, P. M.; Lipscomb, J. D.; Münck, E.; Debrunner, P.; Gunsalus, I. C. *Biochemistry* **1975**, *14*, 4151.

[99] Sharrock, M.; Debrunner, P. G.; Schulz, C.; Lipscomb, J. D.; Marshall, V.; Gunsalus, I. C. *Biochim. Biophys. Acta* **1976**, *420*, 8.

[100] Schappacher, M.; Ricard, L.; Weiss, R.; Montiel-Montoya, R.; Gonser, U.; Bill, E.; Trautwein, A. *Inorg. Chim. Acta* **1983**, *78*, L9.

[101] Cramer, S. P.; Hodgson, K. O. *Prog. Inorg. Chem.* **1979**, *25*, 1.

[102] Scheidt, W. R.; Reed, C. A. *Chem. Rev.* **1981**, *81*, 543.

[103] Cramer, S. P.; Dawson, J. H.; Hodgson, K. O.; Hager, L. P. *J. Am. Chem. Soc.* **1978**, *100*, 7282.

[104] Dawson, J. H.; Davis, I. M.; Andersson, L. A.; Hahn, J. E. *In* "Biochemistry, Biophysics and Regulation of Cytochrome P-450"; Gustafsson, J.-A.; Carlstedt-Duke, J.; Mode, A.; Rafter, J., Eds.; Elsevier: Amsterdam, 1980, p. 565.

[105] Dawson, J. H.; Andersson, L. A.; Hodgson, K. O.; Hahn, J. E. *In* "Cytochrome P-450, Biochemistry, Biophysics and Environmental Implications"; Hietanen, E.; Laitinen, M.; Hänninen, O., eds.; Elsevier: Amsterdam, 1982, p. 523.

[106] Hahn, J. E.; Hodgson, K. O.; Andersson, L. A.; Dawson, J. H. *J. Biol. Chem.* **1982**, *257*, 10934.

[107] Andersson, L. A.; Sono, M.; Dawson, J. H. *Biochim. Biophys. Acta* **1983**, *748*, 341.

[108] Dooley, D. M.; Dawson, J. H. *Coord. Chem. Rev.* **1984**, *60*, 1.

[109] Dawson, J. H.; Dooley, D. M. *In* "Iron Porphyrins"; Part III; Lever, A. B. P.; Gray, H. B., Eds.; Addison-Wesley: Reading, Mass., 1985, in press.

[110] Shimizu, T.; Nozawa, T.; Hatano, M.; Imai, Y.; Sato, R. *Biochemistry* **1975**, *14*, 4172.

[111] Shimizu, T.; Nozawa, T.; Hatano, M. *Biochim. Biophys. Acta* **1976**, *434*, 126.

[112] Dawson, J. H.; Andersson, L. A.; Sono, M.; Gadecki, S. E.; Davis, I. M.; Nardo, J. V.; Svastits, E. W. *In* "Coordination Chemistry of Metalloenzymes"; Bertini, I., Drago, R. S.; Luchinat, C., Eds.; Reidell: Dordrecht, 1983, p. 369.

[113] Shimizu, T.; Iizuka, T.; Shimada, H.; Ishimura, Y.; Nozawa, T.; Hatano, M. *Biochim. Biophys. Acta* **1981**, *670*, 341.

[114] Vickery, L.; Salmon, A.; Sauer, K. *Biochim. Biophys. Acta* **1975**, *386*, 87.

[115] Dawson, J. H.; Trudell, J. R.; Linder, R. E.; Barth, G.; Bunnenberg, E.; Djerassi, C. *Biochemistry* **1978**, *17*, 33.

[116] Sorrell, T. N.; Ph.D. thesis, Stanford University, 1977; Dawson, J. H.; Sorrell, T. N.; Collman, J. P. Unpublished results.

[117] Vickery, L.; Nozawa, T.; Sauer, K. *J. Am. Chem. Soc.* **1976**, *98*, 345.

[118] O'Keeffe, D. H.; Ebel, R. E.; Peterson, J. A.; Maxwell, J. C.; Caughey, W. S. *Biochemistry* **1978**, *17*, 5845.

[119] Böhm, S.; Rein, H.; Jänig, G.-R.; Ruckpaul, K. *Acta Biol. Med. Ger.* **1976**, *35*, K27.

[120] Rein, H.; Böhm, S.; Jänig, G.-R.; Ruckpaul, K. *Croat. Chem. Acta* **1977**, *49*, 333.

[121] Collman, J. P.; Brauman, J. I.; Halbert, T. R.; Suslick, K. S. *Proc. Natl. Acad. Sci. U.S.A.* **1976**, *73*, 3333.

[122] Spiro, T. G. *In* "Iron Porphyrins," Part II; Lever, A. B. P.; Gray, H. B., Eds.; Addison-Wesley: Reading, Mass, 1983, p. 89.

[123] Ozaki, Y.; Kitagawa, T.; Kyogoku, Y.; Shimada, H.; Iizuka, T.; Ishimura, Y. *J. Biochem.* **1976**, *80*, 1447.

[124] Champion, P. M.; Gunsalus, I. C. *J. Am. Chem. Soc.* **1977**, *99*, 2000.

[125] Champion, P. M.; Gunsalus, I. C.; Wagner, G. C. *J. Am. Chem. Soc.* **1978**, *100*, 3743.

[126] Ozaki, Y.; Kitagawa, T.; Kyogoku, Y.; Imai, Y.; Hashimoto-Yutsudo, C.; Sato, R. *Biochemistry* **1978**, *17*, 5826.

[127] Anzenbacher, P.; Sípal, Z.; Chlumsky, J.; Strauch, B. *Stud. Biophys.* **1980**, *78*, 73.

[128] Shimizu, T.; Kitagawa, T.; Mitani, F.; Iizuka, T.; Ishimura, Y. *Biochim. Biophys. Acta* **1981**, *670*, 236.

[129] Champion, P. M.; Stallard, B. R.; Wagner, G. C.; Gunsalus, I. C. *J. Am. Chem. Soc.* **1982**, *104*, 5469.

[130] Anzenbacher, P.; Sípal, Z.; Strauch, B.; Twardowski, J.; Proniewicz, L. M. *J. Am. Chem. Soc.* **1981**, *103*, 5928.

[131] Anzenbacher, P.; Sípal, Z.; Strauch, B.; Twardowski, J. *In* "Cytochrome P-450, Biochemistry, Biophysics and Environmental Implications"; Hietanen, E.; Laitinen, M.; Hänninen, O., Eds.; Elsevier: Amsterdam, 1982, p. 735.

[132] Anzenbacher, P.; Sípal, Z.; Strauch, B.; Twardowski, J. *In* "Raman Spectroscopy: Linear and Nonlinear"; Lascombe, J.; Huong, P. V., Eds.; Wiley: New York, 1982, p. 735.

[133a] Chottard, G.; Schappacher, M.; Ricard, L.; Weiss, R. *Inorg. Chim. Acta* **1983**, *79*, 103.

[133b] Chottard, G.; Schappacher, M.; Ricard, L.; Weiss, R. *Inorg. Chem.* **1984**, *23*, 4557.

[134] Burke, J. M.; Kincaid, J. R.; Peters, S.; Gagne, R. R.; Collman, J. P.; Spiro, T. G. *J. Am. Chem. Soc.* **1978**, *100*, 6083.

[135] Chottard, G.; Battioni, P.; Battioni, J.-P.; Lange, M.; Mansuy, D. *Inorg. Chem.* **1981**, *20*, 1718.

[136] Choi, S.; Spiro, T. G.; Langry, K. C.; Smith, K. M.; Budd, D. L.; La Mar, G. N. *J. Am. Chem. Soc.* **1982**, *104*, 4345.

[137] Spiro, T. G.; Burke, J. M. *J. Am. Chem. Soc.* **1976**, *98*, 5483.

[138] Rimai, L.; Salmeen, I. T.; Petering, D. H. *Biochemistry* **1975**, *14*, 378.

[139] Sono, M.; Dawson, J. H. *J. Biol. Chem.* **1982**, *257*, 5496.

[140] Dawson, J. H.; Trudell, J. R.; Barth, G.; Linder, R. E.; Bunnenberg, E.; Djerassi, C.; Chiang, R.; Hager, L. P. *J. Am. Chem. Soc.* **1976**, *98*, 3709.

[141] Dawson, J. H.; Sono, M.; Hager, L. P. *Inorg. Chim. Acta: Bioinorg. Sect.* **1983**, *79*, 184.

[142] Sono, M.; Dawson, J. H.; Hager, L. P. *J. Biol. Chem.* **1984**, *259*, 13209.

[143] Desbois, A.: Lutz, M.; Banerjee, R. *Biochemistry* **1979**, *18*, 1510.

[144] Remmer, H; Schenkman, J.; Estabrook, R. W.; Sasame, H.; Gillette, J.; Narasimhulu, S.; Cooper, D. Y.; Rosenthal, O. *Mol. Pharmacol.* **1966**, *2*, 187.

[145] Schenkman, J. B.; Remmer, H.; Estabrook, R. W. *Mol Pharmacol.* **1967**, *3*, 113.

[146] Gunsalus, I. C. *Hoppe-Seylers Z. Physiol. Chem.* **1968**, *349*, 1610.

[147] Lu, A. Y. H.; Junk, K. W.; Coon, M. J. *J. Biol. Chem.* **1969**, *244*, 3714.

[148] Hildebrandt, A.; Remmer, H.; Estabrook, R. W. *Biochem. Biophys. Res. Commun.* **1968**, *30*, 607.

[149] Cammer, W.; Schenkman, J. B.; Estabrook, R. W. *Biochem. Biophys. Res. Commun.* **1966**, *23*, 264.

[150] Whysner, J. A.; Ramseyer, J.; Kazmi, G. M.; Harding, B. W. *Biochem. Biophys. Res. Commun.* **1969**, *36*, 795.

[151] Mitani, F.; Horie, S. *J. Biochem.* **1969**, *66*, 139.

[152] Tsai, R.; Yu, C. A.; Gunsalus, I. C.; Peisach, J.; Blumberg, W.; Orme-Johnson, W. H.; Beinert, H. *Proc. Natl. Acad. Sci. U.S.A.* **1970**, *66*, 1157.

[153] Peisach, J.; Blumberg, W. E. *Proc. Natl. Acad. Sci. U.S.A.* **1970**, *67*, 172.

[154] Peisach, J.; Blumberg, W. E. *Adv. Chem. Ser.* **1971**, *100*, 271.

[155] Blumberg, W. E.; Peisach, J. *In* "Probes of Structure and Function of Macromolecules and Membranes"; Chance, B.; Lee, C. P.; Yonetani, T., Eds.; Academic Press: New York, 1971, p. 215.

[156] Ebel, R. E.; O'Keeffe, D. H.; Peterson, J. A. *Arch. Biochem. Biophys.* **1977**, *183*, 317.

[157] Sato, M.; Kon, H. *Inorg. Chem.* **1975**, *14*, 2016

[158] Sato, M.; Kon, H.; Kumaki, K.; Nebert, D. W. *Biochim. Biophys. Acta* **1977**, *498*, 403.

[159] Miyake, Y.; Gaylor, J. L.; Mason, H. S. *J. Biol. Chem.* **1968**, *243*, 5788.

[160] O'Keeffe, D. H.; Ebel, R. E.; Peterson, J. A. *J. Biol. Chem.* **1978**, *253*, 3509.
[161] LoBrutto, R.; Scholes, C. P.; Wagner, G. C.; Gunsalus, I. C.; Debrunner, P. G. *J. Am. Chem. Soc.* **1980**, *102*, 1167.
[162] Peisach, J.; Mims, W. B.; Davis, J. L. *J. Biol. Chem.* **1979**, *254*, 12379.
[163] Groh, S. E.; Nagahisa, A.; Tan, S. L.; Orme-Johnson, W. H. *J. Am. Chem. Soc.* **1983**, *105*, 7445.
[164] Ristau, O.; Rein, H.; Jänig, G.-R.; Ruckpaul, K. *Biochim. Biophys. Acta* **1978**, *536*, 226.
[165] Griffin, B. W.; Peterson, J. A. *J. Biol. Chem.* **1975**, *250*, 6445.
[166] Philson, S. B.; Debrunner, P. G.; Schmidt, P. G.; Gunsalus, I. C. *J. Biol. Chem.* **1979**, *254*, 10173.
[167] Grasdalen, H.; Eriksson, L. E. G.; Ehrenberg, A.; Bäckström, D. *Biochim. Biophys. Acta* **1978**, *541*, 521.
[168] Keller, R. M.; Wüthrich, K.; Debrunner, P. G. *Proc. Natl. Acad. Sci. U.S.A.* **1972**, *69*, 2073.
[169] Parmely, R. C.; Goff, H. M. *J. Inorg. Biochem.* **1980**, *12*, 269.
[170] Berzinis, A. P.; Traylor, T. G. *Biochem. Biophys. Res. Commun.* **1979**, *87*, 229.
[171] Matwiyoff, N. A.; Philson, S. B.; Gunsalus, I. C.; Debrunner, P. R. Unpublished results cited in footnote 18 of ref. [170].
[172] Champion, P. M.; Münck, E.; Debrunner, P. G.; Moss, T. H.; Lipscomb, J. D.; Gunsalus, I. C. *Biochim. Biophys. Acta* **1975**, *376*, 579.
[173] Jänig, G.-R.; Dettmer, R.; Usanov, S. A.; Ruckpaul, K. *FEBS Lett.* **1983**, *159*, 58.
[174] Jänig, G.-R.; Makower, A.; Rabe, H.; Bernhardt, R.; Ruckpaul, K. *Biochim. Biophys. Acta* **1984**, *787*, 8.
[175] Dus, K. M. *In* "From Cyclotrons to Cytochromes"; Kaplan, N. O.; Robinson, A. B., Eds.; Academic Press: New York, 1982, p. 231.
[176] Hanson, L. K.; Sligar, S. G.; Gunsalus, I. C. *Croat. Chem. Acta* **1977**, *49*, 237.
[177] Hanson, L. K. *Int. J. Quant. Chem. Quant. Biol. Symp.* **1979**, *6*, 73.
[178] Jung, C.; Ristau, O. *Chem. Phys. Lett.* **1977**, *49*, 103.
[179] Jung, C.; Friedrich, J.; Ristau, O. *Acta Biol. Med Ger.* **1979**, *38*, 363.
[180] Loew, G. H.; Rohmer, M.-M. *J. Am. Chem. Soc.* **1980**, *102*, 3655.
[181] Loew, G. H. *In* "Iron Porphyrins", Part I; Lever, A. B. P.; Gray, H. B., Eds.; Addison-Wesley: Reading, Mass., 1983, p. 1.
[182] Makinen, M. W.; Churg, A. K. *In* "Iron Porphyrins", Part I; Lever, A. B. P.; Gray, H. B., Eds.; Addison-Wesley: Reading, Mass., 1983, p. 141.
[183] Wagner, G. C.; Gunsalus, I. C.; Wang, M.-Y. R.; Hoffman, B. M. *J. Biol. Chem.* **1981**, *256*, 6266.
[184] Doppelt, O.; Weiss, R. *Nouv. J. Chim.* **1983**, *7*, 341.
[185] Doppelt, P.; Fischer, J.; Weiss, R. *J. Am. Chem. Soc.* **1984**, *106*, 5188.
[186] Schenkman, J. B.; Sligar, S. G.; Cinti, D. L. *Pharmacol. Ther.* **1981**, *12*, 43.
[187] Gibson, G. G.; Tamburini, P. P. *Xenobiotica* **1984**, *14*, 27.
[188] Lewis, B. A.; Sligar, S. G. *J. Biol. Chem.* **1983**, *258*, 3599.
[189] Sligar, S. G. *Biochemistry* **1976**, *15*, 5399.
[190] Sligar, S. G.; Gunsalus, I. C. *Proc. Natl. Acad. Sci. U.S.A.* **1976**, *73*, 1078.
[191] Cole, P. E.; Sligar, S. G. *FEBS Lett.* **1981**, *133*, 252.
[192] Ziegler, M.; Blanck, J.; Ruckpaul K. *FEBS Lett.* **1982**, *150*, 219.
[193] Lange, R.; Hui Bon Hoa, G.; Debey, P.; Gunsalus, I. C. *Eur. J. Biochem.* **1977**, *77*, 479.
[194] Lange, R.; Debey, P. *Eur, J. Biochem.* **1979**, *94*, 485.
[195] Lange, R.; Hui Bon Hoa, G.; Debey, P.; Gunsalus, I. C. *Eur. J. Biochem.* **1979**, *94*, 491.
[196] Lange, R.; Pierre, J.; Debey, P. *Eur. J. Biochem.* **1980**, *107*, 441.
[197] Marden, M. C.; Hui Bon Hoa, G. *Eur. J. Biochem.* **1982**, *129*, 111.
[198] Hui Bon Hoa, G.; Marden, M. C. *Eur. J. Biochem.* **1982**, *124*, 311.

[199] Sligar, S. G.; Cinti, D. L.; Gibson, G.; Schenkman, J. B. *Biochem. Biophys. Res. Commun.* **1979**, *90*, 925.

[200] Sligar, S. G.; Debrunner, P. G.; Lipscomb, J. D.; Namtvedt, M. J.; Gunsalus, I. C. *Proc. Natl. Acad. Sci. U.S.A.* **1974**, *71*, 3906.

[201] Pederson, T. C.; Austin, R. H.; Gunsalus, I. C. *In* "Microsomes and Drug Oxidations"; Ullrich, V.; Roots, I.; Hildebrandt, A.; Estabrook, R. W.; Conney, A. H., Eds.; Pergamon: Oxford, 1977, p. 275.

[202] Peterson, J. A.; Mock, D. M. *Acta Biol. Med. Ger.* **1979**, *38*, 153.

[203] Hintz, M. J.; Peterson, J. A. *J. Biol. Chem.* **1981**, *256*, 6721. Hintz, M. J.; Mock, D. M.; Peterson, L. L.; Tuttle, K.; Peterson, J. A. *J. Biol. Chem.* **1982**, *257*, 14324.

[204] Pederson, T. C.; Gunsalus, I. C. *Fed Proc. Fed. Am. Soc. Exp. Biol.* **1977**, *36*, 663.

[205] Backes, W. L.; Sligar, S. G.; Schenkman, J. B. *Biochem. Biophys. Res. Commun.* **1980**, *97*, 860.

[206] Backes, W. L.; Sligar, S. G.; Schenkman, J. B. *Biochemistry* **1982**, *21*, 1324.

[207] Tamburini, P. P.; Gibson, G. G.; Backes, W. L.; Sligar, S. G.; Schenkman, J. B. *Biochemistry* **1984**, *23*, 4526.

[208] Gould, P. V.; Gelb, M. H.; Sligar, S. G. *J. Biol. Chem.* **1981**, *256*, 6686.

[209] Hui Bon Hoa, G.; Begard, E.; Debey, P.; Gunsalus, I. C. *Biochemistry* **1978**, *17*, 2835.

[210] Hintz, M. J.; Peterson, J. A. *J. Biol. Chem.* **1980**, *255*, 7317.

[211] Bazin, M.; Pierre, J.; Debey, P.; Santus, R. *Eur J. Biochem.* **1981**, *124*, 539.

[212] Estabrook, R. W.; Hildebrandt, A. G.; Baron, J.; Netter, K.; Labman, K. *Biochem. Biophys. Res. Commun.* **1971**, *42*, 132.

[213] Gunsalus, I. C. Paper presented at the Wenner-Gren Symposium on The Structure and Function of Oxidation Reduction Enzymes, Stockholm, 1970; cited in reference 11.

[214] Peterson, J. A.; Ishimura, Y.; Griffin, B. W. *Arch. Biochem. Biophys.* **1972**, *149*, 197.

[215] Peterson, J. A.; Ishimura, Y. *Chem. Biol. Interact.* **1971**, *3*, 300.

[216] Debey, P.; Balny, C.; Douzou, P. *FEBS Lett.* **1976**, *69*, 231.

[217] Balny, C.; Debey, P.; Douzou, P. *FEBS Lett.* **1976**, *69*, 236.

[218] Bonfils, C.; Debey, P.; Maurel, P. *Biochem. Biophys. Res. Commun.* **1979**, *88*, 1301.

[219] Larroque, C.; VanLier, J. E. *FEBS Lett.* **1980**, *115*, 175.

[220] Oprian, D. D.; Gorsky, L. D.; Coon, M. J. *J. Biol. Chem.* **1983**, *258*, 8684.

[221] Sligar, S. G.; Lipscomb, J. D.; Debrunner, P. G.; Gunsalus, I. C. *Biochem. Biophys. Res. Commun.* **1974**, *61*, 290.

[222] Bonfils, C.; Andersson, K. K.; Maurel, P.; Debey, P. *J. Mol. Catal.* **1980**, *7*, 299.

[223] Estabrook, R. W.; Kawano, S.; Werringloer, J.; Kuthan, H.; Tsuji, H.; Graf, H.; Ullrich, V. *Acta Biol. Med. Ger.* **1979**, *38*, 423.

[224] Kuthan, H.; Ullrich, V. *Eur. J. Biochem.* **1982**, *126*, 583.

[225] Tyson, C. A.; Lipscomb, J. D.; Gunsalus, I. C. *J. Biol. Chem.* **1972**, *247*, 5777.

[226] McCandlish, E.; Miksztal, A. R.; Nappa, M.; Sprenger, A. Q.; Valentine, J. S.; Stong, J. D.; Spiro, T. G. *J. Am. Chem. Soc.* **1980**, *102*, 4268.

[227] Welborn, C.; Dolphin, D.; James, B. R. *J. Am. Chem. Soc.* **1981**, *103*, 2869. The potentials reported by Welborn, Dolphin, and James [227] were measured vs. the Ag/AgCl electrode and have been converted to potentials vs. the normal hydrogen electrode at pH 7.0 (E_0') by addition of 0.20 V (Bard, A. J.; Faulkner, L. R. "Electrochemical Methods"; Wiley: New York, 1980, Fig. E.1).

[228] Reed, C. A. *Adv. Chem. Ser.* **1982**, *201*, 333.

[229] Lipscomb, J. D.; Sligar, S. G.; Namtvedt, M. J.; Gunsalus, I. C. *J. Biol. Chem.* **1976**, *251*, 1116.

[230] Gunsalus, I. C.; Meeks, J. R.; Lipscomb, J. D. *Ann N.Y. Acad. Sci.* **1973**, *212*, 107.

[231] Eble, K. S.; Dawson, J. H. *Biochemistry* **1984**, *23*, 2068.

[232] Nordbloom, G. D.; Coon, M. J. *Arch. Biochem. Biophys.* **1977**, *180*, 343.

[233] Gorsky, L. D.; Koop, D. R.; Coon, M. J. *J. Biol. Chem.* **1984**, *259*, 6812.

[234] Hrycay, E. G.; O'Brien, P. J. *Arch. Biochem. Biophys.* **1972**, *153*, 480.

[235] Sligar, S. G.; Kennedy, K. A.; Pearson, D. C. *Proc. Natl. Acad. Sci U.S.A.* **1980**, *77*, 1240.

[236] Wagner, G. C.; Palcic, M.M.; Dunford, H. B. *FEBS Lett.* **1983**, *156*, 244.

[237] Lichtenberger, F.; Ullrich, V. *In* "Microsomes and Drug Oxidations"; Ullrich, V.; Roots, I.; Hildebrandt, A.; Estabrook, R. W.; Conney, A. H., Eds.; Pergamon: Oxford, 1977, p. 218. Blake, R. C., II; Coon, M. J. *J. Biol. Chem.* **1980**, *255*, 4100. Blake, R. C., II; Coon, M. J. *J. Biol. Chem.* **1981**, *256*, 5755. Coon, M. J.; White, R. E.; Blake, R. C., II. *In* "Oxidases and Related Redox Systems"; King, T. E.; Morrison, M.; Mason, H. S., Eds.; Pergamon: Oxford, 1982, p. 857.

[238] Sligar, S. G.; Shastry, B. S.; Gunsalus, I. C. *In* "Microsomes and Drug Oxidations"; Ullrich, V.; Roots, I.; Hildebrandt, A.; Estabrook, R. W.; Cooney, A. H., Eds.; Pergamon: Oxford, 1977, p. 202.

[239] McCarthy, M.-B.; White R. E. *J. Biol. Chem.* **1983**, *258*, 9153.

[240] White, R. E.; Sligar, S. G.; Coon, M. J. *J. Biol. Chem.* **1980**, *255*, 11108.

[241] McCarthy, M.-B.; White, R. E. *J. Biol. Chem.* **1983**, *258*, 11610.

[242] Guengerich, F. P.; Ballou, D. P.; Coon, M. J. *Biochem. Biophys. Res. Comm.* **1976**, *70*, 951.

[243] Hamilton, G. *Adv. Enzymol.* **1969**, *32*, 55.

[244] Sono, M.; Andersson, L. A.; Dawson, J. H. *J. Biol. Chem.* **1982**, *257*, 8308.

[245] Sligar, S. G.; Gelb, M. H.; Heimbrook, D. C. *Xenobiotica* **1984**, *14*, 63.

[246] Heimbrook, D. C.; Sligar, S. G. *Biochem. Biophys. Res. Commun.* **1981**, *99*, 530.

[247] Hjelmeland, L. M.; Aronow, L.; Trudell, J. R. *Biochem. Biophys. Res. Commun.* **1977**, *76*, 541.

[248] Daly, J. *Handb. Exp. Pharmacol.* **1971**, *28*, 285.

[249] Groves, J. T.; McClusky, G. A.; White, R. E.; Coon, M. J. *Biochem. Biophys. Res. Commun.* **1978**, *81*, 154.

[250] White, R. E.; Groves, J. T.; McClusky, G. A. *Acta Biol. Med. Ger.* **1979**, *38*, 475.

[251] Groves, J. T.; Subramanian, D. S. *J. Am. Chem. Soc.* **1984**, *106*, 2177; and references therein.

[252] Smegal, J. A.; Hill, C. L. *J. Am. Chem. Soc.* **1983**, *105*, 3515. Suslick, K. S.; Cook, B. R.; Cook, B. R.; Fox, M. M. *J. Chem. Soc. Chem. Comm.* **1985**, 580.

[253] Gelb, M. H.; Heimbrook, D. C.; Mälkönen, P.; Sligar, S. G. *Biochemistry* **1982**, *21*, 370.

[254] Gelb, M. H.; Mälkönen, P.; Sligar, S. G. *Biochem. Biophys. Res. Commun.* **1982**, *104*, 853.

[255] Eble, K. S.; Dawson, J. H. *J. Biol. Chem.* **1984**, *259*, 14389.

[256] White, R. E.; McCarthy, M.-B.; Egeberg, K. D.; Sligar, S. G. *Arch. Biochem. Biophys.* **1984**, *228*, 493.

Reaction Mechanisms of the Halogens and Oxohalogen Species in Acidic, Aqueous Solution

Richard C. Thompson

Department of Chemistry
University of Missouri
Columbia, Missouri, USA

I. INTRODUCTION

The heavier halogens and several salts of oxohalogen anions have been widely used by chemists for over a century. Some of the early kinetic studies, notably by Bray's group at Berkeley, established the rich chemistry associated with the electron transfer reactions of these reagents. More recently, our knowledge of the mechanistic features has been increased greatly not only by the expanded data base but also through the use of additional experimental techniques such as rapid kinetics instrumentation and both radioactive and stable isotopic tracer studies. There has been a decline in emphasis on main group chemistry during the past two decades, especially in the United States. However, the discovery of a number of oscillating reactions in which the halogens and oxohalogen species play a fundamental role has revived interest on Group VII chemistry enormously.[1] Even aqueous fluorine chemistry is receiving considerable attention, largely due to its potential as a rapid and simple fluorinating agent in nuclear medicine.

It therefore appears appropriate to review the mechanistic information on the redox chemistry of the halogens and oxohalogen species. The scope is very large, and the discussion is restricted to reactions in acidic, aqueous solutions. The review is not comprehensive, in that not all relevant contributions will be discussed, and it is necessary to apologize in advance to those investigators whose work is not referenced, either as a result of our ignorance or prejudice or because of space restrictions. Wherever possible general features have been emphasized, but it must be recognized that these correlations may not stand the test of time.

II. EXPERIMENTAL TECHNIQUES

A. Synthesis and handling procedures

The synthesis of those Group VIIA compounds that are not available in high purity from commercial sources is summarized here. In addition, some safety precautions for the highly reactive or unstable species, as well as some useful handling techniques, are noted.

F_2. Proper safety precautions should be observed, and the user should carefully read the data sheets supplied by the distributor. However, elaborate equipment and unusual ventillation are not required when using diluted ($\leq 20\%$) fluorine, a concentration adequate for many purposes. A specially designed regulator for the fluorine cylinder must be used, but the rest of the plumbing can be fabricated from copper and/or fluorinated plastic tubing and fittings. Glass reaction vessels may be used. The commercial gas contains small amounts of hydrogen fluoride which will etch glass flowmeters.

HOF.   The synthesis and properties of this compound have been reviewed by Appelman.[2] It appears to have been prepared only in his laboratory at Argonne. Small quantities (≤ 1 mM) pose no unusual safety hazards but should be used soon after preparation. We have found that thawing and delivery of HOF by means of an inert gas flow should be down below $0°C$; occasional detonations occurred at room temperature.

Cl_2 solutions.   It is difficult to transfer reproducibly chlorine solutions due to its volatility. Also, the concentration of a stock solution contained in a stoppered vessel will diminish as aliquots are withdrawn due to the increasing gas space. The "shrinking bottle" technique[3] overcomes these difficulties and is suitable for many applications. Solutions introduced into the syringe assembly of most stopped-flow instruments will retain a constant concentration, but that concentration must be determined *in situ*. We have found[4] that storage and quantitative delivery of chlorine solutions are facilitated by the use of equimolar mixtures of OCl^- and Cl^- in excess base. These solutions may be prepared by bubbling Cl_2 into cold, excess aqueous base and are stable for a day or so when refrigerated. The desired chlorine concentration is formed very rapidly when aliquots are injected into acidic reaction solutions; the reaction vessel (often a spectrophotometer cell) should then be stoppered and should contain little gas space.

HOCl, HOBr, and H_2OI^+.   These hypohalous acids have limited stability in acidic aqueous solutions. Their synthesis has been described recently,[5] and a number of useful, earlier references are cited. Hypochlorous acid solutions may be prepared as required by extraction of Cl_2O in CCl_4. The preparation of the latter solution has been described[6]; they may be stored indefinitely at low temperatures.

Salts and solutions of ClO_2^- and BrO_2^-.   Crystalline $NaClO_2$ is available commercially, and may be purified by proper recrystallization.[7] Crystalline $NaBrO_2$ is apparently no longer available commercially. The synthesis is rather difficult if pure salts are desired.[8a,b] The preparation of fairly pure $Ba(BrO_2)_2$ has been described recently.[9] Basic solutions of BrO_2^- and especially ClO_2^- are reasonably stable and may be used as a source of $HBrO_2$ or $HClO_2$ by injection into an acidic reaction solution. The instability of the halous acids due to disproportionation will be discussed in Section VII,A.

Solutions of IO^+, H_2OIO^+, and HIO_2.   Moderately stable solutions of IO^+ may be prepared by the reaction of I_2 and IO_3^- in concentrated H_2SO_4.[9] Substantial dilution yields H_2OIO^+ and HIO_2; their limited stability has been studied.[9]

ClO_2 and BrO_2. Several syntheses of ClO_2 have been described.[10-12] Acidic solutions are quite stable if protected from light. Volatility of ClO_2 from stock solutions and during transfer is only a minor problem due to its substantial solubility. A successful synthesis of pure BrO_2 has not been described and attempts are to be discouraged since aqueous solutions are quite unstable. However, these solutions may be prepared *in situ* by radiolysis or photolysis of bromate[13-16] or by the reaction of bromate ion and bromous acid.[8,17] The results of some studies using these techniques will be discussed later.

$KBrO_4$. The synthetic method of choice is the oxidation of bromate ion in strongly basic solution by fluorine. The isolation procedure for pure, crystalline $KBrO_4$ has been fully described.[18] The method is tedious but has been used successfully in a number of laboratories. The solid is stable indefinitely.

TABLE 1

Absorption maxima and molar extinction coefficients of halogen and oxohalogen species in acidic solution

Ion or molecule	λ_{max} (nm)	$\varepsilon\,(M^{-1}\,cm^{-1})$	Ref.
Cl_2	325	175	[19]
Cl_2^-	340	10^4	[20]
Cl_3^-	325	193	[19]
HOCl	325^a	11.0	[21]
	235	100	[21]
Br_2	395	177	[22]
Br_2^-	360	10^4	[20]
Br_3^-	265	36,100	[23]
Br_2, Br_3^- isosbestic	457	98	[24]
HOBr	265	87	[22]
I_2	460	975	[25]
	270	17,200	[25]
I_2^-	380	10^4	[20]
I_3^-	353	26,400	[25]
	288	40,000	[25]
I_2, I_3^- isosbestic	469	730	[26]
ClO_2^-	260	176	[27]
ClO_2	359	1,250	[10]
BrO_2^b	475	10^3	[13, 14]

^{a} Not an absorption maximum.
^{b} Br_2O_4 transparent at this wavelength.

B. Absorption spectra

Spectrophotometric techniques have been used to monitor most of the reactions we shall discuss. The absorption maxima and molar extinction coefficients for those halogen and oxohalogen species that exhibit useful spectral features in acidic solutions are summarized in Table 1. Readers are cautioned that other values of ε have been reported; for careful work the values should be determined by the investigator.

III. COMPLEXATION OF OXOHALOGEN SPECIES

A. Inner-sphere metal ion complexes

Very few inner-sphere metal ion complexes containing oxohalogen ligands have been observed in aqueous solution. This is not surprising since the perhalate and halate ions are poor donors, and the halite and hypohalite ions are extensively protonated and have limited stability in acidic solution. Substitution at the halogen center by an aquo ligand on the metal ion has been observed only for the labile iodato ion.[28] A full kinetic study of the iodato complexes cis-(en)$_2$Co(OH$_2$)(OIO$_2$)$^{2+}$ and (NH$_3$)$_5$Cr(OIO$_2$)$^{2+}$ has been reported.[29] The formation constants at 25°C, $I = 1.0$ M are 2.3 and 11 M^{-1}, respectively. The rapid substitution rates were monitored by the temperature-jump technique.

A few authentic perchlorato complexes are listed in Table 2. They are all relatively labile. Salts of (NH$_3$)$_5$Co(OClO$_3$)$^{2+}$ have been isolated and are formed in concentrated HClO$_4$ by nitrosation of (NH$_3$)$_5$Co(N$_3$)$^{2+}$ with a mixture of dry NO and NO$_2$.[30] An earlier report of the synthesis of (NH$_3$)$_5$Co(OClO$_3$)$^{2+}$ is apparently erroneous.[30,32,33]

The synthesis of one chlorato complex has been described.[37] The ion (H$_2$O)$_5$Cr(OClO$_2$)$^{2+}$ is formed in good yield by the chromium(VI) oxidation of chlorous acid and can be isolated in aqueous solution by ion exchange techniques. The aquation rate ($k = 4.3 \times 10^{-5}$ s^{-1} + 1.8 $\times 10^{-6}$ M s^{-1}/[H$^+$] at 25°C and $I = 2.0$ M) is nearly identical to that of (H$_2$O)$_5$Cr(ONO$_2$)$^{2+}$. The chlorato complex is probably formed by a two-equivalent oxidation of chlorous acid by a chromium(V) intermediate.

Salts of (NH$_3$)$_5$Co(OClO)$^{2+}$ are formed by oxidation of cobalt(II) in an ammonia–ammonium ion buffer with gaseous ClO$_2$.[38] The chlorito complex

TABLE 2

Perchlorato complexes studied in aqueous solution

Complex	Aquation rate	Temp. (°C)	Ref.
$(NH_3)_5Co(OClO_3)^{2+}$	0.1 s^{-1} (complex)	25	[30]
$(H_2O)_5Cr(OClO_3)^{2+}$	$4.8 \times 10^{-3} \, M^{-1} \text{s}^{-1} \times$ (complex) a_{H_2O}	20	[34]
$(H_2O)_4Cr\overset{\displaystyle OH}{\underset{\displaystyle OH}{<\!\!>}}Cr(OH_2)_3(OClO_3)^{3+}$	a		[35]
$(en)_2Co\overset{\displaystyle NH_2}{\underset{\displaystyle ClO_4}{<\!\!>}}Co(en)_2^{4+}$	$0.11 \, M^{-1} \text{s}^{-1}$ (complex) $\times$ $a_{H_2O}{}^b$	1.5	[36]

a Not measured; assignment of perchlorato complex tentative.
b For ring opening to form the μ-amido complex.

decomposes in acidic solution ($k = 8.4 \times 10^{-6} \text{ s}^{-1}$) via an internal electron transfer reaction to form Co^{2+} and ClO_2.

A weak, inner-sphere, labile complex between $ClO_2{}^-$ and $UO_2{}^{2+}$ has been observed spectrophotometrically.[39] Although it was not possible to uncouple the molar extinction coefficient and the formation constant, the latter was estimated to be approximately $0.02 \, M^{-1}$ at 25°C.

A number of weak, ion pair-type complexes have been reported; the reader is referred to a recent compilation for the thermodynamic data.[40]

B. Ion–molecule complexes

The brown color observed in rather concentrated solutions of chlorous acid during disproportionation is undoubtedly due to the complexation of $ClO_2{}^-$ by ClO_2. This $Cl_2O_4{}^-$ complex is stable in neutral solution and has a formation constant of $1.6 \, M^{-1}$ at 25°C.[41] Its absorption maximum is at 270 nm, with a molar extinction coefficient virtually identical to that of $ClO_2{}^-$. The complex is far less absorbing at 360 nm than is ClO_2.

A complex formulated as $ClO_2 \cdot I^-$ is rapidly formed when solutions containing ClO_2 and I^- are mixed.[42] It has a half-life of 7 s at 25°C; the decomposition products are $ClO_2{}^-$ and I_2. The complex is most conveniently observed ($\varepsilon = 565 \, M^{-1} \text{ cm}^{-1}$) at the $I_2, I_3{}^-$ isosbestic point at 468 nm.

The trihalide ions, $X_2X'^-$, are well known and their potential role in kinetic studies of the halogens in halide-containing solutions must be considered. The formation constants (M^{-1} at 25°C) are 769, I_3^-[43]; 17, Br_3^-[44]; 0.18, Cl_3^-[19]; and 1.14, Br_2Cl^-.[45] For $BrCl_2^-$, a value of $K = [BrCl_2^-][Br^-]/[Br_2][Cl^-]^2 = 7.2 \times 10^{-3}\ M^{-1}$ has been determined.[45]

IV. STANDARD REDUCTION POTENTIALS

Standard reduction potentials are useful guides to the aqueous chemistry of the elements in their various oxidation states. Reasonable values of these potentials have been summarized for the halogens in Table 3. Where feasible, estimates have been included for radical species that are likely intermediates in a number of electron transfer reactions to be discussed.

TABLE 3

Standard reduction potentials (volts) in acidic solution

$$F_2 \xrightarrow{\ 3.0_6\ } 2HF$$

$$OF_2 \xrightarrow{\ 2.3\ } 2HF$$

$$ClO_4^- \xrightarrow{1.22} ClO_3^- \xrightarrow{1.16} HClO_2 \xrightarrow{1.67} HOCl \xrightarrow{1.62} Cl_2 \xrightarrow{1.36} Cl^-$$

with branches: $ClO_3^- \xrightarrow{1.1_3} ClO_2 \xrightarrow{1.1_8} HClO_2$; $HOCl \xrightarrow{0.4} Cl\cdot \xrightarrow{2.6} Cl^-$; $HOCl \xrightarrow{0.4_2} Cl_2^- \xrightarrow{2.3} Cl^-$

$$BrO_4^- \xrightarrow{1.74} BrO_3^- \xrightarrow{1.24} HBrO_2 \xrightarrow{1.74} HOBr \xrightarrow{1.57} Br_2 \xrightarrow{1.09} Br^-$$

with branches: $BrO_3^- \xrightarrow{1.1_5} BrO_2 \xrightarrow{1.3_3} HBrO_2$; $HOBr \xrightarrow{0.6} Br\cdot \xrightarrow{2.1} Br^-$; $HOBr \xrightarrow{0.4_6} Br_2^- \xrightarrow{1.7} Br^-$

$$H_5IO_6 \xrightarrow{1.6_5} IO_3^- \xrightarrow{1.17} H_2OI^+ \xrightarrow{1.27} I_2 \xrightarrow{0.62} I^-$$

with branches: $H_2OI^+ \xrightarrow{0.5} I\cdot \xrightarrow{1.4} I^-$; $I_2 \xrightarrow{0.1_3} I_2^- \xrightarrow{1.1} I^-$

Values of the fluorine couples were taken from Latimer.[46] Aqueous free energies for the chlorine and iodine species with singlet ground states and for chlorine dioxide are from a recent NBS compilation.[47] The bromine potentials with the exception of BrO_4^-,[48] $Br\cdot$, and Br_2^- are those of Noyes.[49,50] We have defined the standard state for the halogens as a hypothetical, ideal 1 M solution. Iodine(I) has been formulated as H_2OI^+ since this form is thought to predominate over HOI at acidities greater than 0.1 M.[5,51,52] The X_2^- and $X\cdot$ values lie within the range estimated by several groups.[53-56]

V.   HALOGENS

A.   Hydrolysis

1. Cl_2, Br_2, and I_2

The thermodynamic and kinetic parameters associated with the hydrolysis (1) of the heavier halogens have been determined and are summarized in Table 4.

$$X_2 + H_2O \underset{k_r}{\overset{k_f}{\rightleftharpoons}} HOX + H^+ + X^- \quad (K_h) \tag{1}$$

The mechanism (2) for halogen hydrolysis in acidic solution proposed by Eigen and Kustin[59] is, taking chlorine as an example,

$$Cl_2 + H_2O \underset{k_{31}}{\overset{k_{13}}{\rightleftharpoons}} HOCl_2^- + H^+ \underset{k_{43}}{\overset{k_{34}}{\rightleftharpoons}} HOCl + H^+ + Cl^- \tag{2}$$

Proton transfer to the intermediate $HOCl_2^-$ is considered to be very rapid.[60]

A different conclusion has been drawn more recently.[61,62] Hurst proposed that proton transfer to $HOCl_2^-$ is relatively slow, and that H_2OCl_2 is instead the most reactive intermediate in the hydrolysis reaction (1) and its reverse. He identified H_2OCl_2 as the reactive oxidant toward the added substrates

TABLE 4

Hydrolysis constants and kinetic parameters for the heavier halogens

Halogen	K_h (M^2), 25°C	Ref.	k_f (s^{-1}), 20°C	Ref.
Cl_2	3.9×10^{-4}	[57]	11.0	[59]
Br_2	5.8×10^{-9}	[58]	110	[59]
I_2	5.4×10^{-13}	[52]^a	3.0	[59]

^{a} $K = 4.5 \times 10^{-2}$ M at 25°C for $H_2OI^+ \rightleftharpoons HOI + H^+$.

hydrogen peroxide and 2,5-dimethylfuran and proposed oxygen exchange between HOCl and H_2O[63] to occur by solvent displacement of OH^- on $HOCl_2{}^-$ as in (3).

$$\text{Cl}_2 + \text{H}_2\text{O} \underset{k_{31}}{\overset{k_{13}}{\rightleftharpoons}} \text{H}_2\text{OCl}_2 \rightleftharpoons \text{HOCl}_2{}^- + \text{H}^+ \underset{k_{43}}{\overset{k_{34}}{\rightleftharpoons}} \text{HOCl} + \text{H}^+ + \text{Cl}^- \tag{3}$$

with branches:
$\nearrow \text{H}_2\text{O} + \text{O}_2 + 2\text{H}^+ + 2\text{Cl}^-$ (via H_2O_2)
$\searrow$ DMFU oxid. products
$+\text{H}_2\text{O} \Updownarrow -\text{H}_2\text{O}$ oxygen exchange

The mechanistic details of chlorine hydrolysis are an open question at the present time. In any event, it is clear that in kinetic studies of oxidations by either Cl_2 or HOCl, both species as well as intermediates formed during the hydrolysis reaction must be considered as potential reactive forms of the oxidant. A similar problem may arise with Br_2 and I_2, although here the hydrolysis equilibrium is far less favorable.

2. F_2: The role of HOF as an intermediate

Moissan reported in 1900 that fluorine reacts vigorously with water.[64] A rate constant of 10^5 s^{-1} at 1°C for the irreversible hydrolysis has recently been determined.[65]

$$\text{F}_2 + \text{H}_2\text{O} \xrightarrow{k_h} \text{HOF} + \text{HF} \tag{4}$$

followed by

$$\text{HOF} + \text{H}_2\text{O} \xrightarrow{k_5} \text{H}_2\text{O}_2 + \text{HF} \tag{5}$$

Reaction (5) is also quite rapid, but the value of k_5 is not known. The experimental technique consisted of measurements of the rate of absorption of gaseous fluorine into aqueous solution. The value of the hydrolysis rate constant k_h was estimated from these data by means of the theory of mass transfer with chemical reaction.[66,67] This approach has been employed successfully by chemical engineers for years,[68,69] but has been little appreciated by chemists. The appreciable uncertainty in k_h arises largely from the required estimate of the solubility of fluorine in water.

Equations (4)–(7) summarize the principal reactions that occur when fluorine is continuously passed into neutral or acidic solutions free of oxidizable substrates.

$$\text{F}_2 + \text{H}_2\text{O}_2 \xrightarrow{k_6} \text{O}_2 + 2\text{HF} \tag{6}$$

$$\text{HOF} + \text{H}_2\text{O}_2 \xrightarrow{k_7} \text{O}_2 + \text{HF} + \text{H}_2\text{O} \tag{7}$$

At 1°C, $k_h/k_6 = 0.24$ M and $k_5/k_7 = 0.86$ M.[65] Oxygen transfer from HOF to H_2O is quantitative in reaction (5) and allows the synthesis of specifically labeled $H-O-O^*-H$ and $O-O^*$.[70] Substantial oxygen transfer from HOF to the dioxygen product in reaction (7) is found, a unique observation in the oxidation of hydrogen peroxide.[65]

While the hydrolysis reactions of F_2 and HOF are rapid, they are by no means "instantaneous." Therefore, in the presence of oxidizable substrates competition between hydrolysis and direct reaction with F_2, HOF, or both should be possible. In fact, it has been recognized for some time that fluorination of a number of substrates can be achieved in aqueous solution by treatment with fluorine.[71–73] Direct experiments with HOF have revealed that its principal mode of reaction is oxygen atom transfer, either to water or to an appropriate substrate.[70]

The possible reactions that can occur when fluorine is passed through an aqueous solution of an oxidizable substrate R are numerous. In addition to reactions (4)–(7), we must at least consider reactions (8)–(11).

$$F_2 + R \rightarrow R' \tag{8}$$

$$HOF + R \rightarrow R'' \tag{9}$$

$$R' + H_2O_2 \rightarrow R + O_2 \tag{10}$$

$$R'' + H_2O_2 \rightarrow R + O_2 \tag{11}$$

The oxidation products R' and R'' may be the same, in which case difficulty is encountered in establishing whether reaction (8), reaction (9), or both occurred. If reactions (10) and (11) are rapid, it may even be falsely concluded that the substrate is unreactive to both F_2 and HOF. Clearly, considerable caution must be exercised in mechanistic interpretations of this type of experiment.

Rather definitive results were obtained with a simple azido complex ion, $(H_2O)_5CrN_3^{2+}$.[74] When F_2 is passed through fairly concentrated solutions of $(H_2O)_5CrN_3^{2+}$, the principal chromium-containing products are $(H_2O)_6Cr^{3+}$ and $(H_2O)_5CrF^{2+}$. With lower substrate concentrations, the nitrosyl complex $(H_2O)_5CrNO^{2+}$ is also formed; it is the major product when HOF is used instead of F_2. Since none of the three chromium(III) products are reactive toward H_2O_2, F_2, or HOF, mechanistic assignments are possible. Direct reaction of F_2 with $(H_2O)_5CrN_3^{2+}$ yields $(H_2O)_6Cr^{3+}$ and the fluorinated product $(H_2O)_5CrF^{2+}$. With lower substrate concentrations, hydrolysis of F_2 is competitive, and the hydrolysis product HOF undergoes either hydrolysis or competitive direct reaction by oxygen transfer to form $(H_2O)_5CrNO^{2+}$. Note that the direct experiments with HOF provide important confirmation of this scheme.

Based on the limited data available, it appears that reactions of aqueous F_2 with substrates yield fluorinated or oxidized products. In contrast, the principal mode of HOF reactions is oxygen atom transfer rather than fluorination. These features may be applicable to nonaqueous systems as well.

B. Reduction of Cl_2, Br_2, and I_2

As we have discussed, aqueous solutions of the halogens are in rapid equilibrium with the corresponding hypohalous acid. It has been observed that in some cases the reactive form of the oxidant is the hypohalous acid or an intermediate in the hydrolysis reaction. In principle this potential complication can be tested by a comparison of the kinetic results when the hydrolysis is suppressed. However, this approach presents some difficulties. If the hydrolysis is suppressed by increasing the acidity, one must worry about an observed (H^+) dependence arising from the halogen–substrate reaction and not the hydrolysis. If added halide ion is used instead to suppress the hydrolysis, complications from halide catalysis or the X_3^- formed may be encountered. An alternative approach is to measure the electron transfer rate with the hypohalous acid directly. If it is slower than the rate found with the corresponding halogen, autocatalysis will usually be observed and one can safely conclude that hydrolysis is unimportant in the halogen system. If the HOX reaction is more rapid, considerable care will be required in making mechanistic assignments for the X_2 system.

Examples where direct reaction with the halogen has been demonstrated will be considered in this section. Since these reactions often involve successive one-equivalent steps, it is worthwhile to consider whether X_2^- or $X\cdot$ and X^- are the likely intermediates. Reactions of X_2^- species have been investigated by flash photolysis or radiolysis techniques. In either case the $X\cdot$ formed by oxidation of X^- undergoes reaction (12) very rapidly.[20]

$$X\cdot + X^- \rightleftharpoons X_2^-, \quad k_{\text{forward}} \approx 10^{10} \ M^{-1} \ s^{-1} \tag{12}$$

$$K \geq 10^5 \ M^{-1}$$

The subsequent reaction with the added substrate must be rapid under the experimental conditions; otherwise disproporitionation of X_2^- is observed:

$$X_2^- + X_2^- \rightarrow X_3^- + X^-, \quad k \approx 5 \times 10^9 \ M^{-1} \ s^{-1} \tag{13}$$

It is therefore reasonable to consider X_2^- as the intermediate in one-equivalent reductions of halogens or oxidations of halide ions. Under most experimental conditions any halogen atoms that may be formed will effectively be converted by reaction (12) to the dihalogen radical anion.

Vanadium(III) is oxidized much more rapidly by Cl_2 than by HOCl.[75] Suppression of Cl_2 hydrolysis by the addition of Cl^- did not affect the kinetic results. With excess V(III), VO^{2+} was the only oxidation product. These observations indicate that the chlorine reaction is direct and proceeds by one-equivalent steps. The proposed Cl_2^- intermediate was found to competitively oxidize added VO^{2+} or Co^{2+}. The rate-determining step in the V(III)–Cl_2 reaction is likely outer-sphere in that it occurs more rapidly than substitution rates normally observed for vanadium(III). The hydrogen ion dependence of the rate expression is predominantly reciprocal first order and was attributed to hydrolysis of V^{3+} rather than Cl_2. The second-order rate constant at 25°C in 1.0 M $HClO_4$ is 20 M^{-1} s^{-1}.

In contrast to V(III), oxovanadium(IV), VO^{2+}, is oxidized more rapidly by HOCl than by Cl_2.[76] Even with 1.0 M Cl^- and 1.0 M H^+, some 35% of the observed reaction rate is due to oxidation by HOCl. The remainder was attributed to the Cl_2–VO^{2+} reaction. At 25°C and $I = 1.0$ M, $-d[Cl_2]/dt = 7 \times 10^{-3}$ M $s^{-1}[V(IV)][Cl_2]/[H^+]^2$. The hydrogen ion dependence is awkward to rationalize; the authors prefer equilibrium (14) prior to electron transfer.

$$VO^{2+} + Cl_2 + H_2O \rightleftharpoons (VO_2 \cdot Cl_2) + 2H^+ \tag{14}$$

Another possibility is the reaction of $VO(OH)^+$ with $HOCl_2^-$ in the rate-determining step.

The substitution-inert complex ion tris(1,10-phenanthroline)iron(II) is about as weak a reductant as VO^{2+}, yet it is oxidized by Cl_2 approximately 10^2 faster than by HOCl.[77,78] Clearly, no single factor determines the relative reactivity of Cl_2 and HOCl. We shall return to this point in Section VI.

The rapid Cr^{2+}–Cl_2 reaction proceeds by predominantly if not exclusively inner-sphere, one-equivalent steps.[79] Reduction of Cl_2 by V^{2+} is sufficiently rapid to be classified as outer-sphere and involves primarily one-equivalent steps.[80] The Fe^{2+}–Cl_2 reaction has a rate constant of 80 M^{-1} s^{-1} at 25°C and the high yields of $FeCl^{+2}$ ($>70\%$) establish a predominantly inner-sphere attack in the rate-determining step.[81] The Fe^{2+}–Cl_2^- reaction has been studied independently.[82] Both inner-sphere ($k = 4.0 \times 10^6$ M^{-1} s^{-1}) and outer-sphere ($k = 1.0 \times 10^7$ M^{-1} s^{-1}) paths were observed. The inner-sphere reaction appears to be limited by substitution on iron(II).

It is likely that a two-equivalent process occurs in the Cl_2–H_2O_2 reaction; there is no evidence for a free-radical reaction.[61] The intermediate HOOCl [reactions (15) and (16)] has been proposed by several investigators.[83,84]

$$H_2O_2 + Cl_2 \rightleftharpoons HOOCl + H^+ + Cl^- \tag{15}$$

$$HOOCl \rightarrow O_2 + H^+ + Cl^- \tag{16}$$

As mentioned earlier, H_2OCl_2 has been proposed as the reactive form of the oxidant.[61] The formation of HOOCl by nucleophilic displacement of Cl^-

from Cl_2 (or H_2OCl_2) is unfavorable if the peroxide moiety is complexed to vanadium(V) or titanium(IV). Instead, the reaction rate is limited by loss of the peroxo ligand from $VO(O_2)^+$ and $Ti(O_2)^{2+}$.[4,85]

Relatively few studies of the reduction of the heavier halogens have been reported; seldom have the corresponding rates of HOBr and HOI (or H_2OI^+) been measured. The hydrolysis constants are much smaller for these halogens than for Cl_2, and examples in studies of the reduction of Br_2 and I_2 where a significant portion of the reaction rate was attributed to the hypohalous acid are rare. However, the formation of Br_3^- and I_3^- is much more favorable than for Cl_3^-, and their kinetic importance is often detected.

Reductions of I_2, I_3^-, and Br_2 by V^{2+} are quite rapid. At 25°C and $I = 1.0\ M$, the rate constants ($M^{-1}\ s^{-1}$) are 7.5×10^3, 9.7×10^2, and 3.0×10^4, respectively.[80] One-equivalent, outersphere pathways predominate.

The reduction of Br_2 and Br_3^- by Fe^{2+} has been thoroughly investigated.[23] The mechanism proposed is given by Eqs. (17)–(21).

$$Fe^{2+} + Br_2 \underset{k_2}{\overset{k_1}{\rightleftharpoons}} Fe^{3+} + Br_2^- \tag{17}$$

$$Fe^{2+} + Br_3^- \underset{k'_2}{\overset{k_3}{\rightleftharpoons}} Fe^{3+} + Br_2^- + Br^- \tag{18}$$

$$Fe^{2+} + Br_2^- \overset{k_4}{\longrightarrow} Fe^{3+} + 2Br^- \tag{19}$$

$$Br_2 \rightleftharpoons HOBr + H^+ + Br^- \tag{20}$$

$$Fe^{2+} + HOBr \overset{k_5}{\longrightarrow} products \tag{21}$$

At 29.8°C and $I = 1.0\ M$, $k_1 = 0.76\ M^{-1}\ s^{-1}$, $k_3 = 34\ M^{-1}\ s^{-1}$, $k_5 = 2.1 \times 10^4\ M^{-1}\ s^{-1}$, and $k'_2[Br^-]/k_4 = 0.055$. The value of k_4 is now known to be $4.5 \times 10^6\ M^{-1}\ s^{-1}$ at 30°C and $I = 0.2\ M$.[82] Therefore, the rate constant k_2 for the $Fe^{3+}-Br_2^-$ reaction must be quite large. Unfortunately, the experimental procedures used in studies of X_2^- are best suited for determining reduction rates, so a direct measurement of k_2 is not available.

Hydrolyzed forms of the metal ions are proposed as the reactive reductants in the Ti(III)–I_2, I_3^-,[86] V(III)–I_2,[87] U(IV)–I_2,[88] and U(IV)–Br_2[88,89] reactions. Only in the last two systems has the possibility of a two-equivalent process been suggested.

VI. REDUCTION OF HYPOHALOUS ACIDS

These reactions are normally complicated by halogen formation as the reaction proceeds. In some cases the halogen is unreactive on the time scale of the kinetic experiment, but with other reductants the halogen is more reactive than the hypohalous acid. The ability to predict whether Cl_2 or HOCl, for which substantial kinetic data are available, will be more reactive toward a given substance, is limited. However, several tentative guidelines can be suggested:

1. If the reducing agent is a metal ion and its reaction with both HOCl and Cl_2 is inner-sphere, then the redox reaction with HOCl will be faster. While this is an empirical correlation based on limited available data, it is intuitively reasonable in that substitution at the oxygen of HOCl is favored over substitution at a halogen atom of Cl_2. It can be further asserted that in general for a two-equivalent process, the reaction is necessarily inner-sphere regardless of the chemical features of the reactants. This notion is not new,[78,90,91] and to date this author is not aware of an exception.

2. If the reducing agent is not a metal ion and a two-equivalent process occurs, then the electrophile (Cl_2 or HOCl)–nucleophile (reductant) association is critical. This guideline is based on extensive data for H_2O_2.[61,62] In acidic solution, the reaction rate is much more rapid with Cl_2 than with HOCl. However, in mildly basic solution it has been proposed that both Cl_2 and HOCl (but not OCl^-) react very rapidly with HO_2^-.

3. If the reaction is outer-sphere or if the reductant is relatively inert to substitution, Cl_2 will normally be more reactive. This is an empirical correlation. It is noted that the structural changes are probably less for a one-equivalent reduction of Cl_2 than HOCl, while the $E°$ values for these two processes are similar.

These guidelines may well be applicable to the heavier halogens, but the experimental data are not available. Some studies of the reduction of hypohalous acids, mainly HOCl, are considered next.

The full rate expression for the U(IV)–HOCl reaction at 25°C and $I = 3.0$ M is $-d[U(IV)]/dt = 0.01$ M $s^{-1}[Cl_2][H^+]^2 + 1.1$ $s^{-1}[HOCl][U^{4+}]/[H^+]$.[91,92] No interference from the accumulating Cl_2 product was observed under most of the experimental conditions. The assignment of a two-equivalent process for the HOCl reaction is based on kinetic arguments and some indirect but convincing oxygen-18 tracer experiments.[91] The different $[H^+]$ dependence for the Cl_2 and HOCl reactions is likely due to the formation of the O–U–O configuration as a prerequisite to oxidation of uranium(IV).[91,93] With Cl_2, doubly hydrolyzed U(IV) is necessary, whereas the combination of singly hydrolyzed U(IV) and inner-sphere attack of HOCl achieves a comparable configuration.

That V(III) reacts more slowly with HOCl than Cl_2 is consistent with the last guideline, in that the latter reaction is probably outer-sphere.[75] Based on initial rate measurements, an upper limit of 3 M^{-1} s^{-1} for the reaction was estimated (25°C, 1 M $HClO_4$). With vanadium(IV), the reverse reactivity order is observed.[76] Inner-sphere pathways are proposed, in harmony with the first guideline. The rate expression at 25°C and $I = 1.0 M$ is $-\frac{1}{2}d[V(IV)]/dt = 40 M^{-1} s^{-1} [V(IV)][HOCl]/(1 + 6.0 M^{-1}[H^+])$.

The substitution-inert ion, $(phen)_3Fe^{2+}$, reacts more slowly with HOCl than with Cl_2.[77,78] The HOCl reaction was studied over the pH range 5–9 in order to avoid the formation of Cl_2.[78] The electron transfer rate is limited by dissociation of one 1,10-phenanthroline ligand from the iron(II) center. The $(phen)_2Fe^{2+}$ intermediate was found to either re-form the tris complex or reduce HOCl and OCl^-. Although a definitive assignment of a one- or two-equivalent process was not made, the former is preferable. It is interesting to note that an outer-sphere reduction of HOCl has not been demonstrated. An uninvestigated possibility is the V(II)–HOCl reaction. However, a slower rate would be anticipated than with Cl_2, but it may be more rapid than substitution on V(II).

The Fe^{2+} reduction of several oxidants, including Cl_2 and HOCl, has been examined primarily to determine whether the dimeric iron(III) product, presumably $Fe(OH)_2Fe^{4+}$, was formed.[81] As mentioned, with Cl_2 the predominant product is $FeCl^{2+}$. With HOCl, approximately 15% of the dimeric product is found in 0.1 M $HClO_4$; the yield is decreased with increasing acidity. The interesting question is whether one- or two-equivalent processes are responsible for its formation. It can be asserted that a two-equivalent oxidation of iron(II) is very unlikely, with one possible exception. In the present case a sequence of inner-sphere, one-equivalent steps can account for the observations. The possible exception is the Fe(VI)–Fe(II) reaction (which does not appear to have been investigated). Here, an iron(IV) intermediate is likely, and it would be of interest to examine the yield of $Fe(OH)_2Fe^{2+}$. The possible two-equivalent oxidation of iron(II), however, would involve an iron(V) intermediate. The Cr(VI)–Cr(II) reaction has been investigated by the use of a radioactive chromium tracer.[94] The reaction scheme proposed is summarized in Eqs. (22)–(25).

$$*Cr(VI) + Cr(II) \longrightarrow *Cr(V) + Cr^{3+} \tag{22}$$

$$*Cr(V) + Cr(II) \begin{cases} \xrightarrow{\sim 80\%} *Cr(IV) + Cr^{3+} \quad (a) \\ \xrightarrow{\sim 20\%} *Cr^{3+} + Cr(IV) \quad (b) \end{cases} \tag{23}$$

$$*Cr(IV) + Cr(II) \longrightarrow *Cr(OH)_2Cr^{4+} \tag{24}$$

$$Cr(IV) + Cr(II) \longrightarrow Cr(OH)_2Cr^{4+} \tag{25}$$

The two-equivalent oxidation of Cr(II) occurs to a limited extent in reaction (23b), and an analogous pathway may exist in the Fe(VI)–Fe(II) reaction.

Substantial amounts of a dimeric chromium(III) product, presumably $Cr(OH)_2Cr^{4+}$, are formed in the rapid Cr^{2+}–HOCl reaction.[79] This does not arise from a $Cr(IV)$–$Cr(II)$ reaction, nor in other oxidations of $Cr(II)$ except by $Cr(VI)$. Instead, a sequence of inner-sphere one-equivalent steps is proposed.[37,79]

Hypobromous acid is fairly stable in acidic solution.[22] Very few kinetic studies of the reduction of HOBr have been reported; it has often been considered a nuisance in bromate reactions and occasionally in bromine systems. The observation that cerium(III) in sulfuric acid solutions is unable to reduce HOBr at a significant rate[95] is important in mechanistic interpretations of the Belousov–Zhabotinsky oscillating reaction.[49,50,96]

VII. HALOUS ACIDS

A. Disproportionation

There is general agreement that the primary reaction in the disproportionation of halous acids is Eq. (26):

$$2HXO_2 \rightarrow HOX + XO_3^- + H^+ \tag{26}$$

The reactions may be complicated by interactions of the hypohalous acid with the halous acid. This is clearly the case in the chlorine system, where reaction (26) is considerably slower than the proposed sequence of reactions (27)–(29).[97]

$$HOCl + HClO_2 \rightarrow Cl_2O_2 + H_2O \tag{27}$$

$$H_2O + Cl_2O_2 \rightarrow Cl^- + ClO_3^- + 2H^+ \tag{28}$$

$$2Cl_2O_2 \rightarrow Cl_2 + 2ClO_2 \tag{29}$$

The Cl_2 formed in Eq. (29) reacts even more rapidly with $HClO_2$ (or ClO_2^-) than does HOCl.

$$Cl_2 + ClO_2^- \rightarrow Cl_2O_2 + Cl^- \tag{30}$$

The chloride ion formed in Eqs. (28) and (30) is known to react rapidly with hypochlorous acid [Eq. (31)] and at higher concentrations it can react with chlorous acid.[59,10]

$$HOCl + H^+ + Cl^- \rightleftharpoons Cl_2 + H_2O \tag{31}$$

Clearly, both the stoichiometry and the kinetics associated with the disproportionation of chlorous acid are complex.

It is possible that HOBr and $HBrO_2$ do not interact significantly during the much more rapid disproportionation of bromous acid. However, this interaction can assume importance depending on the experimental conditions, and certainly plays an important role in the isotopic exchange of the bromine–bromate–hypobromous acid system.[98] Bromide ion formed or inadvertently present in the reaction mixtures rapidly reduces both $HBrO_2$ and HOBr [Eqs. (32) and (33)]:

$$HBrO_2 + H^+ + Br^- \rightleftharpoons 2HOBr \tag{32}$$

$$HOBr + H^+ + Br^- \rightleftharpoons Br_2 + H_2O \tag{33}$$

The H_2OI^+ formed during the disproportionation of HIO_2 causes autocatalysis of this reaction.[9] Not only is reaction (34) rapid, but the iodide ion formed reacts rapidly with both HIO_2 and H_2OI^+ (or HOI).

$$H_2OI^+ + HIO_2 \rightarrow I^- + IO_3^- + 3H^+ \tag{34}$$

The kinetics of the disproportionation reactions of the halous acids have been unambiguously determined only in the case of chlorous acid.[10,99] The full rate expression is summarized in Eq. (35):

$$-d[HClO_2]/dt = k_1[HClO_2]^2 + k_2[HClO_2][Cl^-]^2/(K + [Cl^-])$$
$$+ k_3[HClO_2][ClO_2^-] \tag{35}$$

The last term in this expression has not been studied extensively and is unimportant at $[H^+] \geq 0.1$ M, but it clearly plays a significant role at pH values near the pK_a of chlorous acid (~ 2).[10,99] At 25°C and 2.0 M H^+, $k_1 = 1.2 \times 10^{-2}$ M^{-1} s^{-1}, $k_2 = 3.0 \times 10^{-2}$ M^{-1} s^{-1}, and $K = 1.2 \times 10^{-3}$ M.[10] The second term in the rate expression summarizes the catalysis by chloride ion. The overall stoichiometry under many experimental conditions is approximately given by Eq. (36):

$$4HClO_2 \rightarrow 2ClO_2 + ClO_3^- + Cl^- + 2H^+ + H_2O \tag{36}$$

The rate of disproportionation of bromous acid is much faster, but the value of the second-order rate constant in highly acidic solution is controversial . A lively debate has arisen from both experimental difficulties and the importance of this reaction in the mechanistic interpretations and numerical simulations of the Belousov–Zhabotinsky oscillating reaction. Values of the second-order rate constant, including both experimental and estimated, currently range from 2×10^3 to 2×10^9 M^{-1} s^{-1}![100]

Noyes estimated a value of 4×10^7 M^{-1} s^{-1} in his classic paper on the mechanism of the Belousov–Zhabotinsky oscillating reaction.[49] This value has been used in most of the numerical simulations.[100,101] These simulations successfully correlate many experimental observations on the oscillating reaction, although a large number of reactions and estimates of their rate constants are required.

Sullivan and Thompson concluded that the disproportionation rate constant for $HBrO_2$ must be considerably less than 4×10^7 M^{-1} s^{-1}, based on kinetic studies of the $Ce(IV)-HBrO_2$ reaction in sulfuric acid solutions.[102] Schelly and co-workers determined a value of 8×10^3 M^{-1} s^{-1} at 25°C in 1.5 M H_2SO_4 by electrochemical methods. However, experiments by Försterling's group on the $BrO_3^- - HBrO_2$ reaction indicate a much larger value, similar to that used in the numerical simulations.[8,17] Clearly, additional experimental work is needed.

Schelly and co-workers have studied the disproportionation of iodous acid in 0.05 to 0.15 M H_2SO_4 at 24°C.[9] The reaction is autocatalytic due to the reaction of H_2OI^+, a product of the disproportionation, and HIO_2. They were able to assign upper limits for k_{HIO_2} of 2.0 M^{-1} s^{-1} (0.15 M H_2SO_4) and 5.4 M^{-1} s^{-1} (0.05 M H_2SO_4). Much larger estimates for k_{HIO_2} ranging from 45 to 6×10^5 M^{-1} s^{-1} have been used in numerical simulations of the Briggs–Rauscher oscillating reaction.[103,104]

B. Reduction of chlorous acid

These systems may be complicated due to interactions of intermediates such as $HOCl$ and Cl_2 with $HClO_2$ as shown in Eqs. (27) and (30). However, the reactions can often be simplified by the removal of troublesome intermediates with appropriate scavengers. The primary reaction between uranium(IV) and chlorous acid was shown to be either Eq. (37) or Eq. (38) by a combination of kinetic and oxygen-18 tracer studies.[105]

$$U^{4+} + ClO_2^- + H_2O^* \rightarrow {}^*OUO^{2+} + HOCl + H^+ \tag{37}$$

or

$$U^*OH^{3+} + HClO_2 \rightarrow {}^*OUO^{2+} + HOCl + H^+ \tag{38}$$

Reaction (38) is appealing in that the OUO core is formed during the rate-determining step. Phenol was used as the scavenger. Neither phenol nor uranium(IV) compete effectively with chlorous acid for the HOCl intermediate. However, the phenol apparently does effectively remove the Cl_2O_2 produced in Eq. (27), as summarized in Eq. (39).

$$Cl_2O_2 + C_6H_5OH \rightarrow ClC_6H_4OH + H^+ + Cl(III) \tag{39}$$

The resulting stoichiometric ratio $\Delta[U(IV)]/\Delta[Cl(III)]$ is unity, and the rate expression at 25°C and $I = 2.0$ M is given in Eq. (40):

$$\frac{-d[U(IV)]}{dt} = \frac{360 \ M^{-1} \ s^{-1} [U(IV)][Cl(III)]}{(1 + 3.9 \ M^{-1}[Cl(III)])(1 + 26 \ M^{-1}[H^+])} \tag{40}$$

The reduction of chlorous acid by iron(II) is proposed to involve a one-equivalent pathway in the rate-determining step.[106] The chlorine(II) intermediate thus formed rapidly oxidizes another equivalent of the iron(II) reactant present in large excess. The kinetics were simplified by the presence of two scavengers—Cl^- to convert the HOCl intermediate to Cl_2 and phenol to remove the Cl_2. The resulting net stoichiometry is $\Delta[Fe(II)]/[Cl(III)] = 2.0$, and the rate expression at 25°C and $I = 2.0\ M$ is given in Eq. (41):

$$-d[Fe(II)]/dt = (1.9 \times 10^3\ M^{-1}\ s^{-1} + 58\ s^{-1}/[H^+])[Fe(II)][Cl(III)] \quad (41)$$

Only traces of dimeric iron(III), presumably $Fe(OH)_2Fe^{4+}$, were found as an immediate product, in contrast to the Fe^{2+}–HOCl reaction.[81]

The presence of phenol in the V(III)–$HClO_2$ reaction causes a 1:1 stoichiometry, and the second-order rate constant is $75\ M^{-1}\ s^{-1}$ at 25°C in $1.0\ M\ HClO_4$.[75] In this system the phenol is proposed to scavenge a chlorine(II) intermediate, in contrast to the iron(II) reaction.

C. Reduction of chlorite in solutions of pH 3–7

The predominant form of chlorine(III) in this pH region is ClO_2^-. A number of reductions of chlorine(III) have been investigated under these conditions. In no case has it been necessary to invoke the reaction of an HOCl intermediate with the oxidant, although this interaction is often troublesome in more acidic solutions. It is not clear whether this simplification comes from a slower reaction of HOCl with ClO_2^- than with $HClO_2$, or is due to alternative rapid reactions of HOCl in the systems examined.

The reduction of chlorite ion by iodide ion has been thoroughly investigated and with excess ClO_2^- constitutes one type of the well-known "iodine clock reaction."[26,107–109] The rate law at 25°C and $I = 0.5\ M$ is given by Eq. (42).[26]

$$d[(I_2) + (I_3^-)]/dt = 9.2 \times 10^2\ M^{-1}\ s^{-1}[ClO_2^-][I^-][H^+]$$
$$+ 5.1 \times 10^{-3}\ s^{-1}[ClO_2^-][I_2]/[I^-] \quad (42)$$

The two terms in Eq. (42) could be studied separately by adjustment of the $[I^-]$, especially since at high $[I^-]$ the I_2 is largely converted to I_3^-, an unreactive component. The first term arises from reaction (43), followed by rapid reduction of the products by the excess iodide ion.

$$I^- + HClO_2 \rightarrow IO^- + HOCl \quad (43)$$

The formation of iodine is initially slow under the experimental conditions of the clock reaction. However as the $[I_2]$ increases, the autocatalytic second term rapidly becomes dominant due to reactions (44) and (45):

$$I_2 + H_2O \rightleftharpoons HOI + H^+ + I^-, \quad \text{rapid } K_h \tag{44}$$

$$HOI + HClO_2 \rightarrow HIO_2 + HOCl, \quad \text{rate determining} \tag{45}$$

With sufficient $[I^-]$, the products of reaction (45) are rapidly reduced to a net yield of two I_2 molecules per chlorite ion. We note that the reactive form of chlorine(III) is $HClO_2$ in both rate terms according to the proposed mechanism.[26] The rapid, autocatalytic formation is abruptly terminated if excess $[ClO_2^-]$ is present, and the iodine is then consumed very rapidly. These features may be accentuated for demonstration purposes by the addition of starch indicator.

It has been shown that the sudden consumption of iodine is due to reaction (46).[110]

$$2I_2 + 5ClO_2^- + 2H_2O \rightarrow 4IO_3^- + 5Cl^- + 4H^+ \tag{46}$$

In the kinetic study the $[I^-]$ was maintained at very low levels by the addition of IO_3^- to the reactant I_2 solutions. This procedure minimizes the formation of I_2 by chlorine- and iodine-containing intermediates and removes "spurious" effects observed in the absence of added IO_3^-.[110] The rate expression summarized in Eq. (47) was determined at 22°C, pH = 2.8 to 4.6, and a variable ionic strength $\leq 0.1\ M$.

$$-\tfrac{1}{2} d[I_2]/dt = (11\ M^{-1}\ s^{-1}[ClO_2^-] + 1.1 \times 10^{-2}\ s^{-1}[ClO_2^-]/[H^+] + 0.54\ s^{-1})[I_2] \tag{47}$$

The three pathways are attributed to the formation of an $IClO_2$ intermediate by reactions between ClO_2^- and I_2, I_2OH^-, and H_2OI^+, respectively. It should be noted that the third term is similar to the autocatalytic pathway observed under different conditions by Kern and Kim.[26] However, these authors proposed a rate-determining reaction between $HClO_2$ and HOI rather than between ClO_2^- and H_2OI^+.

The reaction between nitrous acid and chlorite has been studied.[111] An interesting feature of the proposed mechanism, Eqs. (48)–(50), is the formation of peroxonitrite, $ONOO^-$, as an intermediate.

$$HNO_2 + ClO_2^- \rightleftharpoons ONOO^- + HOCl \tag{48}$$

$$ONOO^- \rightarrow NO_3^- \tag{49}$$

$$HNO_2 + HOCl \rightarrow NO_3^- + 2H^+ + Cl^- \tag{50}$$

A chloride ion-catalyzed pathway was detected and attributed to reactions (51) and (52).

$$HOCl + Cl^- \rightleftharpoons HOCl_2^- \qquad (51)$$

$$HNO_2 + HOCl_2^- \rightarrow NO_3^- + 2H^+ + 2Cl^- \qquad (52)$$

Again note the recurring evidence that the intermediate species H_2OCl_2 and $HOCl_2^-$ may play important roles in the chemistry of Cl_2 and $HOCl$.

The rate of reduction of excess chlorite ion by tris(1,10-phenan-throline)iron(II) in slightly acidic solutions is equal to the dissociation rate of one ligand from the iron(II) center.[112] It is likely but not certain that the electron transfer reaction is inner-sphere.

The outer-sphere reaction between $Fe(CN)_6^{-4}$ and ClO_2^- is proposed to be a one-equivalent, reversible process.[113] A detailed investigation disclosed, in addition to the anticipated homogeneous reaction, and interfacial pathway whose velocity was dependent on the surface material of the reaction vessel. We can only hope that this feature was not present but undetected in other studies discussed in this review.

A number of chemical oscillators in which chlorite is a critical component have recently been discovered.[114–116] Given the intense activity in this field, we may anticipate numerous reports on these oscillators and increased interest in the redox chemistry of chlorite in the near future.

D. Oxidation of chlorous acid and chlorite

We see from the potential diagram that one- and two-equivalent oxidations of chlorous acid have similar free energy changes. However, although a two-equivalent transformation with concomitant oxygen transfer from the oxidant would appear to be an attractive possibility, it is not often observed. Instead, chlorine dioxide is normally an important intermediate or the final oxidation product if an excess of chlorous acid is used.

The following reaction scheme has been proposed for the bromate–chlorous acid reaction.[117]

$$BrO_3^- + HClO_2 + H^+ \xrightarrow{k_1} BrO_2 + ClO_2 + H_2O, \quad \text{rate determining} \qquad (53)$$

$$BrO_2 + HClO_2 \xrightarrow{k_2} HBrO_2 + ClO_2, \quad \text{rapid} \qquad (54)$$

$$BrO_2 + ClO_2 + H_2O \xrightarrow{k_3} HBrO_2 + ClO_3^- + H^+, \quad \text{rapid} \qquad (55)$$

$$HBrO_2 + HClO_2 \xrightarrow{k_4} HOBr + ClO_3^- + H^+, \quad \text{rapid} \qquad (56)$$

The HOBr formed in Eq. (56) was scavenged with allyl alcohol to simplify the study. At 25°C and $I = 2.0\ M$, values of $k_1 = 1.7\ M^{-2}\ s^{-1}$ and the ratio $k_2/k_3 = 1.9$ (in 2.0 M $HClO_4$) were determined. The stoichiometry is variable and depends on the ratios $[HClO_2]/[ClO_2]$ and $[HClO_2]_0/[BrO_3^-]_0$.

With excess BrO_3^-, the formation of ClO_2 reaches a maximum and then is consumed after an induction period. We note that a two-equivalent oxidation of $HClO_2$ is proposed in Eq. (56). An oxygen-18 tracer study in which the source of the ClO_3^- oxygen atoms are determined would be of interest.

In a later study the BrO_3^-–$HClO_2$ reaction in sulfuric acid solution was examined in a flow reactor.[118] Numerical simulations led to the suggestion that the allyl alcohol used in the earlier study might scavenge oxobromine intermediates such as $HBrO_2$ in addition to HOBr, though to a limited extent.

The formation of ClO_2 as the initial oxidation product was also observed in the MnO_4^-–ClO_2^- reaction over the pH range 1.4–5.0.[119] Parallel paths involving the reduction of MnO_4^- by both ClO_2^- (k_1) and $HClO_2(k_2)$ were found; at 25°C and $I = 1.0$ M, $k_1 = 24$ M^{-1} s^{-1} and $k_2 = 92$ M^{-1} s^{-1} if the best fit value of 2.7×10^{-3} M for the dissociation constant of $HClO_2$ is used. The results indicate that ClO_2 is also formed in the oxidation of chlorine(III) by intermediate manganese species.

Stanbury has begun studies of outer-sphere electron transfer reactions of the chlorite/chlorine dioxide couple as part of a program designed to determine self-exchange rates of small molecules in aqueous solution.[11,120] The oxidant $IrCl_6^{2-}$ converts ClO_2^- to ClO_2 with a rate constant of 1.1×10^4 M^{-1} s^{-1} if the best fit value of $K_a = 1.6 \times 10^{-2}$ M for chlorous acid is used.[120] The corresponding rate constant for $IrBr_6^{2-}$ is 1.9×10^4 M^{-1} s^{-1}. The current value for the self-exchange constant for the ClO_2^-/ClO_2 couple derived from these studies is 1.6×10^2 M^{-1} s^{-1}. A direct measurement using isotopic exchange techniques was attempted by Dodgen and Taube, and a lower limit of 4×10^2 M^{-1} s^{-1} was determined.[121] This apparent agreement may be coincidental since ClO_2 and ClO_2^- are known to form a weak complex.[41] As a result the true self-exchange rate may be considerably greater depending on the degree of overlap. Readers interested in this area should note the cautionary point raised recently by Espenson.[122] The oxidation of chlorous acid to chlorine dioxide by cerium(IV) is rapid even in sulfuric acid solution ($k = 5.6 \times 10^5$ M^{-1} s^{-1} in 1.5 M H_2SO_4 plus 1.5 M $NaClO_4$ at 25°C).[102] As is usually the case with cerium(IV), the reaction is even more rapid in perchloric acid solution.

Reaction (57) has been studied over the pH range 4.6–7.0.[27]

$$HSO_5^- + ClO_2^- \rightarrow SO_4^{2-} + ClO_3^- + H^+ \tag{57}$$

The key mechanistic feature proposed is a nucleophilic attack by the chlorine in ClO_2^- at the terminal oxygen in peroxosulfate. An oxygen-18 tracer experiment would provide a crucial test of this reasonable proposal. The synthesis and use of specifically labeled $O_3SOO^*H^-$ have been reported.[70,123,124]

E. Redox reactions of chloritopentaamminecobalt(III)

The principal mode of decomposition of this chlorito complex ion in acidic solution is the internal redox reaction (58).[38]

$$(NH_3)_5Co(OClO)^{2+} + 5H^+ \rightarrow 5NH_4^+ + Co^{2+} + ClO_2 \tag{58}$$

The rate constant at 25°C and $I = 0.50$ M is 8.0×10^{-6} M^{-1} s^{-1}, a value within the range observed for the acid hydrolysis of simple penta-amminecobalt(III) complex.[125] Coordination to the metal center rather than to a proton greatly reduces the rate of oxidation of chlorite, a feature documented in detail for the external oxidant cobalt(III). While electro-static effects may account for some of the rate diminution, it is likely that bond lengthening in the chlorito complex in contrast to bond shortening in the chlorine dioxide product is an important factor.

Oxygen-18 tracer experiments demonstrate that the cobalt–oxygen bond in the chlorito complex is largely broken ($\sim 66\%$) during oxidation, but substantially retained (84–99%) by reduction with Fe^{2+}, VO^{2+}, and HSO_3^-. The kinetics of the reduction reactions appear to be complex.

F. Oxidation of bromous and iodous acid

Kinetic studies of bromous acid are complicated by its rapid disproportionation. However, the expedient of mixing at least slightly basic solutions of bromite, which are moderately stable, with the desired reactant in excess acid has allowed the investigation of a few reactions.

The kinetic profiles of mixtures of cerium(IV) and bromous acid in sulfuric acid solution have been examined by stopped-flow techniques.[102] The rate of disappearance of cerium(IV) was found to be extremely dependent on the distribution of the sulfatocerium(IV) species $Ce(SO_4)_3^{2-}$, $Ce(SO_4)_2$, and $CeSO_4^{2+}$ present. This distribution can be systematically varied by the use of appropriate mixtures of Na_2SO_4 and $HClO_4$ or by altering the concentration of H_2SO_4.[126,127] Two important conclusions drawn from this study are (1) $Ce(SO_4)_3^{2-}$, the overwhelmingly dominant form of cerium(IV) in 1.5 M H_2SO_4 (a common medium for studies of the Belousov–Zhabotinsky oscillating reaction), is quite unreactive toward $HBrO_2$, and (2) the second-order rate constant for the disproportionation of $HBrO_2$ must be consider-ably smaller than the commonly used value of 4×10^7 M^{-1} s^{-1}.[102] These conclusions have sparked considerable controversy.[8,9,17,100] The same experimental method has allowed studies of reactions (59) and (60), and indirectly the disproportionation of $HBrO_2$ as discussed earlier.[8,17]

$$BrO_3^- + HBrO_2 + H^+ \rightarrow Br_2O_4 + H_2O \tag{59}$$

$$Br_2O_4 \rightleftharpoons 2BrO_2 \tag{60}$$

The only experimental value of a rate constant for the oxidation of HIO_2 is that for the H_2OI^+ reaction: $k = 130\ M^{-1}\ s^{-1}$ at 24°C in 0.15 M H_2SO_4.[9] Values for several other oxidations of HIO_2 have been estimated for numerical simulations of the Briggs–Rauscher oscillating reaction, but these are quite speculative.[103,104] The finding that HIO_2 in low concentration is moderately stable should allow direct investigations of its redox chemistry in the future.[9]

VIII. HALOGEN DIOXIDES

A. Reduction

Several metal ions have been shown to reduce chlorine dioxide to chlorite or chlorous acid in the rate-determining step. Values of k_1 at 25°C, appropriate to the rate expression $-d[ClO_2]/dt = k_1[ClO_2][\text{reductant}]$, are summarized in Table 5. Reaction (61) proceeds in two stages, which could be well separated over the pH range 5.0–8.5.[42]

$$2ClO_2 + 2I^- \rightarrow 2ClO_2^- + I_2 \tag{61}$$

The first corresponds to the formation of a spectrophotometrically observable $ClO_2 \cdot I^-$ complex, with the rate expression $d[ClO_2 \cdot I^-]/dt = 3.0 \times 10^3\ M^{-1}\ s^{-1}\ [ClO_2][I^-]$ at 25°C and $I = 0.75\ M$. No pH dependence on the rate was detected, but below pH 5.0 complications from the $ClO_2^- - I^-$ reaction become important. The slower decomposition of the complex is cleanly first order with a rate constant of 0.10 s^{-1} at 25°C.

Oxidations by bromime dioxide are of interest currently because of their important role in bromate-driven oscillators. However, direct studies are

TABLE 5

Rate constants for the reduction of chlorine dioxide at 25°C

Reductant	$I\ (M)$	$k_1\ (M^{-1}\ s^{-1})$	Mechanism	Ref.
V^{3+}	1.0	22	Inner-sphere or outer-sphere	[75]
VO^{2+}	1.0	5	Inner-sphere or outer-sphere	[128]
$Fe(phen)_3^{2+}$	0.10	4.5×10^4	Outer-sphere	[11][a]
$Co(terpy)_2^{+2}$	0.10	2.1×10^7	Outer-sphere	[120]

[a] An earlier report of this rate constant (Ref. [112]) is apparently in error.

experimentally difficult for several reasons. First, BrO_2 is unstable with respect to disproportionation over the entire pH range.[14] Consequently, the BrO_2 must be prepared *in situ* by such techniques as pulse radiolysis or flash photolysis of BrO_3^-. The thermal reaction between bromate ion and the reductant of interest must therefore be quite slow. This condition is fulfilled by $Fe(CN)_6^{4-}$ at a pH $\geq$ 6.0. The second-order rate constant for the BrO_2–$Fe(CN)_6^{4-}$ reaction is quite large—$2 \times 10^9\ M^{-1}\ s^{-1}$ at 20°C.[13] If the electron transfer rate with BrO_2 is not very rapid, a second complication due to dimerization of BrO_2 occurs, as shown in Eq. (62).[129]

$$2BrO_2 \rightleftharpoons Br_2O_4, \qquad K = 1.4 \times 10^4\ M^{-1} \tag{62}$$

Just this problem was encountered in a study of the slower BrO_2–Mn(II) reaction.[13]

Reaction of IO_2 is undoubtedly important in iodate oscillators. However, no direct studies of this oxidant have been reported.

B. Oxidation of chlorine dioxide

A kinetic study of the Co^{3+}–ClO_2 reaction appears to be the only example of the oxidation of a halogen dioxide.[12] The principal reactive form of the cobalt(III) is $CoOH^{2+}$, as is often the case with this oxidant.[130] The rate expression at 25°C and $I = 2.1\ M$ is $-d[ClO_2]/dt = 1.1 \times 10^2\ s^{-1}$ $[Co(III)][ClO_2]/[H^+]$.

IX. HALATES

A. Reduction

The halate ions are much less basic than the halite or hypohalite ions. Since halogen–oxygen bond breaking normally accompanies reduction, we might anticipate positive hydrogen dependencies in the rate expressions, although the chemistry of the reducing agent must also be considered. Experimentally, these dependencies are *usually* less than or equal to the number of hydrogen ions that appear in the net process (that is, the proposed reaction in the rate-determining step). This feature is illustrated for a few examples listed in Table 6. The parallel path with a first-order hydrogen ion dependence found for the ClO_3^-–VO^{2+} reaction would not be anticipated by this empirical correlation. Nonetheless, the idea is useful in that it does embody a large set of experimental data. It is the duty of the experimentalist to consider carefully the

TABLE 6

Some hydrogen ion dependencies for reduction of halates

System	Net process	[H$^+$] dependence	Ref.
$BrO_3^- + Br^-$	$BrO_3^- + Br^- + 2H^+ = HBrO_2 + HOBr$	Second order	[131]
$BrO_3^- + HN_3$	$BrO_3^- + HN_3 + H^+ = HBrO_2 + NOH + N_2$	First order	[132]
$ClO_3^- + V^{2+}$	$ClO_3^- + V^{2+} + 2H^+ = ClO_2 + V^{3+} + H_2O$	Zero order	[133]
$ClO_3^- + VO^{2+}$	$ClO_3^- + VO^{2+} = ClO_2 + VO_2^+$	Zero and first order	[128]

possibility that medium effects are responsible for *minor* pathways in the determination of hydrogen ion dependencies.

The halate ions often are ultimately reduced to the corresponding halide ions. Thus, many intermediates will be formed and determining which ones are important can be viewed as either an exciting challenge or a hopeless task. It will be seen that the latter viewpoint is unduly pessimistic, at least in some cases. We might anticipate that interactions between intermediates will be minimal, due to their low and often steady-state concentrations during halate reductions, but this anticipation is not invariably realized.

As a final introductory comment, note that perhaps the simplest scheme for the reduction of XO_3^- to X^- would involve a sequence of oxygen atom transfers from the oxidant to the reductant, steps that may be considered as two-equivalent processes. Such a scheme is undoubtedly operative in some systems, although it is often difficult to establish experimentally. A nice example is the oxygen-18 tracer study of the $ClO_3^- - H_2SO_3$ reaction.[134] It was found that 2.5 oxygen atoms from ClO_3^- were transferred, 1.5_5 from $HClO_2$, and approximately 0.36 from HOCl. The rate expression had been shown previously to be $k[ClO_3^-][H_2SO_3]$.[135] The results are consistent with the two-equivalent, oxygen transfer scheme outlined above, with $HClO_2$ and HOCl as intermediates and with incomplete oxygen transfer at the HOCl stage. We note that the rate law by itself is consistent with innumerable mechanisms. A generalization may be made for halate reactions: the simpler the rate expression, the less mechanistic information that may be gleaned from the kinetic study.

1. Chlorate

Kinetic results and the experimental conditions for the reduction of chlorate ion by several metal ions are summarized in Table 7. With V^{2+} in excess the reaction products are V^{3+} and Cl^-. Apparently the reduction of

TABLE 7

Kinetic data for the reduction of chlorate by metal ions at 25°C

Metal ion	$I\,(M)$	[H$^+$] dependence	k	Ref.
V^{2+}	1.6	Zero order	4.4 $M^{-1}\,s^{-1}$	[133]
V^{3+}	1.0	Zero order	5.4 $M^{-1}\,s^{-1}$	[75]
VO^{2+}	2.0	Zero and first order	$9.5 \times 10^{-3}\,M^{-1}\,s^{-1}$	[128]
			$3.2 \times 10^{-3}\,M^{-2}\,s^{-1}$	
Cr^{2+}	2.0	First order	17 $M^{-2}\,s^{-1}$	[79]
Cr$^{2+\,a}$	2.0	Zero order	1.3 $M^{-1}\,s^{-1}$	[37]

^{a} The oxidant is $(H_2O)_5Cr(OClO_2)^{2+}$.

chlorine-containing intermediates is quite rapid, although this has been demonstrated only for the potential intermediate Cl_2.[80] The rate-determining reaction may be inner-sphere, but the subsequent fast steps are clearly outer-sphere. The rates of reduction of ClO_3^- and ClO_2 by V^{3+} are similar, and with large excesses of ClO_3^- the reduction product is found to be ClO_2. The reaction was suggested to be outer-sphere.

The stoichiometric ratio $\Delta[VO^{2+}]/[ClO_3^-]_0$ with excess VO^{2+} is 5.0 in the absence of Cl$^-$ but 4.0 in its presence. The rate of reduction of intermediates is much more rapid than for ClO_3^-. However, the reduction of Cl_2 is a slower process, and it is the reduction product. Added chloride ion replaces one equivalent of vanadium(IV) as the reductant as the HOCl stage.

Considerable experimental data are available for the Cr^{2+}–ClO_3^- reaction. The rate constants for reduction of the potential intermediates ClO_2, $HClO_2$, HOCl, and Cl_2 are several orders of magnitude larger than for the ClO_3^- reaction. The three chromium(III) products $(H_2O)_5CrCl^{2+}$, $(H_2O)_6Cr^{3+}$, and a dimeric species formulated as $(H_2O)_4Cr(OH)_2Cr(H_2O)_4^{4+}$ are formed in variable yields depending on the experimental conditions. This product distribution was determined for the ClO_3^-, ClO_2, $HClO_2$, HOCl, and Cl_2 reactions. The chlorate reaction is predominantly, if not exclusively, inner-sphere. Nearly all of the Cl$^-$ produced is contained in the $(H_2O)_5CrCl^{2+}$, and 0.32 oxygen atoms from the chlorate ion are transferred to each $(H_2O)_6Cr^{3+}$. The basic conclusion drawn from these results is that oxochlorine complexes of chromium(III) are produced as intermediates by inner-shpere attacks, and these intermediates are unstable to further attach by chromium(II). The formation of the dimeric chromium(III) product does *not* arise via a chromium(IV) intermediate, but instead through reactions generically represented by Eq. (63).

$$Cr\!-\!O\!-\!Cl\!-\!O_x \longrightarrow Cr(OH)_2Cr^{4+} + ClO_{x-1} \qquad (63)$$
$$\overset{\displaystyle |}{\underset{\displaystyle Cr^{2+}}{\vdots}}$$

A study of the Cr^{2+} reduction of the $(H_2O)_5Cr(OClO_2)^{2+}$ complex has fortified these suggestions. Chromium-51 tracer experiments demonstrate that the chromium(III) originally in the complex predominantly appears in the dimeric product. We note from the rate expressions in Table 6 that the chromium(III) in the complex formally replaces the proton present in the activated complex for the $Cr^{2+}-ClO_3^-$ reaction.

The $Cl^--ClO_3^-$ system has been reviewed in depth.[99] Although an oxygen atom transfer to form $HClO_2$ and $HOCl$ might be anticipated, reaction (64) is observed.

$$2Cl^- + 2ClO_3^- + 4H^+ = Cl_2 + 2ClO_2 + 2H_2O \qquad (64)$$

Relatively high concentrations of reactants are required to overcome the unfavorable equilibrium. Equation (64) is exactly what would be expected if Cl_2O_2 is an important intermediate at fairly high concentrations, as discussed in Section VII,A. The chlorine tracer results of Dodgen and Taube are consistent with this interpretation if it is assumed that the Cl_2O_2 intermediate is unsymmetrical, as in $Cl-ClO_2$. Aside from some disagreement on just how the intermediate is formed, the principal experimental observation that is at odds with this scheme is the reported lack of induced exchange of oxygen between chlorate ion and water during the partial (10%) reduction of ClO_3^- by Cl^-.[136] While it is not intended to resolve this issue, it should be noted that the experimental details were not given in Ref. [136]. If high concentrations of Cl^- and H^+ were used, then the bimolecular decomposition of Cl_2O_2 would be favored, especially at $100°C$. However, only the unimolecular decomposition would lead to induced exchange.

2. Bromate

These reactions have been studied more extensively than those of chlorate ion. This is undoubtedly due to the faster rates usually observed with bromate ion and the ongoing interest in bromate oscillators.

a. Two-equivalent processes. Some results of kinetic studies of the reduction of bromate ion by potentially two-equivalent reductants are summarized in Table 8. It can be seen that in each system two protons are formally present in the activated complex, including that for oxygen exchange with solvent. It is very unlikely that this arises from the prior formation of BrO_2^+.[137] Instead, a reasonable picture is that protonation of oxo ligands on the bromate is accompanied by substitution of the reducing agent. The concept of replacement as a prerequisite to electron transfer in reductions of oxoanions has been developed in detail by Edwards.[142] Attack by HSO_3^- or H_2SO_3 has been demonstrated to occur at a bromate oxygen.[134] The

TABLE 8

Kinetic results for the reduction of BrO_3^- by potentially two-equivalent reductants at 25°C

Rate law $(-d[BrO_3^-]/dt)$	Reductant	I (M)	Rate constant	Ref.
$k[BrO_3^-][R][H^+]^2$	I^-	1.0	$49\ M^{-3}\ s^{-1}$	[137]
	Br^-	0.11	3.3	[131]
	Cl^-	1.2	6.5×10^{-3}	[138]
Oxygen exchange	H_2O	1.0	$4.0 \times 10^{-3}\ M^{-2}\ s^{-1}$	[139, 140]
$k[BrO_3^-][R][H^+]$	H_2O_2	0.06	$5.7 \times 10^{-4}\ M^{-2}\ s^{-1}$	[131]
	HN_3	2.0	6.2×10^{-2}	[132]
	HSO_3^-	2.1	4.3×10^3	[141]

intermediates $HBrO_2$ and $HOBr$, formed by successive two-equivalent reductions, have been proposed for all the redox reactions listed in Table 8. It is worthwhile to consider the evidence for these conclusions.

The evidence for HSO_3^- and H_2SO_3 is very convincing. In addition to the kinetic and oxygen-18 tracer results, the absence of dithionate ion as significant reaction product[141] virtually eliminates the possibility of one-equivalent pathways.[143]

The stoichiometry [Eq. (65)] for the HN_3 system is consistent only with two-equivalent reductions of BrO_3^- and $HBrO_2$. The $HOBr$ product is relatively unreactive and was scavenged for simplicity.

$$BrO_3^- + 2H-N^*-N-N^* \rightarrow 2N \cdot N^* + N^*-N^*-O \qquad (65)$$

The nitrogen-15 tracer results (the asterisks indicate the labeling) are very difficult to reconcile with alternative schemes.

If one-equivalent pathways were significant in the H_2O_2 reaction, it is quite likely that complex chain reactions due to the HO_2^- intermediate would occur. Since no such complications were detected in the kinetic studies, a sequential two-equivalent scheme is most probable. The O_2 product contains only oxygen derived from the peroxide moiety,[144] a nearly inevitable result in oxidations of hydrogen peroxide.[145,146] Presumably, attack by H_2O_2 at the bromine in BrO_3^- occurs prior to electron transfer in the rate-determining step.

Two-equivalent paths are often written for the reduction of bromate ion by halide ions. There is no firm evidence that this scheme is correct. Indeed, as we have discussed, there is substantial evidence for the formation of Cl_2O_2 in the "analogous" $Cl^--ClO_3^-$ reaction. One question is clear: is Br_2O_2 or, $HBrO_2$ and $HOBr$, formed in the rate-determining step of the $Br^--BrO_3^-$ reaction? Unfortunately, the answer is not known. Noyes has wrestled with related questions in his classic paper on the Belousov–Zhabotinsky oscillating reaction.[49] In a thorough mechanistic analysis of the $Br_2-BrO_3^-$ isotopic exchange data,[98] he proposed the Br_2O_2 intermediate as a means of rapidly equilibrating $HBrO_2$ with lower oxidation states of bromine [Eq. (66)].

$$HOBr + HBrO_2 \rightleftharpoons Br_2O_2 + H_2O \tag{66}$$

The only similarity of reaction (66) to that of chlorine chemistry is the forward reaction. The stoichiometry of the reverse reaction is different from that for the unimolecular decomposition of Cl_2O_2. Also, the Br_2O_2 is proposed to be a symmetrical molecule, whereas Cl_2O_2 is almost certainly unsymmetrical.

It is concluded that the mechanistic details of the reduction of bromate ion by halide ions are uncertain. Edward's proposal of substitution as a prerequisite to electron transfer is quite likely correct. However, it is an open question whether the attack by the halide ion occurs at the bromine or oxygen site of the bromate ion.[137] The former possibility is preferred. The challenge to the experimentalist is to design a definitive experiment.

b. One-equivalent processes. A number of these systems have been investigated and a rich variety of chemistry has emerged. Those reducing agents with oxidation potentials in the range -0.67 to -1.02 V will be discussed first and are referred as "moderately weak." The general rate Equation (67) may be written for this class of reductants (R), although the relative importance of the terms varies widely and additional features are important.

$$d[BrO_3^-]/dt = (k_1 + k_2[H^+]^2)[BrO_3^-][R] + k_3[H^+]^2[BrO_3^-][Br^-] \tag{67}$$

Autocatalysis is observed when the last term is significant. Numerical results at $25°C$ and experimental details for seven reductants are summarized in Table 9, which is a modified version of that presented in Ref. [147].

It is thought that the moderately weak reducing agents listed are capable of reacting with BrO_3^- in a one-equivalent process to form BrO_2. We had noted earlier that the $E°$ values of the $BrO_3^--BrO_2$ couple is approximately 1.15 V. Mercury(0) is exceptional in that it is proposed to reduce BrO_3^- by a two-equivalent reaction. It is included in the table because it derives from Hg_2^{2+}, the predominant form of the reductant. It is interesting that while Hg_2^{2+} could in principle reduce BrO_3^- to BrO_2, the system apparently finds an alternative two-equivalent route that is not available for the other species listed.

Whether autocatalysis is observed depends in part on whether the product Br_2 accumulates or is competitively reduced during the BrO_3^- reaction. We note that the weaker reductants Hg^0, VO^{2+}, and $IrCl_6^{3-}$ do not show

TABLE 9

Kinetic parameters (25°C) and experimental details for the reduction of bromate by moderately weak, one-equivalent agents

Reductant	$E°$ (V)[a]	k_1 (M^{-1} s^{-1})	k_2 (M^{-3} s^{-1})	k_4 (M^{-3} s^{-1})	I (M)	Ref.
Fe(bpy)(CN)$_4$$^{2-}$	0.55[b]	6.2×10^{-3}	0.23	2.9	0.5	[149]
Fe(CN)$_6$$^{4-}$	0.40	1.3×10^{-3}	0.19	1.9	0.5	[150]
Fe^{2+}	0.77	0.37	[c]	[d]	0.5	[151]
Fe(byp)$_2$(CN)$_2$	0.81[e]	3.0×10^{-2}	0.76	2.9	0.5	[149]
Hg^0[f]	0.92	3.7	2.0		1.0	[152]
VO^{2+}	1.0	4.9			2.0	[153]
IrCl$_6$$^{3-}$	0.89[g]		10		0.5	[147]

[a] Values listed are standard reduction potentials for the oxidized form and are taken from Latimer[46] unless otherwise indicated.
[b] Value from Ref. [199].
[c] A more complex term was observed.
[d] Autocatalysis not observed with excess BrO$_3$$^-$.
[e] Value from Ref. [148].
[f] As derived from a study of the BrO$_3$$^-$–Hg$_2$$^{2+}$ reaction.
[g] Value from Ref. [200].

autocatalysis. Of course, autocatalysis will also not be observed if the rate of reduction of BrO$_3$$^-$ greatly exceeds that of the BrO$_3$$^-$–Br$^-$ reaction. Detailed kinetic studies with strong, one-equivalent reducing agents have not been reported.

Reasonable arguments have been presented for either outer-sphere or inner-sphere mechanisms for the k_2 path. The form of the rate expression for this path is identical to that seen for the halide reductions of bromate ion. I propose that the three iron(II) complexes and IrCl$_6$$^{3-}$ attack H$_2$BrO$_3$$^+$ at the central bromine site, just as has been suggested for the halide reactions. I further suggest that the less competitive k_1 path is also inner-sphere for these reductants, with a similar attack on singly protonated BrO$_3$$^-$.

It is likely that the Fe^{2+} and VO^{2+} reactions are inner-sphere as well, but with electrophilic attack at a bromate oxygen. While a k_2 term was not observed for Fe^{2+}, the data were consistent with a parallel pathway involving FeOBrO$_2$$^+$ and H$^+$. The absence of a [H$^+$] dependence for VO^{2+} is not necessarily inconsistent with an inner-sphere mechanism. It may be due to a cancellation of effects; addition of two protons is optimal for the BrO$_3$$^-$– BrO$_2$ transformation, whereas loss of two protons is optimal for the VO^{2+}– VO$_2$$^+$ transformation. The VO^{2+} system is unique among the moderately weak reductants in that disproportionation of a bromous acid intermediate was proposed based on kinetic evidence. This complication was not found for the other reductants.

Although the reduction potentials may be a useful guide in thinking about one-equivalent reductions of BrO_3^-, they are nearly useless in predicting gross reduction rates. In fact, the results in Table 9 are more consistent with an inverse correlation and suggest that the other factors such as have been discussed are much more important in this regard.

The reactions of weak, one-equivalent reductants are considered next. A "weak reductant" is defined as one with an oxidation potential more negative than approximately -1.15 V, but not so negative that it is unreactive toward bromate ion. There are undoubtedly many such species,[154] but we shall consider primarily Ce(III), Mn(II), and Np(V) in sulfuric acid solution.[95,96] There is compelling kinetic evidence that these metal ions do not react directly with BrO_3^-. Instead, after a poorly defined induction period, the observed rate law at 25°C in 3 M H_2SO_4 is Eq. (68) (provided large excesses of Ce(III) or Mn(II) are present):

$$-d[BrO_3^-]/dt = 6.0 \ M^{-1} \ s^{-1} \ [BrO_3^-]^2 \tag{68}$$

The reaction rate is independent of the (excess) metal ion concentration and even the identity of the reductant. A further unusual feature is that HOBr is the sole reduction product. The proposed mechanism is given in Eqs. (69)–(71), and is a critical component of mechanistic interpretations of the Belousov–Zhabotinsky oscillating reaction.

$$BrO_3^- + HBrO_2 + H^+ \xrightarrow{k_a} 2BrO_2 \tag{69}$$

$$BrO_2 + R + H^+ \xrightarrow{k_b} HBrO_2 + Ox \tag{70}$$

$$2HBrO_2 \xrightarrow{k_c} BrO_3^- + HOBr + H^+ \tag{71}$$

The reducing agent is designated as R, and Ox represents its oxidized form (Ce(IV), Mn(III), or Np(VI)). This scheme leads to the kinetic expression given in Eq. (72) if the steady-state approximation is applied to the intermediates BrO_2 and $HBrO_2$.

$$-d[BrO_3^-]/dt = (k_a^2/4k_c)[BrO_3^-]^2[H^+]^2 \tag{72}$$

It is quite probable that Br_2O_4 is the immediate product of reaction (69), but it is known to dissociate rapidly.[141] Reaction (69) is reversible if BrO_2 is not rapidly depleted by use of a large excess of Ce(III) or Mn(II) or by using the more reactive Np(V). The kinetic behavior becomes quite complex with Mn(II) or Ce(III) if excess BrO_3^- is present. Under this condition reaction (70) may be reversible and oxidation of BrO_2 by Mn(III) or Ce(IV) may be encountered. The hydrogen ion dependence shown in Eq. (72) was not demonstrated in the original study[95] but was partially confirmed in a later study of the BrO_3^-–NpO_2^+ reaction in perchloric acid solution.[154]

3. Iodate

Iodate is much more basic than bromate and chlorate, and its oxygen exchange rate with water is quite rapid in acidic solutions. The protonation constant for IO_3^- is $2.0\ M^{-1}$ at $25°C$ and $I = 1.0\ M$.[155-157] The rate constant for oxygen exchange in acidic solutions at $5°C$ and $I = 1.0\ M$ is $3.2 \times 10^3\ M^{-1}\ s^{-1}\ [H^+] + 1.4 \times 10^5\ M^{-2}\ s^{-1}\ [H^+]^2$.[158] Both of these features must be considered when reaction mechanisms for the reduction of iodate are formulated. The kinetically observed reduction product is usually I_2.

The reduction of iodate by vanadium(II) proceeds by one-equivalent steps, but the kinetics are complicated by side reactions.[159] The best estimate of the rate law at $25°C$ and $I = 1.0\ M$ is Eq. (73).

$$-d[I(V)]\,dt = 360\ M^{-1}\ s^{-1}\ [HIO_3][V^{2+}] \tag{73}$$

The rate is too rapid for substitution on V^{2+} to precede electron transfer, but the reaction may be inner-sphere by substitution on HIO_3.

The vanadium(III) reduction probably occurs by one-equivalent steps also, although the V(III)–V(V) reaction is sufficiently rapid[160] to make this conclusion less certain.[159] Under the conditions of excess V(III), $25°C$, and $I = 1.0\ M$, the rate law is Eq. (74).

$$-d[I(V)]/dt = [190\ M^{-1}\ s^{-1}/(1 + 2.0\ M^{-1}\ [H^+])][I(V)][V(III)] \tag{74}$$

This expression is consistent with parallel pathways involving $IO_3^- + V^{3+}$ and $HIO_3 + VOH^{2+}$. With large excesses of iodate, the observed rate constant varies with the ratio $[V(III)]_0/[I(V)]_0$. An analogous variation was also seen in the VO^{2+}–BrO_3^- system[153] and in the present case may be rationalized as involving competition between the disproportionation of an intermediate, probably HIO_2, and its reduction by V^{3+}. The effect is significant at relatively high acidities but disappears when $[H^+] \leq 0.2\ M$.

The kinetics of the Fe(II)–I(V) reaction are very complicated but may be simplified if the iodine(I) intermediate is scavenged by allyl alcohol.[161] The preferred mechanism identifies HIO_3 as the reactive form of iodate and the intermediate $(FeHIO_3)^{2+}$, which competitively undergoes electron transfer either internally or by reaction with Fe^{2+}.

The reduction of iodate by the substitution-inert complex ions $Fe(CN)_6^{4-}$ and $IrCl_6^{3-}$ has been studied.[162,163] The rate laws are complex when the $[H^+]$ dependence is included and may be interpreted mechanistically in a number of ways. One formulation invokes substitution of the reductant into the coordination sphere of iodate and electron transfer within the binuclear complex via parallel pathways with first- and second-order $[H^+]$ dependencies.

The iodate–iodide reaction in acidic solutions (the Dushman reaction) is quite rapid; the kinetics have been studied with very low iodide ion concentrations either by measuring the rate of exchange between radioactive I_2 and IO_3^- [164,165] or by addition of Ag^+.[166,167] The rate law at 25°C, $I = 0.7$ M is Eq. (75).

$$-d[IO_3^-]/dt = 1.4 \times 10^3 \ M^{-3} \ s^{-1} \ [IO_3^-][I^-][H^+]^2 \tag{75}$$

The $(H_2O)_5CrI^{2+}$–IO_3^- system has been investigated.[168] The stoichiometry [Eq. (76)] is analogous to that observed with free I^-

$$5CrI^{2+} + IO_3^- + 6H^+ = 5Cr^{3+} + 3I_2 + 3H_2O \tag{76}$$

The net reaction occurs much more rapidly than the thermal aquation of CrI^{2+}. The reaction is autocatalytic, and at 20°C, $I = 1.0$ M and $[H^+] = 0.95$ M, the rate expression is Eq. (77).

$$-d[CrI^{2+}]/dt = 8 \ M^{-1} \ s^{-1} \ [CrI^{2+}][I_2] \tag{77}$$

The intriguing mechanism proposed is a rate-determining formation of CrI_3^{2+} at low concentration, followed by a relatively rapid reaction with IO_3^-.

The iodate ion is an important component in the Bray–Liebhafsky[169,170] and Briggs–Rauscher[171] oscillating reactions. Given the current interest in these systems, increased attention to the kinetics of reduction of iodate is anticipated in the near future.

B. Oxidation

It may seem bizarre to consider the halates as reducing agents, but in fact they can be partially oxidized to perhalates by sufficiently potent oxidants. One such intriguing oxidant to consider is XeF_2. The standard reduction potential for XeF_2 [Eq. (78)] has been estimated to be 2.64 V.[172]

$$XeF_2 + 2H^+ + 2e^- = Xe(g) + 2HF \tag{78}$$

In spite of this formidable potential, the half-life of XeF_2 in water is approximately 27 min at 25°C.[173] It forms an intermediate of uncertain composition that has been proposed as the reactive species in the oxidation of ClO_3^- and BrO_3^-.[174] The complete scheme as outlined by Appelman is given in Eqs. (79)–(84):

$$XeF_2 + H_2O \xrightarrow{k_1} XeO + 2HF \tag{79}$$

$$XeO + H_2O \xrightarrow{k_2} Xe + H_2O_2 \tag{80}$$

$$XeO + ZO_3^- \xrightarrow{k_3} Xe + ZO_4^-* \tag{81}$$

$$ZO_4^-* \xrightarrow{k_4} ZO_4^- \tag{82}$$

$$ZO_4^-* + H_2O \xrightarrow{k_5} ZO_3^- + H_2O_2 \tag{83}$$

$$XeF_2 + H_2O_2 \xrightarrow{k_6} Xe + O_2 + 2HF \tag{84}$$

In this scheme the reactive intermediate formed from XeF_2 is represented by XeO; ZO_3^- is either ClO_3^- or BrO_3^-; and ZO_4^-* is a second reactive intermediate that either froms the perhalate [Eq. (82)] or oxidizes water [Eq. (83)]. Under the experimental condition of excess halate, application of the steady-state approximation for XeO, ZO_4^-*, and H_2O_2 leads to Eq. (85), where

$$[XeF_2]_0/[ZO_4^-]_\infty = K_1 + K_2/[ZO_3^-] \tag{85}$$

$K_1 = 2k_5/k_4$ and $K_2 = 2k_2(k_4 + k_5)/k_3k_4$. In slightly acidic solution at 25°C, the experimental values are $K_1 = 1.07$ (for ClO_3^-) and 8.1 (for BrO_3^-), and $K_2 = 0.23$ (for ClO_3^-) and 0.93 (for BrO_3^-). The residual BrO_3^- gains oxygen-18 from the solvent to an extent that appears to be consistent with four equivalent oxygens in BrO_4^-*. The iodate system is more complicated due to a direct XeF_2–IO_3^- reaction and reduction of IO_4^- by H_2O_2.

While the mechanism of these reactions is interesting, XeF_2 is not a practical reagent for the synthesis of BrO_4^- due to the low yields observed.

X. REDUCTION OF PERHALATES

A. Perchlorate

Perchlorate salts are used to maintain constant ionic strength in a variety of studies. The perchlorate ion forms very weak complexes with metal ions and is usually kinetically inert as an oxidizing agent in spite of its formidable reduction potential. However, a few metal ions have been found to reduce ClO_4^- at a significant rate. Some well-documented examples are listed in Table 10. Notable exceptions are Cr^{2+} and Eu^{2+} which do not normally react with ClO_4^-. Some of these metal ions can function as catalysts. For example, Ru^{2+} catalyzes the reduction of ClO_4^- by Cr^{2+}.[180] Additional examples of catalysts are shown in Table 11, although under certain conditions some of the "catalyst" is consumed.

We anticipate that a number of additional examples of the reduction of perchlorate ion will be discovered as chemists study the unusual oxidation states of the less familiar elements. The renewed interest of Taube's group in

TABLE 10

Kinetic results for the reduction of perchlorate by metal ions

Metal ion	Reduction product	Oxidation product	I (M)	T $(°C)$	k $(M^{-1} s^{-1})^a$	Ref.
Ru^{2+}	$ClO_3^{-\ b}$	Ru^{3+b}	0.3	25	3.2×10^{-3}	[175]
$Ru(NH_3)_5OH_2^{2+}$	Cl^-	$(NH_3)_5RuCl^{2+\ c}$	0.15	25	3.3×10^{-3}	[176]
Ti^{3+}	Cl^-	TiO^{2+}	1.0	25	$2.4 \times 10^{-5\ d}$	[177, 178]
$Ru(NH_3)_6^{2+}$	Cl^-	$Ru(NH_3)_6^{3+\ e}$	0.62	25	3.8×10^{-5}	[176]
V^{2+}	Cl^-	f	2.5	50	2.8×10^{-6}	[179]
V^{3+}	Cl^-	VO^{2+}	2.5	50	5×10^{-7}	[179]

^{a} $-d[ClO_4^-]/dt = k[ClO_4^-][\text{metal ion}]$.
^{b} With excess $[ClO_4^-]$, less than 2% dimeric Ru(III) and only traces of $RuCl^{2+}$ were detected.
^{c} 0.14 M Cl^- present in reaction solutions.
^{d} In addition, a term $= 1.6 \times 10^{-5}$ M^{-2} s^{-1} $[H^+]$ was determined. Numerical values listed are from Ref. [177].
^{e} Competitive spontaneous decomposition of $Ru(NH_3)_6^{2+}$ was observed.
^{f} The V(III) product competitively reduces ClO_4^-.

TABLE 11

Some catalysts for the reduction of perchlorate ion

Catalyst	Reductant	Medium	Ref.
Solid MnO_2	Cr^{2+}	1 M $HClO_4$	[181]
Nb(V) or W(VI)	Sn(II)	11 M HCl	[182]
Mo(VI)	Sn(II)	3 M H_2SO_4	[183]

the reactivity of ClO_4^- should lead to new developments shortly.[184] As a cautionary note, the possibility of catalysis must be carefully considered when reduction of ClO_4^- is observed.

The mechanistic features are at best only partially understood. Certainly replacement of an oxo ligand by a nucleophile does not precede electron transfer because of the almost unbelievably slow rate of oxygen exchange between ClO_4^- and H_2O.[136] An outer-sphere reduction of ClO_4^- is unlikely since the strength of the observed reducing agents is nearly immaterial. Even the hydrated electron is inert toward ClO_4^-, and the upper limit of 10^5 M^{-1} s^{-1} is dictated by the lifetime of e_{aq}^-. The true, though hypothetical, rate constant may be many orders of magnitude smaller. Although $Ru(NH_3)_6^{3+}$ is the oxidation product of the $Ru(NH_3)_6^{2+}$–ClO_4^- reaction, attack at the octahedral edge of the Ru(II) center has been suggested.

We are left with substitution at the reducing agent as the likely event prior to electron transfer. This has been convincingly proposed for Ru^{2+}, since the activation parameters for ClO_4^- reduction are strikingly similar to those for substitution by Cl^-, Br^-, and I^-. The presumed inner-sphere perchlorato complex survives long enough for electron transfer to occur due to the substitution inertness of the metal center. The reluctance of ClO_4^- to become involved in electron transfer may be understood in part by the scheme shown in Eqs. (78)–(80):

$$ClO_4^- + M^{n+} \underset{k_2}{\overset{k_1}{\rightleftharpoons}} M(OClO_3)^{n-1} \tag{86}$$

$$M(OClO_3)^{n-1} \overset{k_3}{\longrightarrow} \text{products} \tag{87}$$

$$-d[ClO_4^-]/dt = [k_1 k_3/(k_2 + k_3)][ClO_4^-][M^{n+}] \tag{88}$$

The ratio k_1/k_2 will be small for most, if not all, metal ions. However, the factors that affect the magnitude of k_3 are poorly understood, yet this is probably the crucial point. Oxygen atom transfer to the reductant is an attractive proposal but has not been demonstrated in a single case at the time of this writing.

Only for Ti^{3+} has a significant term dependent on $[H^+]$ been reported, in contrast to many reductions of oxoanions with the halogen in lower oxidation states. While the role of the proton in the Ti^{3+} system is not understood, it is useful nonetheless to suggest that *any* authentic $[H^+]$ dependent observed for the reduction of ClO_4^- will be associated with the chemistry of the reductant and not with ClO_4^- due to its extremely weak basicity.

B. Perbromate

The perbromate ion is comparable to the perchlorate ion in several ways. It is an extremely weak base, tetrahedral with no tendency to form octahedral species, thermodynamically a very potent oxidant, and extremely sluggish with regard to oxygen exchange with water.[72] Nevertheless, it shows considerably enchanced reactivity toward electron transfer relative to ClO_4^-. It is therefore surprising that so few kinetic studies of the reduction of BrO_4^- have appeared. This may be due in part to the unavailability of perbromate salts from commercial sources, but the synthesis has been fully described.[18]

Oxygen atom transfer from BrO_4^- to sulfur(IV) and arsenic(III) has been demonstrated.[185] The reactivity order of some potentially two-equivalent reducing agents is $N_3^- < ClO_2 \sim SeO_3^{2-} \sim NO_2^- < P(III) \ll S(IV) \leq As(III) < Sb(III)$. The factors that govern these reactivities are poorly understood. However, the increasing rate with increasing size of Group VA elements may reflect increased ease of oxygen transfer due to the increasing polarizability of the reductant.[186] Any $[H^+]$ dependence observed in BrO_4^- reactions to data may be ascribed solely to the chemistry associated with the reductant and *not* the oxidant.

Kinetic studies of the reduction of BrO_4^- by one-equivalent species are virtually nonexistent. We note, however, that the hydrated electron reacts very rapidly with BrO_4^- (and BrO_3^-) and IO_4^- (and IO_3^-), but not with ClO_4^- (or ClO_3^-).[187] This illustrates that while oxygen lability is often an important factor in oxohalogen reductions, it is by no means the only factor.

C. Periodate

The periodate ion is unique among the perhalates in that it readily forms octahedral species. Species IO_4^-, $H_4IO_6^-$, and H_5IO_6 are in rapid equilibrium in acidic solutions, with H_5IO_6 the predominant form in strongly acidic solution.[188] The rate of oxygen exchange between periodate and water is rapid.[189] It is therefore not unexpected that inner-sphere mechanisms dominate the reduction of periodate, by substitution either on the reductant (Fe^{2+},[190] VO^{2+},[191] and $Co(edta)^{2-}$,[192]) or on the oxidant ($Fe(CN)_6^{4-}$,[193] Cr^{3+},[194,195] and I^- [196–198]). In addition, $[H^+]$-dependent terms in the rate laws are often observed and may be associated with the oxidant, in sharp contrast to reduction of ClO_4^- and BrO_4^-.

References

[1] For a recent, readable review, see Epstein, I. R.; Kustin, K.; DeKepper, P.; Orban, M. *Sci. Am.* **1983**, *248*, 112.
[2] Appelman, E. H. *Acc. Chem. Res.* **1973**, *6*, 113.
[3] Silverman, R. A.; Gordon, G. *Anal. Chem.* **1974**, *46*, 178.
[4] Thompson, R. C. *Inorg. Chem.* **1983**, *22*, 584.
[5] Noszticzius, Z.; Noszticzius, E.; Schelly, Z. A. *J. Am. Chem. Soc.* **1982**, *104*, 6194.
[6] Cady, G. H. *Inorg. Synth.* **1957**, *5*, 160.
[7] Mellor, J. W. "Inorganic and Theoretical Chemistry"; Longmans, Green: London, 1956, Suppl. II, Part I, p. 570.
[8] a) Forsterling, H. D.; Lamberz, H. J.; Schreiber, H. *Z. Naturforsch. A.* **1980**, *35*, 1354.
   b) Lamberz, H. J. Ph.D. Thesis, Phillipps University, Marburg, 1982, to be published.
[9] Noszticzius, Z.; Noszticzius, E.; Schelly, Z. A. *J. Phys. Chem.* **1983**, *87*, 510.
[10] Kieffer, R. G.; Gordon, G. *Inorg. Chem.* **1968**, *7*, 235,239.
[11] Lednicky, L. A.; Stanbury, D. M. *J. Am. Chem. Soc.* **1983**, *105*, 3098.
[12] Thompson, R. C. *J. Phys. Chem.* **1968**, *72*, 2642.
[13] Field, R. J.; Raghavan, N. V.; Brummer, J. C. *J. Phys. Chem.* **1982**, *86*, 2443.
[14] Buxton, G. V.; Dainton, F. S. *Proc. R. Soc. London, Ser. A* **1968**, *304*, 427.
[15] Amichai, O.; Czapski, G.; Treinin, A. *Isr. J. Chem.* **1969**, *7*, 351.
[16] Amichai, O.; Treinin, A. *J. Phys. Chem.* **1970**, *74*, 3670.

[17] Forsterling, H. D.; Lamberz, H. J.; Schreiber, H.; Zittlau., W. *Acta Chim. Acad. Sci. Hung.* **1982,** *110,* 251.

[18] Appelman, E. H. *Inorg. Synth.* **1972,** *13,* 1.

[19] Zimmerman, G.; Strong, F. C. *J. Am. Chem. Soc.* **1957,** *79,* 2063.

[20] Ross, A. R.,; Neta, P. "Rate Constants for Reactions of Inorganic Radicals in Aqueous Solution"; *N5RDS-NBS Publication No. 65,* **1979,** p. 3.

[21] Silverman, R. A.; Gordon, G. *J. Phys. Chem.* **1980,** *84,* 625.

[22] Betts, R. H.; Mackenzie, A. N. *Can. J. Chem.* **1951,** *29,* 666.

[23] Carter, P. R.; Davidson, N. *J. Phys. Chem.* **1952,** *56,* 877.

[24] Woodruff, W. H.; Margerum, D. W. *Inorg. Chem.* **1973,** *12,* 958.

[25] Awtrey, A. D.; Connick, R. E. *J. Am. Chem. Soc.* **1951,** *73,* 1842.

[26] Kern, D. M.; Kim, C. *J. Am. Chem. Soc.* **1965,** *87,* 5309.

[27] Johnson, R. W.; Edwards, J. O. *Inorg. Chem.* **1966,** *5,* 2073.

[28] Oxygen exchange rates of oxohalogen species are discussed in a recent review. See: Gamsjäger, H.; Murmann, R. K. *Adv. Inorg. Bioinorg. Mech.* **1983;** *2,* 317.

[29] Wharton, R. K.; Taylor, R. S.; Sykes, A. G. *Inorg. Chem.* **1975,** *14,* 33.

[30] Harrowfield, J. M.; Sargenson, A. M.; Singh, B.; Sullivan, J. C. *Inorg. Chem.* **1975,** *14,* 2864.

[31] Duval, R. *Ann. Chim. (Paris)* **1932,** *18,* 241.

[32] Jones, W. E.; Swaddle, T. W. *Can. J. Chem.* **1967,** *45,* 2647.

[33] Ramasami, T.; Wharton, R. K.; Sykes, A. G. *Inorg. Chem.* **1975,** *14,* 359.

[34] Jones, K. M.; Bjerrum, J. *Acta Chim. Scand.* **1965,** *19,* 974.

[35] Ostrich, I.; Leffler, A. J. *Inorg. Chem.* **1983,** *22,* 921.

[36] Taylor, R. S.; Sykes, A. G. *Chem. Commun.* **1969,** 1137.

[37] Thompson, R. C. *Inorg. Chem.* **1975,** *14,* 1279.

[38] Thompson, R. C. *Inorg. Chem.* **1979,** *18,* 2379.

[39] Gordon, G.; Kern, D. M. H. *Inorg. Chem.* **1964,** *3,* 1055.

[40] Martell, A. E.; Smith, R. M. "Critical Stability Constants", Vol. 4; Plenum: New York, 1976.

[41] Gordon, G.; Emmenegger, F. *Inorg. Nucl. Chem. Lett.* **1966,** *2,* 395.

[42] Fukutomi, H.; Gordon, G. *J. Am. Chem. Soc.* **1967,** *89,* 1362.

[43] Gwnne, E.; Davies, M. *J. Am. Chem. Soc.* **1952,** *74,* 2748.

[44] Griffith, R. O.; McKeown, A.; Winn, A. G. *Trans. Faraday Soc.* **1932,** *28,* 101.

[45] Bell, R. P.; Pring, M. *J. Chem. Soc. (A)* **1966,** 1607.

[46] Latimer, W. M. "Oxidation Potentials"; 2nd ed., Prentice-Hall: New York, 1952.

[47] *J. Phys. Chem. Ref. Data* **1982, II** (Suppl. 2).

[48] Schreiner, F.; Osborne, D. W.; Pocius, A. V.; Appelman, E. H. *Inorg. Chem.* **1970,** *10,* 2320.

[49] Field, R. J.; Körös, E.; Noyes, R. M. *J. Am. Chem. Soc.* **1972,** *94,* 8649.

[50] Noyes, R. M. *J. Am. Chem. Soc.* **1980,** *102,* 4644.

[51] Bell, R. P.; Gelles, E. *J. Chem. Soc.* **1951,** 2734.

[52] Allen, T. L.; Keefer, A. M. *J. Am. Chem. Soc.* **1955,** *77,* 2957.

[53] Henglein, A. *Radiat. Phys. Chem.* **1980,** *15,* 151.

[54] Thornton, A. T.; Laurence, G. S. *J. Chem. Soc., Dalton Trans.* **1972,** 1632.

[55] Woodruff, W. H.; Margerum, D. W. *Inorg. Chem.* **1972,** *12,* 962.

[56] Raycheba, J. M. T.; Margerum, D. W. *Inorg. Chem.* **1981,** *20,* 45.

[57] Connick, R. E.; Chia, Y. *J. Am. Chem. Soc.* **1959,** *81,* 1280.

[58] Liebhafsky, H. A. *J. Am. Chem. Soc.* **1939,** *61,* 3513.

[59] Eigen, M.; Kustin, K. *J. Am. Chem. Soc.* **1962,** *84,* 1355.

[60] Eigen, M. *Angew. Chem. Int. Ed. Engl.* **1964,** *3,* 1.

[61] Held, A. M.; Halko, D. J.; Hurst, J. K. *J. Am. Chem. Soc.* **1978,** *100,* 5732.

[62] Hurst, J. K.; Carr, P. A. G.; Hovis, F. E.; Richardson, R. J. *Inorg. Chem.* **1981,** *20,* 2435.

[63] Anbar, M.; Taube, H. *J. Am. Chem. Soc.* **1958,** *80,* 1073.

[64] Moissan, H. "Le Fluor et Ces Composes"; Paris, 1900.

[65] Appelman, E. H.; Thompson, R. C. *J. Am. Chem. Soc.* **1984,** *106,* 4167.

[66] Astarita, G. "Mass Transfer with Chemical Reaction"; Elsevier: Amsterdam, 1967.

[67] Danckwerts, P. V. "Gas Liquid Reactions"; McGraw-Hill: New York, 1970.

[68] Spalding, C. W. *A.I.Ch.E.J.* **1962,** *8,* 685.

[69] Hikita, H.; Asai, S.; Himukashi, Y.; Takatsuka, T. *J. Chem. Eng.* **1973,** *5,* 77.

[70] Appelman, E. H.; Thompson, R. C.; Engelkemeir, A. G. *Inorg. Chem.* **1979,** *18,* 909.

[71] Knoth, W. H.; Miller, H. C.; Sauer, J. C.; Balthis, J. H.; Chia, Y. T.; Muetterties, E. L. *Inorg. Chem.* **1964,** *3,* 159.

[72] Appelman, E. H. *Inorg. Chem.* **1969,** *8,* 223.

[73] Appelman, E. H.; Basile, L. J.; Thompson, R. C. *J. Am. Chem. Soc.* **1979,** *101,* 3384.

[74] Thompson, R. C.; Appelman, E. H.; Sullivan, J. C. *Inorg. Chem.* **1977,** *16,* 2921.

[75] Cornelius, R. D.; Gordon, G. *Inorg. Chem.* **1976,** *15,* 997, 1002.

[76] Dreyer, K.; Gordon, G. *Inorg. Chem.* **1972,** *11,* 1174.

[77] Shakhashiri, B. Z.; Gordon, G. *Inorg. Chem.* **1968,** *7,* 2454.

[78] Ondrus, M. G.; Gordon, G. *Inorg. Chem.* **1971,** *10,* 474.

[79] Thompson, R. C.; Gordon, G. *Inorg. Chem.* **1966,** *5,* 557, 562.

[80] Malin, J. M.; Swinehart, J. H. *Inorg. Chem.* **1969,** *8,* 1407.

[81] Conocchioli, T. J.; Hamilton, J. H.; Sutin, N. *J. Am. Chem. Soc.* **1965,** *87,* 926.

[82] Thornton, A. T.; Laurence, G. S. *J. Chem. Soc., Dalton Trans.* **1973,** 804.

[83] Markower, B.; Bray, W. C. *J. Am. Chem. Soc.* **1933,** *55,* 4765.

[84] Connick, R. E. *J. Am. Chem. Soc.* **1947,** *69,* 1309.

[85] Thompson, R. C. *Inorg. Chem.* **1984,** *23,* 1794.

[86] Adegite, A.; Iyun, J. F. *Inorg. Chem.* **1979,** *18,* 3602.

[87] Ramsey, J. B.; Heldman, M. J. *J. Am. Chem. Soc.* **1936,** *58,* 1153.

[88] Adegite, A.; Ford-Smith, M. H. *J. Chem. Soc. Dalton Trans.* **1973,** 134.

[89] Andrewes, A.; Gordon, G. *Inorg. Chem.* **1964,** *3,* 1733.

[90] Gordon, G. *Proc. Int. Conf. Coord. Chem., 13th* **1970,** 11.

[91] Silverman, R. A.; Gordon, G. *Inorg. Chem.* **1976,** *15,* 35.

[92] Adegite, A.; Ford-Smith, M. H. *J. Chem. Soc., Dalton Trans.* **1973,** *138.*

[93] Newton, T. W.; Baker, F. B. *Adv. Chem. Ser.* **1967,** (71), 268.

[94] Hegedus, L. S.; Haim, A. *Inorg. Chem.* **1967,** *4,* 664.

[95] Thompson, R. C. *J. Am. Chem. Soc.* **1971,** *93,* 7315.

[96] Noyes, R. M.; Field, R. J.; Thompson, R. C. *J. Am. Chem. Soc.* **1971,** *93,* 7315.

[97] Emmenegger, F.; Gordon, G. *Inorg. Chem.* **1967,** *6,* 633.

[98] Betts, R. H.; MacKenzie, A. N. *Can. J. Chem.* **1951,** *29,* 655.

[99] Gordon, G.; Kieffer, R. G.; Rosenblatt, D. H. *Prog. Inorg. Chem.* **1972,** *15,* 201. This article contains a more extensive discussion of the disproportionation of chlorous acid and also reviews the reported values of the dissociation constant for chlorous acid.

[100] Bar-Eli, K.; Ronkins, J. *J. Phys. Chem.* **1984,** *88,* 2844.

[101] Edelson, D.; Field, R. J.; Noyes, R. M. *Int. J. Chem. Kinet.* **1975,** *7,* 417.

[102] Sullivan, J. C.; Thompson, R. C. *Inorg. Chem.* **1979,** *18,* 2375.

[103] Noyes, R. M.; Furrow, S. D. *J. Am. Chem. Soc.* **1982,** *104,* 45.

[104] Epstein, I. R.; DeKepper, P. *J. Am. Chem. Soc.* **1982,** *104,* 49.

[105] Buchacek, R.; Gordon, G. *Inorg. Chem.* **1972,** *11,* 2154.

[106] Ondrus, M. G.; Gordon, G. *Inorg. Chem.* **1972,** *11,* 985.

[107] DeMerus, J.; Sigalla, J. *Chim. Phys. Phys.-Chim. Biol.* **1966,** *63,* 453.

[108] Indelli, A. *J. Phys. Chem.* **1964,** *68,* 3027.

[109] Bray, W. C. *Z. Phys. Chem.* **1906,** *54,* 741.

[110] Grant, J. L.; DeKepper, P.; Epstein, I. R.; Kustin, K.; Orbán, M. *Inorg. Chem.* **1982,** *21,* 2192.

[111] Emeish, S. S.; Howlett, K. E. *Can. J. Chem.* **1980**, *58*, 159.
[112] Shakhashiri, B. Z.; Gordon, G. *J. Am. Chem. Soc.* **1969**, *91*, 1103.
[113] Higginson, W. C. E.; Khan, A. H. *J. Chem. Soc., Dalton Trans.* **1981**, 2537.
[114] Epstein, I. R. *J. Phys. Chem.* **1984**, *88*, 187.
[115] Alamgir, M.; Epstein, I. R. *J. Phys. Chem.* **1984**, *88*, 2848.
[116] Orbán, M.; Dateo, C.; DeKepper, P.; Epstein, I. R. *J. Am. Chem. Soc.* **1982**, *104*, 5911.
[117] Thompson, R. C. *Inorg. Chem.* **1973**, *12*, 1905.
[118] Orbán, M.; Epstein, I. R. *J. Phys. Chem.* **1983**, *87*, 3212.
[119] Ahlstrom, C.; Boyd, D. W.; Epstein, I. R.; Kustin, K. *Inorg. Chem.* **1984**, *23*, 2185.
[120] Stanbury, D. M.; Lednicky, L. A. *J. Am. Chem. Soc.* **1984**, *106*, 2847.
[121] Dodgen, H.; Taube, H. *J. Am. Chem. Soc.* **1949**, *71*, 2501.
[122] McDowell, M. S.; Espenson, J. H.; Bakac, A. *Inorg. Chem.* **1984**, *23*, 2232.
[123] Thompson, R. C. *Inorg. Chem.* **1981**, *20*, 1005.
[124] Thompson, R. C.; Wieland, P.; Appelman, E. H. *Inorg. Chem.* **1979**, *18*, 1974.
[125] Basolo, F.; Pearson, R. G. "Mechanisms of Inorganic Reactions"; 2nd ed., Wiley, New York, 1968, Ch. 3.
[126] Hanna, S. B.; Kessler, R. R.; Merbach, A.; Ruzicka, S. *J. Chem. Educ.* **1976**, *53*, 524.
[127] Hanna, S. B.; Sarac, S. A. *J. Org. Chem.* **1977**, *42*, 2063.
[128] Melvin, W. S.; Gordon, G. *Inorg. Chem.* **1972**, *11*, 1912.
[129] The numerical value for the dimerization constant of BrO_2 given in Ref. [14] is incorrect, as pointed out in Refs. [8 and 13].
[130] Davies, C.; Warnqvist, B. *Coord. Chem. Rev.* **1970**, *5*, 1970.
[131] Young, H. A.; Bray, W. C. *J. Am. Chem. Soc.* **1932**, *54*, 4284.
[132] Thompson, R. C. *Inorg. Chem.* **1969**, *8*, 1891.
[133] Gordon, G.; Tewari, P. *J. Phys. Chem.* **1966**, *70*, 200.
[134] Halperin, J.; Taube, H. *J. Am. Chem. Soc.* **1952**, *74*, 375.
[135] Nixon, A. C.; Krauskopf, K. B. *J. Am. Chem. Soc.* **1932**, *54*, 4606.
[136] Hoering, T. C.; Ishimori, F. T.; McDonald, H. O. *J. Am. Chem. Soc.* **1958**, *80*, 3876.
[137] Barton, A. F. M.; Wright, G. A. *J. Chem. Soc. (A)* **1968**, 1747.
[138] Sigalla, J. *J. Chim. Phys.* **1958**, *55*, 758.
[139] Hoering, T. C.; Butler, R. C.; McDonald, H. O. *J. Am. Chem. Soc.* **1956**, *78*, 4829.
[140] Gamsjäger, H.; Grütter, A.; Baertschi, P. *Helv. Chim. Acta* **1972**, *55*, 781.
[141] Williamson, F. S.; King, E. L. *J. Am. Chem. Soc.* **1957**, *79*, 5397.
[142] Chaffee, E.; Edwards, J. O. *Prog. Inorg. Chem.* **1970**, *13*, 205.
[143] Higginson, W. C. E.; Marshall, J. *J. Chem. Soc.* **1957**, 447.
[144] O'Connor, C. J. *Int. J. Appl. Radiat. Isotop.* **1967**, *18*, 700.
[145] Cahill, A. E.; Taube, H. *J. Am. Chem. Soc.* **1952**, *74*, 2312.
[146] Anbar, M. *J. Am. Chem. Soc.* **1961**, *83*, 2031.
[147] Birk, J. P. *Inorg. Chem.* **1978**, *17*, 504.
[148] Campion, R. J.; Purdie, N.; Sutin, N. *Inorg. Chem.* **1964**, *3*, 1091.
[149] Birk, J. P.; Kozub, S. G. *Inorg. Chem.* **1978**, *17*, 1186.
[150] Birk, J. P.; Kozub, S. G. *Inorg. Chem.* **1973**, *12*, 2460.
[151] Birk, J. P. *Inorg. Chem.* **1973**, *12*, 2468.
[152] Davies, R.; Kipling, B.; Sykes, A. G. *J. Am. Chem. Soc.* **1973**, *95*, 7250.
[153] Thompson, R. C. *Inorg. Chem.* **1971**, *10*, 1892.
[154] Knight, G. C.; Thompson, R. C. *Inorg. Chem.* **1973**, *12*, 63.
[155] Naidich, S.; Ricci, J. E. *J. Am. Chem. Soc.* **1939**, *61*, 3268.
[156] Ramette, R. W. *J. Chem. Educ.* **1959**, *36*, 191.
[157] Gamsjäger, H.; Gerber, F.; Antonsen, O. *Chimica (Switz.)* **1973**, *27*, 94.
[158] Von Felton, H.; Gamsjäger, H.; Baertschi, P. *J. Chem. Soc., Dalton Trans.* **1976**, 1683.
[159] Bakać, A.; Thornton, A. T.; Sykes, A. G. *Inorg. Chem.* **1976**, *15*, 274.

[160] Daugherty, N. A.; Newton, T. W. *J. Phys. Chem.* **1964**, *68*, 612.

[161] Higginson, W. C. E.; McCarthy, D. A. *J. Chem. Soc., Dalton Trans.* **1980**, 797.

[162] Sulfab, Y.; Elfaki, H. A. *Can. J. Chem.* **1974**, *52*, 2001.

[163] Birk, J. P. *Inorg. Chem.* **1978**, *17*, 1372.

[164] Myers, O. E.; Kennedy, J. W. *J. Am. Chem. Soc.* **1950**, *73*, 897.

[165] Connick, R. E.; Hugus, Z. Z. Brookhaven Conference Report BNL-C-8, Chemical Conference No. 2, 1948, p. 164.

[166] Furuichi, R.; Matsuzaki, I.; Simic, R.; Liebhafsky, H. A. *Inorg. Chem.* **1972**, *11*, 952.

[167] Furuichi, R.; Liebhafsky, H. A. *Bull. Chem. Soc. Jpn.* **1975**, *48*, 745.

[168] Espenson, J. H. *Inorg. Chem.* **1964**, *3*, 968.

[169] Bray, W. C. *J. Am. Chem. Soc.* **1921**, *43*, 1262.

[170] Liebhafsky, H. A.; McGavock, W. C.; Reyes, R. J.; Roe, G. M.; Wu, L. S. *J. Am. Chem. Soc.* **1978**, *100*, 87.

[171] Briggs, T. S.; Rauscher, W. C. *J. Chem. Educ.* **1973**, *50*, 496.

[172] Malm, J. G.; Appelman, E. H. *At. Energy Rev.* **1969**, *7*, 21.

[173] Appelman, E. H. *Inorg. Chem.* **1967**, *6*, 1305.

[174] Appelman, E. H. *Inorg. Chem.* **1971**, *10*, 1881.

[175] Kallen, T. W.; Earley, J. E. *Inorg. Chem.* **1971**, *10*, 1152.

[176] Endicott, J. F.; Taube, H. *Inorg. Chem.* **1965**, *4*, 437.

[177] Cope, V. W.; Miller, R. G.; Fraser, R. T. M. *J. Chem. Soc. (A)* **1967**, 301.

[178] Duke, F. R.; Quinney, P. D. *J. Am. Chem. Soc.* **1954**, *76*, 3800.

[179] King, W. R.; Garner, C. S. *J. Phys. Chem.* **1954**, *58*, 29.

[180] Seewald, D.; Sutin, N.; Watkins, K. O. *J. Am. Chem. Soc.* **1969**, *91*, 7307.

[181] Zabin, B. A. Taube, H. *Inorg. Chem.* **1964**, *3*, 963.

[182] Haight, G. P.; Swift, A. C.; Scott, R. *Acta Chem. Scand. A* **1979**, *33*, 47.

[183] Haight, G. P.; Sager, W. F. *J. Am. Chem. Soc.* **1952**, *74*, 6056.

[184] See: *Chem. Eng. News* **1984**, *62*, 31.

[185] Appelman, E. H.; Kläning, U. K.; Thompson, R. C. *J. Am. Chem. Soc.* **1979**, *101*, 929.

[186] Edwards, J. O. "Inorganic Reaction Mechanisms"; Benjamin, New York, 1964, Chs. 4 and 5.

[187] Olsen, K. J.; Sehested, K.; Appelman, E. H. *Chem. Phys. Lett.* **1973**, *19*, 213.

[188] Crouthamel, C.; Hayes, A.; Martin, D. *J. Am. Chem. Soc.* **1951**, *73*, 82.

[189] Pecht, I.; Luz, Z. *J. Am. Chem. Soc.* **1965**, *87*, 4068.

[190] El-Eziri, F. R.; Sulfab, Y. *Inorg. Chim. Acta* **1977**, *25*, 15.

[191] Sulfab, Y.; Abu-Shaby, A. I. *Inorg. Chim. Acta* **1977**, *21*, 115.

[192] Kassim, A. Y.; Sulfab, Y. *Inorg. Chim. Acta* **1977**, *24*, 247.

[193] Sulfab, Y. *J. Inorg. Nucl. Chem.* **1976**, *38*, 2271.

[194] Kassim, A. Y.; Sulfab, Y. *Inorg. Chem.* **1981**, *20*, 506.

[195] Sulfab, Y. *Polyhedron* **1983**, *2*, 679.

[196] Indelli, A.; Ferranti, F.; Secco, F. *J. Phys. Chem.* **1966**, *70*, 631.

[197] Marques, C.; Hasty, R. A. *J. Chem. Soc., Dalton Trans.* **1980**, 1266.

[198] Ferranti, F.; Indelli, A. *J. Chem. Soc., Dalton Trans.* **1984**, 1773.

[199] Stasiw, R.; Wilkins, R. G. *Inorg. Chem.* **1969**, *8*, 156.

[200] Margerum, D. W.; Chellappa, K. L.; Bossu, F. P.; Burce, G. L. *J. Am. Chem. Soc.* **1975**, *97*, 6894.

Kinetic Studies on edta Complexes of Transition Metals

Hiroshi Ogino and Makoto Shimura

Department of Chemistry
Faculty of Science, Tohoku University
Sendai, Japan

I.   INTRODUCTION

Ethylenediamine-N,N,N',N'-tetraacetic acid (edta) was first reported in 1937.[1] It is a potentially sexidentate ligand which contains two nitrogen donor atoms and four oxygen donor atoms. This particular ligand forms very stable 1:1 complexes with various metal ions at appropriate pH. The edta complexes are much more stable toward base hydrolysis than the corresponding hydrated metal ions. When hydrated metal ions are coordinated to edta, the redox behavior is also altered. Thus, the reduction potentials of metal–edta redox couples always shift to the negative side compared with the potentials of the corresponding hydrated metal redox couples. This trend indicates that, by ligating with edta, the higher oxidation state of metal ions is stabilized relative to the lower oxidation state. These features provide wide applicability of edta and metal–edta complexes not only in analytical

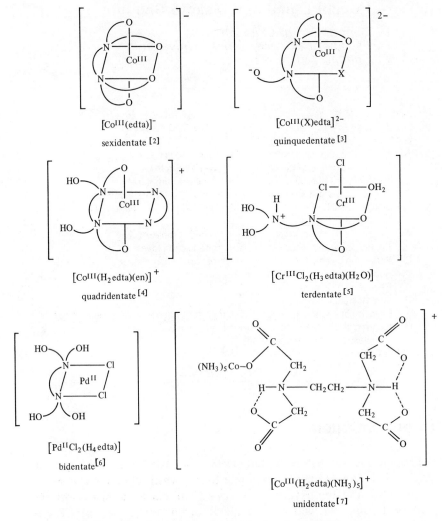

Fig. 1. Examples of various coordination modes of metal–edta complexes.

procedures but also in industrial, medical, biological, and agricultural usage. From these points of view, it is important to have an understanding of kinetic aspects of metal–edta complexes.

Though edta is a sexidentate ligand, it is quite another matter whether all six donor atoms are used upon coordination to a metal ion. Examples of metal complexes in which edta acts as sexidentate to unidentate ligand are shown in Fig. 1. Many other coordination modes are possible, though these are not shown in Fig. 1.

Many aminocarboxylic acid homologs of edta are known. By variation of the amino carboxylic acid, one can change not only the charge, but also reduction potential, hydrophilicity, etc. Therefore, aminopolycarboxylic acids serve as a convenient probe for elucidating factors controlling reactions.

Recently an amino acid, mugineic acid, was isolated from the roots of water-cultured barley (*Hordeum vulgare* L. var. Minorimugi).[8] This is the first

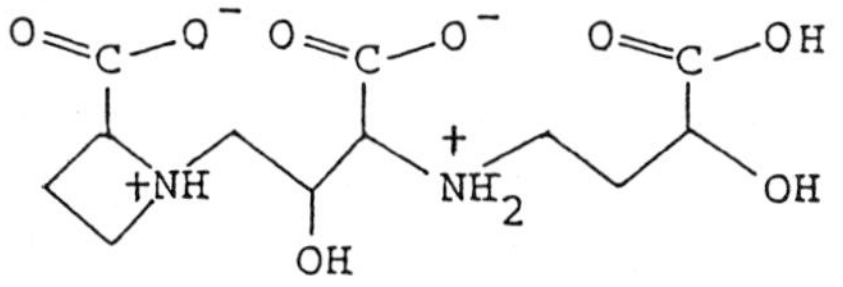

Mugineic acid

phytosiderophore which plays a role in the uptake and transport of iron in higher plants: certain plants release mugineic acid from the roots to collect iron in the soil. It is interesting to note that mugineic acid has a structure similar to edta. Also, several homologs of mugineic acid are now known.[9–11]

TABLE 1

Abbreviations used in this review

bedtra^{3-}	N-Benzylethylenediamine-N,N',N'-triacetate
bpy	2,2'-Bipyridine
cydta^{3-}	Cyclohexanediamine-N,N,N',N'-tetracetate
dtpa^{5-}	Diethylenetriaminepentaacetate
edda^{2-}	Ethylenediamine-N,N'-diacetate
eddadp^{4-}	Ethylenediamine-N,N'-diacetate-N,N'-3-propionate
edta^{4-}	Ethylenediamine-N,N,N',N'-tetraacetate
en	Ethylenediamine
hedtra^{3-}	N-Hydroxyethylethylenediamine-N,N',N'-triacetate
ida^{2-}	Iminodiacetate
medtra^{3-}	N-Methylethylenediamine-N,N',N'-triacetate
mida^{2-}	N-Methyliminodiacetate
nta^{3-}	Nitrilotriacetate
pdta^{4-}	1,2-Propanediamine-N,N,N',N'-tetraacetate
phen	1,10-Phenanthroline
tedta^{4-}	Thiobis(ethylenenitrilo)tetraacetate

Substitution reactions of metal–edta complexes have been referred to previously.[12–14] In this review, two prime subjects, unusually rapid ligand substitution reactions and electron transfer reactions, will be dealt with. Abbreviations used here are summarized in Table 1.

II. UNUSUALLY RAPID SUBSTITUTION REACTIONS OF BOUND WATER IN METAL(III)–edta COMPLEXES

This section concerns kinetic studies of the reaction:

$$[M(III)(H_2O)L] + X \rightleftharpoons [M(III)XL] + H_2O \qquad (1)$$

where L denotes edta or its homologs and X denotes various anions or molecules, mainly unidentate ligands. Investigations of reaction (1) are a rather narrow field, but relate to fundamental problems encountered in solution chemistry of coordination compounds. There are apparently no unequivocal methods for knowing whether transition metal–edta complexes in aqueous solutions contain water molecules in the first coordination sphere or not, and how many donor atoms in edta are coordinated to central metal ions. An exception is the substitution-inert cobalt(III)–edta complexes. Crystallographic structures of various $[Co(edta)]^-$ salts as well as $[Co(H_2O)(Hedta)] \cdot 3H_2O$ have been determined.[15–17] In $[Co(edta)]^-$, $edta^{4-}$ acts as a sexidentate ligand, in which two glycinate rings in the plane of the diamine backbone are sterically strained and two out-of-plane glycinate rings are less strained. In $[Co(H_2O)(Hedta)] \cdot 3H_2O$, the $Hedta^{3-}$ acts as a quinquedentate ligand with one CH_2COOH group free. One water molecule occupies an in-plane position. It would be safe to consider that these structures hold even in aqueous solution owing to their inertness. However, it is usually not possible to specify the structures that substitution-labile metal–edta complexes adopt in solution. Therefore, there have been many conflicting reports on the solution structures of metal–edta complexes. There remains a serious problem even in the case of the solution structure of the chromium(III)–edta complex as will be discussed below. Chromium(III) ions are normally inert. However, one of the coordination sites of chromium(III)–edta has been found to be labile.[18–20] Similar phenomena have been observed for several other metal(III)–edta complexes.

A. The reactions of chromium(III)–edta and related complexes

For present purposes it is assumed that $edta^{4-}$ in the chromium(III)–edta complex acts as a quinquedentate ligand with one water molecule occupying the sixth coordination position.

TABLE 2

Rate constants of the anation reactions (k_f) of
$[Cr(H_2O)(edta)]^-$ and the aquation reactions (k_b) of $[CrX(edta)]^{2-}$ [a]

X	N_3^-	OAc^-	NCS^-	CrO_4^{2-}	MoO_4^{2-}	WO_4^{2-}
k_f (M^{-1} s^{-1})	98 ± 5	3.3 ± 0.4	13.7 ± 0.6	3.2 ± 0.2	21 ± 2	27 ± 4
k_b (s^{-1})	13.4	5.4 ± 0.6	26.8 ± 1.9	0.060	0.46	1.50

[a] From Refs. [20 and 21].

The first example of a rapid ligand substitution reaction[18] is (2):

$$[Cr(H_2O)(edta)]^- + OAc^- \underset{k_b}{\overset{k_f}{\rightleftharpoons}} [Cr(OAc)(edta)]^{2-} + H_2O \qquad (2)$$

The rate of this reaction shows an acid dependency which arises from the following reaction[20,21]:

$$[Cr(H_2O)(edta)]^- + HOAc \rightleftharpoons [Cr(OAc)(edta)]^{2-} + H_3O^+ \qquad (3)$$

Subsequently, many ligands X were found to react rapidly with $[Cr(H_2O)(edta)]^-$. In Table 2, some rate constant values are summarized. All values are so large as to lie in the stopped-flow range. Since the anation rate constant (k_f value) of $[Cr(H_2O)_6]^{3+}$ with NCS^- and the aquation rate constant (k_b value) of $[Cr(NCS)(H_2O)_5]^{2+}$ are 1.8×10^{-6} M^{-1} s^{-1} and 9.2×10^{-9} s^{-1}, respectively,[22] the replacement of five water ligands by $edta^{4-}$ results in an increase in rate in the range 10^7 to 3×10^9. The lability has been attributed to internal associative catalysis by a pendant carboxylate group.[18-20,23] In fact, this interpretation is supported by the results of the kinetic experiments using various $[Cr(H_2O)L]^{n-}$ complexes where L denotes N-substituted ethylenediaminetriacetate[19,21]:

$$L = \begin{array}{c} ^-O_2CCH_2 \\ ^-O_2CCH_2 \end{array} NCH_2CH_2N \begin{array}{c} R \\ CH_2CO_2^- \end{array}$$

Rate constants with respect to the nature of the pendant group R are given in Table 3. The labilizing power of the pendant group R for both the forward and backward reactions is in the order of $CH_2CH_3 < CH_3 < CH_2CH_2CH_2OH < H \ll CH_2CO_2H$, $CH_2CO_2^-$, CH_2CH_2OH, $CH_2CH_2OCOCH_3$, $CH_2CO_2Co(NH_3)_5^{2+}$. If the R groups have coordinating ability, the k_f and k_b values become quite larger. Therefore, it is apparent that the transient coordination of the pendant group has an essential importance for the unexpected rapid substitution reaction of $[Cr(H_2O)L]^{n-}$ complexes. Recently, a trimeric complex (1) was prepared,[24] and substitution reactions of the H_2O with OAc^- and SCN^- were investigated. As expected the reactions turned out to be slow. On the other hand, a dimeric complex

TABLE 3

Rate constants of the anation reactions (k_f) of $[Cr(H_2O)L]^{n-}$ and the aquation reactions (k_b) of $[CrXL]^{(n+m)-}$ [a]

N-subsitutent R	X = OAc⁻		X = NCS⁻	
	k_f (M^{-1} s^{-1})	k_b (s^{-1})	k_f (M^{-1} s^{-1})	k_b (s^{-1})
$CH_2CO_2^-$	3.3 ± 0.4	5.4 ± 0.06	13.7 ± 0.6	26.8 ± 1.9
CH_2CO_2H	—	—	0.77 ± 0.04	$(3.2 \pm 0.3) \times 10^{-2}$
CH_2CH_2OH	7.6 ± 0.6	0.45 ± 0.03	3.3 ± 0.1	0.24 ± 0.02
$CH_2CH_2OCOCH_3$	4.5 ± 0.7	0.38 ± 0.06	—	—
$CH_2CO_2Co(NH_3)_5^{2+}$ [b]	44.5 ± 1.1	2.1 ± 0.4	133 ± 1	4.2 ± 0.1
H	$(2.88 \pm 0.14) \times 10^{-2}$	$(1.6 \pm 0.1) \times 10^{-3}$	$(3.0 \pm 0.1) \times 10^{-2}$	$(2.4 \pm 0.2) \times 10^{-3}$
$CH_2CH_2CH_2OH$ [b]	$(2.1 \pm 0.2) \times 10^{-3}$	$(3.8 \pm 0.8) \times 10^{-4}$	—	—
CH_3	$(7.3 \pm 0.4) \times 10^{-4}$	$(5.9 \pm 0.3) \times 10^{-5}$	—	—
CH_2CH_3 [b]	$(7.3 \pm 0.4) \times 10^{-4}$	$(2.9 \pm 0.1) \times 10^{-5}$	—	—

[a] From Refs. [19 and 21]. L denotes N-substituted ethylenediaminetriacetate.
[b] H. Ogino, A. Masuko, and S. Ito, unpublished results.

(2) shows very large reactivity toward OAc^- and SCN^- (see Table 3). In the former complex, two oxygen donor atoms in the free $CH_2CO_2^-$ group of $[Cr(H_2O)(edta)]^-$ are used for the coordination to two cobalt(III) ions. Therefore, the R group shows no associative assistance by the pendant group. In contrast complex **2** has a free carbonyl oxygen which can exert internal associative catalysis upon substitution.

(1)   (2)

Various unidentate ligands, N_3^-, OAc^-, SCN^-, ONO^-, MO_4^{2-} (M = Cr, Mo, and W), have been found to react with $[Cr(H_2O)L]^{n-}$ to form $[CrXL]^{(n+m)-}$. The formation constants K_x of these species lie in the range of $0.6-110\ M^{-1}$. No formation of $[CrX_2L]^{(n+2m)-}$ has been observed. However, $[Cr(H_2O)(hedtra)]$ was found to react with a bidentate ligand, acetylacetone (Hacac), to form $[Cr(acac)(hedtra)]^-$ in which $acac^-$ acts as a bidentate ligand.[23] The formation constant of the latter complex from $[Cr(H_2O)(hedtra)]$ and $acac^-$ was evaluated to be $(5.9 \pm 0.2) \times 10^5\ M^{-1}$ $(I = 0.20\ M, 25°C)$, so that bidentate $acac^-$ shows the normal chelate effect on the stability constant relative to unidentate anions. $[Cr(acac)(hedtra)]^-$ aquates to the original complex $[Cr(H_2O)(hedtra)]$ with an enhanced rate relative to $[Cr(acac)(edda)]$, which has no pendant group. It is again proposed that the pendant carboxylate group in $[Cr(acac)(hedtra)]^-$ is responsible for the rate enhancement.[23]

The Cr(III)–edta complex undergoes the following acid–base equilibria:

$$[Cr(H_2O)(Hedta)] \underset{pK_1 = 1.8-3.1^{[19,25-27]}}{\rightleftharpoons} [Cr(H_2O)(edta)]^- \underset{pK_2 = 7.39-7.5^{[19,25-28]}}{\rightleftharpoons}$$

$$[Cr(OH)(edta)]^{2-} \underset{pK_3 = 12.2^{[26]}}{\rightleftharpoons} [Cr(OH)_2(edta)]^{3-} \qquad (4)$$

The $[Cr(H_2O)_2(edta)]^-$ complex in which $edta^{4-}$ acts as a quadridentate ligand can be produced by rapid mixing of the solution of $[Cr(OH)_2(edta)]^{3-}$ with appropriate concentration of acid. $[Cr(H_2O)_2(Hedta)]$ and $[Cr(H_2O)_2(H_2edta)]^+$ are also formed depending on pH. Thorneley *et al.* measured the rate of the conversion of quadridentate species to quinquedentate $[Cr(H_2O)(edta)]^-$ and/or $[Cr(H_2O)(Hedta)]$ by the stopped-flow method.[29] Rate constants for the conversion processes (ring closure reactions) are shown in reaction (5):

$$\left.\begin{array}{l} [Cr(H_2O)_2(H_2edta)]^+ \xrightarrow{1.4\,s^{-1}} \\[2mm] [Cr(H_2O)_2(Hedta)] \xrightarrow{\sim 130\,s^{-1}} \\[2mm] [Cr(H_2O)_2(edta)]^- \xrightarrow{330\,s^{-1}} \end{array}\right\} \left\{\begin{array}{l} [Cr(H_2O)(Hedta)] \\[2mm] \Updownarrow \\[2mm] [Cr(H_2O)(edta)]^- \end{array}\right\} \tag{5}$$

The magnitude of these rate constants are again very large when one considers that the reactions involve substitution at chromium(III) centers.

$[Cr(H_2O)(hedtra)]$ reacts with oxalate (ox^{2-}) to give $[Cr(\underline{o}x)(hedtra)]^{2-}$ and $[Cr(\underline{o}xH)(hedtra)]^-$, where $\underline{o}x^{2-}$ and $\underline{o}xH^-$ denote unidentately coordinated ox^{2-} and oxH^-, respectively.[30,31] These reactions are complete within several seconds at 25°C, and are followed by an extremely slow reaction to produce $[Cr(ox)_3]^{3-}$ without formation of any detectable intermediates at pH 2.5–3.9:

$$\left.\begin{array}{l} [Cr(\underline{o}x)(hedtra)]^{2-} \\[2mm] \Updownarrow \\[2mm] [Cr(\underline{o}xH)(hedtra)]^- \end{array}\right\} \xrightarrow{ox^{2-}} [Cr(ox)_3]^{3-} + hedtra^{3-} \tag{6}$$

The second step takes more than a few hundred hours to completion. Therefore, it is not possible in this case to evaluate the lability of the fifth coordination site of the Cr(III)–hedtra complex.

Peerce *et al.* obtained two different geometrical isomers from the reaction of $Cr(ClO_4)_3$ with tedta.[32] The structures of the two isomers were concluded to have cis and trans configurations with respect to the coordinated nitrogen atoms. In the cis isomer, the ligand is quinquedentate and a water molecule is coordinated to chromium(III) ion. This conclusion was derived from the observation that the isomer reacted with azide rapidly.

Another sort of labilizing effect by the edta ligand has been found in the chromium–carbon bond cleavage of $[CrR(edta)]^{2-}$ ($R = CH_2OH^-$, $CH(CH_3)OH^-$, $C(CH_3)_2OH^-$).[33] The reaction takes place by the two paths:

$$[CrR(edta)]^{2-} \xrightarrow{H_2O} [Cr(H_2O)(edta)]^- + \text{organic products} \tag{7}$$

$$[CrR(edta)]^{2-} \xrightarrow{H^+} [Cr(H_2O)(edta)]^- + \text{organic products} \tag{8}$$

Reaction (8) is more than 10^3 times faster than the corresponding reaction of $[CrR(H_2O)_5]^{2+}$, while reaction (7) proceeds in a rate similar to that of the corresponding $[CrR(H_2O)_5]^{2+}$.[34] It has been proposed that the protonation of free CH_2COO^- of the edta ligand is involved in reaction (8). The CH_2COOH group thus formed acts as an electrophile toward the chromium–carbon bond in the same molecule, and results in an accelerated chromium–carbon bond cleavage.

B. Solution structure of the chromium(III)–edta complex

So far, we have been using the formulation $[Cr(H_2O)(edta)]^-$ in which chromium(III) has nearly octahedral geometry and $edta^{4-}$ acts as quinquedentate ligand, so that one water molecule occupies the sixth coordination position. Among chromium(III)–edta complexes prepared so far, terdentate ($[CrX_2(H_2O)(H_3edta)]$ where $X^- = Cl^-$, Br^-, and CN^-),[5,35] quadridentate ($[Cr(H_2O)_2(edta)]^-$),[29] and quinquedentate or sexidentate ($[Cr(H_2O)(edta)]^-$ or $[Cr(edta)]^-$) bonding modes are encountered. The following argument focuses on the distinction between quinquedentate and sexidentate formulations, since this is the most controversial subject in the present topic.

In 1943, Brintzinger *et al.* first isolated the chromium(III)–edta complex, and formulated it as $H[Cr(edta)]$ from analytical results.[36] This suggested that the $edta^{4-}$ acts as a sexidentate ligand. Schwarzenbach and Biedermann reinvestigated the solid complex and presented the $[Cr(H_2O)(Hedta)]$ formulation.[28] They observed that the water molecule in the complex is very firmly coordinated and cannot be removed at 100°C under vacuum. This observation was confirmed by Thorneley *et al.*[29] Dwyer and Garvan confirmed the quinquedentate nature of $Hedta^{3-}$ in the complex from the IR spectrum, which indicated the presence of an uncoordinated COOH group.[37] Recently, the $[Cr(H_2O)(Hedta)]$ formulation has been firmly established by X-ray crystallography, where one water molecule occupies an in-plane position and the CH_2COOH group is free.[38]

However, there is a complicated situation for the uninegatively charged chromium(III)–edta complex. Sawyer and McKinnie measured the IR spectrum of $NaCr(edta) \cdot 2H_2O$, but they could not conclude whether the $edta^{4-}$ had a quinquedentate or a sexidentate nature.[39] Hoard *et al.* found that $NH_4Cr(edta) \cdot 2H_2O$, $RbCr(edta) \cdot 2H_2O$, $NH_4[Co(edta)] \cdot 2H_2O$, and $Rb[Co(edta)] \cdot 2H_2O$ are an isomorphous series in the orthorhombic space group $P2_12_12_1$.[40,41] Since $[Co(edta)]^-$ contains sexidentate $edta^{4-}$, it was concluded that the chromium(III) complexes must be formulated as $NH_4[Cr(edta)] \cdot 2H_2O$ and $Rb[Cr(edta)] \cdot 2H_2O$ and that the $edta^{4-}$ is also sexidentate in the chromium complexes. Hoard *et al.* stated that it is probable that in an intermediate pH range (5–9), both $[Cr(H_2O)(edta)]^-$ and $[Cr(OH)(edta)]^{2-}$ are converted, though perhaps slowly, into sexidentate $[Cr(edta)]^-$. Kushi *et al.* completed the structural determination of $K[Cr(edta)] \cdot 2H_2O$ by X-ray crystallography, and found that the salt belongs to the orthorhombic space group $P2_12_12_1$, and is isomorphous with $K[Co(edta)] \cdot 2H_2O$.[42] Therefore, the chromium complex contains sexidentate $edta^{4-}$ without any water molecule in the first coordination sphere. The structure of $[Cr(edta)]^-$ is similar to that of $[Co(edta)]^-$, but more strained than $[Co(edta)]^-$.

The space group $P2_12_12_1$ demands that each crystal of $K[Cr(edta)] \cdot 2H_2O$ contains a pure enantiomer (occurrence of spontaneous resolution). In fact, when a single crystal of $K[Cr(edta)] \cdot 2H_2O$ was dissolved into water, optical activity was observed.[42] Dwyer and Garvan could not resolve the chromium complex and concluded then that a rapid rearrangement is responsible for the failure to observe any optical activity.[37] It is now evident that their results do not mean the rapid rearrangement in the complex, but a failure to resolve optically the isomers. Jørgensen reported that the electronic spectra of $[Cr(H_2O)(edta)]^-$ and $[Cr(H_2O)(Hedta)]$ are the same as predicted by crystal field theory, where the perturbation from the ligands decreases with about sixth power of the distance.[43] However, the visible spectra of $[Cr(H_2O)(edta)]^-$ and $[Cr(H_2O)(Hedta)]$ are not the same.[21,27] Such large difference in spectra does not seem to be reasonable if the two species are different only by the protonation of the free CH_2COO^- group. This spectral phenomenon may be interpreted in two ways. One possibility is that the uninegatively charged species is not $[Cr(H_2O)(edta)]^-$ but $[Cr(edta)]^-$. The second possibility is that there is hydrogen bonding between the hydrogen of the coordinated water and the oxygen of the free CH_2COO^- group in $[Cr(H_2O)(edta)]^-$, but that such hydrogen bonding is absent in $[Cr(H_2O)(Hedta)]$ and hence the spectral difference results. A hydrogen bonding similar to this is also postulated for $[Ru(H_2O)(edta)]^-$.[44]

Sykes and colleagues measured the rate of interconversion of chromium(III)–edta species present in pH 1–9.[29] For example, if the pH of a chromium(III)–edta solution is changed suddenly from 5 to 1, either reaction (9) or (10) must occur:

$$[Cr(H_2O)(edta)]^- + H^+ \rightarrow [Cr(H_2O)(Hedta)] \tag{9}$$

$$[Cr(edta)]^- + H_3O^+ \rightarrow [Cr(H_2O)(Hedta)] \tag{10}$$

Reaction (9) involves addition of a proton to CH_2COO^- group and should be extremely rapid and difficult to follow, whereas reaction (10) involves a change in degree of chelation of the $edta^{4-}$ from six to five. Therefore, the reaction should be slower and readily observable. The reaction was found to be too rapid to follow by stopped-flow and T-jump methods. This observation favors the [Cr(H$_2$O)(edta)]$^-$ formulation.

Recently, Wheeler and Legg presented a new approach to this problem and claimed that edta^{4-} forms a sexidentate complex with chromium(III) between pH 3.5 and 6.5.[45,46] They prepared [Cr(H$_2$O)(Hedta-α-d_8)] in which eight acetate protons are all deuterated, and measured the ^{2}H NMR spectra of the complex at various pH's. Their conclusion was derived mainly from the observation that the spectrum at pH 4.8 or 6.3 consists of only four resonances indicative of a complex with C$_2$ symmetry. However, their results are also understandable from the standpoint of [Cr(H$_2$O)(edta)]$^-$ formulation, if the site of the H$_2$O exchanges rapidly:

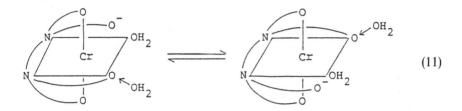

$$(11)$$

As described in Section II,A, Cr(III)–edta complexes can accommodate various X, e.g., N$_3$$^-$, NCS$^-$, OAc$^-$, CrO$_4$$^{2-}$, very easily. It is difficult to understand why only the case with X = H$_2$O should constitute a special case with a rapid interchange as in reaction (11). Considering that the concentration of water is nearly 50 M, it is more difficult to accept that X = H$_2$O is the special case. If [Cr(H$_2$O)(edta)]$^-$ gives four resonances due to the rapid occurrence of reaction (11), the site exchange rates must exceed several thousand per second. This value does not seem to be unreasonable considering the data given in Table 3 and the high concentration of water.

As mentioned above, however, Sykes' experiment on the interconversion of chromium(III)–edta species favors the [Cr(H$_2$O)(edta)]$^-$ formulation.[29] In that experiment, it was assumed that reaction (10) should be slow. If the [Cr(edta)]$^-$ formulation is correct, incorporation of a water molecule into the first coordination sphere must occur so rapidly that the rates are difficult to follow by stopped-flow or T-jump method. This point awaits further studies. Wheeler and Legg also pointed out that the difference of pK$_a$ values between the uninegatively charged species of chromium(III)–edta and [Cr(H$_2$O)(medtra)] (7.39 and 6.25, respectively[19]) are most likely due to the replacement of an acetate arm by hydroxide ion in the former complex, and by the ionization of the coordinated water in the latter complex. However, the difference is also understandable if the presence of hydrogen bonding between water ligand and free acetate arm in $[Cr(H_2O)(edta)]^-$ is considered. As the hydrogen bonding becomes stronger, the pK_a value should increase.

An alternative explanation may be possible which is consistent with many experimental results. The uninegatively charged species, $[Cr(H_2O)(edta)]^-$, could have a seven-coordinate geometry in which the $edta^{4-}$ is sexidentate and one water molecule occupies the seventh coordination position. Coordination of this kind had been found for $[Fe(III)(H_2O)(edta)]^-$ and $[Mn(II)(H_2O)(edta)]^{2-}$.[47-52] The former complex has a nearly pentagonal bipyramidal shape, while the shape of the latter complex is roughly trigonal prism, as shown in Fig. 2. If $[Cr(H_2O)(edta)]^-$ is seven coordinate, its geometry may be approximated as a pentagonal bipyramid as in $[Fe(H_2O)(edta)]^-$. Because an optically active chromium(III)–edta complex having uninegative charge really exists, and the racemization reaction is slow,[42] a trigonal bipyramid structure is not possible. It is interesting to note that, in $Li[Fe(edta)] \cdot 3H_2O$, the $edta^{4-}$ acts as a sexidentate ligand occupying all six sites of the octahedral coordination sphere of iron(III). This indicates that the energy differences between six-coordinate and seven-coordinate structures are slight. Since at least one coordination site of chromium(III)–edta is labile and the ionic radius of chromium(III) ion is larger than that of cobalt(III) ion, it may well be that the chromium(III)–edta species is seven coordinate in aqueous solution. In summary, the solution structure of chromium(III)–edta is still speculative. For elucidation of this problem, further studies must be undertaken.

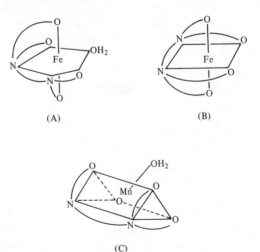

Fig. 2. Idealized structures of edta complexes in the compounds (A) $Li[Fe(H_2O)(edta)] \cdot 2H_2O$,[47] (B) $Li[Fe(edta)] \cdot 3H_2O$,[52] and (C) $[Mn_3(Hedta)_2] \cdot 10H_2O$.[49]

C. The reactions of ruthenim(III)–edta and hedtra complexes

Enhanced ligand substitution reactivity upon coordination of edta is also seen in ruthenium(III) complex. Although it is not clear whether or not the edta^{4-} acts as quinquedentate or sexidentate ligand as is the case of chromium(III) complex, the $[Ru(H_2O)(edta)]^-$ formulation will be used tentatively in this section.

In 1966, Ezerskaya and Solovykh reported a qualitative observation that, in dilute solutions, chloride ions in $[RuCl(Hedta)]^-$ and $[RuCl_2(H_2edta)]^-$ are rapidly replaced by water molecules.[53] In agreement with this, the dilute solutions of $[RuCl(Hedta)]^-$ and $[Ru(H_2O)(Hedta)]$ give almost identical electronic spectra. Matsubara and Creutz reported the occurrence of the unusually rapid substitution reactions of $[Ru(H_2O)(edta)]^-$ with SCN$^-$ and several molecules[44]:

$$[Ru(H_2O)(edta)]^- + X \underset{k_b}{\overset{k_f}{\rightleftharpoons}} [RuX(edta)]^- + H_2O \qquad (12)$$

The data obtained by them are given in Table 4 along with related data. Yoshino *et al.* reported that the reactions of $[Ru(H_2O)(edta)]^-$ with a variety of sulfur-containing ligands (cysteine, thiourea, etc.) are too rapid to be followed, but thiosulfate reacts at a more moderate rate.[54,55] Although there are only scattered data in the literature for ligand substitution reactions of other ruthenium(III) complexes, it is clear that the substitution reactions of $[Ru(H_2O)(edta)]^-$ are much more rapid than those of other ruthenium(III) complexes.[56–59]

Matsubara and Creutz concluded that the substitution reactions of $[Ru(H_2O)(edta)]^-$ proceed through an associative route, because the rate constants (k_1) are sensitive to the equilibrium constants (K) as shown in Table 4. For the unusually rapid substitution of $[Ru(H_2O)(edta)]^-$, they proposed the following mechanism. In the $[Ru(H_2O)(edta)]^-$, there is a hydrogen bonding between coordinated water and the oxygen of the free carboxylate. This hydrogen bonding creates a very open area and may sterically activate $[Ru(H_2O)(edta)]^-$ toward associative substitution. For the reverse reaction, aquation of $[RuX(edta)]^-$, the pendant $CH_2CO_2^-$ group would make a hydrogen bonding to the incoming ligand, water. In accordance with this interpretation, when one oxygen of the pendant carboxylate group is blocked by binding to a graphite electrode, the rate of substitution of the electrode-attached ruthenium(III) is reduced.[60,61] This behavior seems to be different from that of $[Cr(H_2O)(edta)]^-$, which shows high reactivity even after blocking the pendant $CH_2CO_2^-$ group by $[Co(NH_3)_5]^{3+}$ (see Section II,A). However, $[Ru(H_2O)(hedtra)]$ containing the CH_2CH_2OH group as a pendant group also shows high reactivity

TABLE 4

Rate constants for the ligand substitution of $[Ru(H_2O)(edta)]^-$ and $[Ru(H_2O)(hedtra)]$

Ligand	edta system[a]				hedtra system[b]			
	$k_1\ (M^{-1}\,s^{-1})$	$k_{-1}\ (s^{-1})$	$K\ (M^{-1})$	Ref.	$k_1\ (M^{-1}\,s^{-1})$	$k_{-1}\ (s^{-1})$	$K\ (M^{-1})$	Ref.
Br^-	—	—	—	—	0.13	1.9×10^{-2}	6.6	[31]
$S_2O_3^{2-}$	2.9	—	1.0×10^5	[54]	—	—	—	—
CH_3CN	30	3.2	9	[44]	0.48	3.0×10^{-2}	16	[31]
SCN^-	270	0.5	540	[44]	7.5	1.3×10^{-2}	580	[31]
Pyridine	6300	0.06	1×10^5	[44]	16	—	$\gg 3 \times 10^3$	[31]
Isonicotinamide	8300	0.7	1.2×10^4	[44]	—	—	—	—
Pyrazine	20000	—	1×10^4	[44]	50	—	—	[44]

[a] Temperature, 25°C and $I = 0.2\ M$ except for $S_2O_3^{2-}$ (30°C and $I = 0.1\ M$).
[b] Temperature, 25°C and $I = 0.5\ M$ except for pyrazine (25°C and $I = 0.2\ M$).

toward various ligands as shown in Table 4. Therefore, the present authors are not inclined to believe that the activation of $[Ru(H_2O)(edta)]^-$ by the internal hydrogen bonding is an appropriate interpretation. It is very interesting that the substitution rates of $[Ru(III)(H_2O)(edta)]^-$ shown in Table 4 are much higher than those of $[Ru(II)(H_2O)(edta)]^{2-}$.[44,62] Unless $[Ru(III)(H_2O)(edta)]^-$ is removed entirely, attempts to study the kinetics of substitution of the ruthenium(II) complex, $[Ru(II)(H_2O)(edta)]^{2-}$, result in zero-order kinetics due to the following reaction sequence:

$$[Ru(III)(H_2O)(edta)]^- + X \rightleftharpoons [Ru(III)X(edta)]^- + H_2O \qquad (13)$$

$$[Ru(III)X(edta)]^- + [Ru(II)(H_2O)(edta)]^{2-}$$

$$\xrightarrow{fast} [Ru(II)X(edta)]^{2-} + [Ru(III)(H_2O)(edta)]^- \qquad (14)$$

D. The reactions of some other metal(III)–edta (and related) complexes

Following the identification of unusually rapid substitution reactions for chromium(III)– and ruthenium(III)–edta complexes, similar phenomena have been found for titanium(III)–, iron(III)–, and osmium(III)–edta complexes.

Although direct measurement of the kinetics of substitution of titanium(III)–edta has not been made, there is some indirect evidence that titanium(III)–edta is more substitution labile than $[Ti(H_2O)_6]^{3+}$ ions. Thompson and Sykes investigated the electron transfer reaction between $[Co(H_2O)(NH_3)_5]^{3+}$ and $[Ti(H_2O)(Hedta)]$ and estimated that the water exchange rate constant of $[Ti(H_2O)(Hedta)]$ is $> \sim 10^7 \, s^{-1}$.[63] This is much larger than the corresponding value of $[Ti(H_2O)_6]^{3+}$ ($\sim 10^5 \, s^{-1}$).[64] Adegite *et al.* found an increased rate of reduction of $[Ru(OAc)(NH_3)_5]^{2+}$ by titanium(III)–edta relative to that of $[Ti(H_2O)_6]^{3+}$.[65] This may be caused by increased lability of the coordinated water molecule of the reductant, although increased prior association between the oxidant and reductant cannot be ruled out.

$[Fe(H_2O)(hedtra)]$ and $[Fe(H_2O)_2(nta)]$ react very rapidly with SCN^- to give $[Fe(NCS)(hedtra)]^-$ and $[Fe(NCS)(H_2O)(nta)]^-$, respectively. The lower limit of the rate constants were estimated to be $10^7 \, M^{-1} \, s^{-1}$.[21] This is unusual, because the water exchange rate constant of $[Fe(H_2O)_6]^{3+}$ has been known to be $160 \, s^{-1}$.[66,67] Matsubara and Creutz reported briefly in their paper that osmium(III) ions are markedly labilized on complexing with $edta^{4-}$, and undergo substitution 10^6 times more rapidly than $[Os(H_2O)(NH_3)_5]^{3+}$.[44]

TABLE 5

Equilibrium constants and rate constants of the
reactions given in Scheme 1[a]

Equilibrium constant or rate constant	I (M)	Temp. (°C)
$pK_1 = 3.1$	0.1	15
$pK_2 = 8.1$	0.1	15
$K_a = 1.28\ M^{-1}$	1.0	25
$K = 1.2 \times 10^{-3}$ [b]	0.1	15
$k_1 = 2.2 \times 10^{-4}\ s^{-1}$	0.1	25
$k_{-1} = 3.0 \times 10^{-4}\ M^{-1}\ s^{-1}$	1.0	25
$k_2 = 1.8 \times 10^{-3}\ s^{-1}$	0.1	25
$k_3 = 4.3 \times 10^{-5}\ s^{-1}$	0.1	25

[a] From Refs. [73 and 74].
[b] $K = [[Co(H_2O)(edta)]^-]/[[Co(edta)]^-]$.

There are some works on the substitution reactions of manganese(III)–edta and related complexes.[68–72] However, it is still premature to argue the lability of manganese(III) center, because the literature data for ligand substitution in other manganese(III) complexes are sparse.

Higginson and colleagues investigated acid–base equilibria and interconversion between cobalt(III)–edta complexes.[73,74] Their results are summarized in Scheme 1[73,74] and Table 5. Besides those interconversions, there are some studies (Scheme 1[73,74]) on the rates of ring closure of $[CoX(edta)]^{2-}$ ($X^- = Cl^-$ and Br^-) to give $[Co(edta)]^-$. The rate constants at 25°C are $2.5 \times 10^{-6}\ s^{-1}$ for $X^- = Cl^-$ and $(2\text{–}20) \times 10^{-6}\ s^{-1}$ for $X^- = Br^-$.[74–77] When these values are compared with the data given in Table 5, the rate enhancement of the ring closure is seen for $[Co(H_2O)(Hedta)]$ and $[Co(H_2O)(edta)]^-$. Dyke and Higginson, and Wilkins and colleagues have suggested that the pendant CH_2COOH or CH_2COO^- group participates in the transition state in an I_a process for the ring closure of $[Co(H_2O)(Hedta)]$ and $[Co(H_2O)(edta)]^-$, while for the corresponding reactions of $[CoX(edta)]^{2-}$ ($X^- = OH^-$, Cl^-, and Br^-) an I_d type substitution has been considered likely.[74,77]

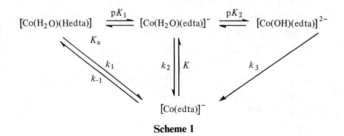

Scheme 1

As has been shown in this section, the extra lability of ligand substitution reactions of metal(III)–edta complexes seems to be a rather general phenomenon, not a unique effect to particular metal ions. In fact, metal ions of most of the first transition series elements, from titanium(III) to cobalt(III), are labilized by coordination of edta. Ligand substitution reactions of vanadium(III)–edta have not been studied to our knowledge, and this complex certainly merits attention in the future.

III. ELECTRON TRANSFER REACTIONS INVOLVING METAL–edta AND RELATED COMPLEXES

As has been mentioned in a previous section, it is often difficult to determine conclusively whether a coordinated water molecule(s) is retained in a metal–edta complex or not, except in the case of substitution inert complexes such as cobalt(III). Coordinated water is therefore omitted from the formulae used in this section unless otherwise specified.

A. Reduction of aminopolycarboxylatocobalt(III) complexes

Electron transfer reactions between cobalt(III) complexes and Cr^{2+} or Fe^{2+} have been fairly extensively studied. Reduction of $[Co(edta)]^-$, $[Co(H_2O)(hedtra)]$, $[CoX(Hedta)]^-$, and $[CoX(hedtra)]^-$ $(X^- = Cl^-, Br^-)$ by Fe^{2+} has been investigated.[78–80] The inner-sphere mechanism was suggested for the reduction of halogeno-cobalt(III) complexes. This was confirmed in the reaction of $[CoCl(Hedta)]^-$ with Fe^{2+} by following the formation of $FeCl^{2+}$ and the subsequent disappearance of the $FeCl^{2+}$.[81]

Reduction of metal–edta complexes by Cr^{2+} sometimes results in the formation of the chromium(III)–edta complex with a novel mode of coordination as shown in Eq. (15).[82,83]

$$[Co(NH_3)_5\{Co(edta)(H_2O)\}]^{2+} + Cr^{2+} + nH^+ \xrightarrow{k_{et}}$$
$$(3)$$

$$[Co(NH_3)_5\{Cr(H_nedta)(H_2O)_5\}]^{(n+2)+} + Co^{2+} \qquad (15)$$
$$(4)$$

Reaction (15) proceeds by an attack of Cr^{2+} toward the CO_2^- group of the edta^{4-} with rate constant $k_{et} = (2.55 \pm 0.15) \times 10^3 \ M^{-1} \ s^{-1}$ at 25°C and $I = 0.2 \ M$ (LiClO$_4$). Complex **3** can be regarded as an analog of $[Co(H_2O)(Hedtra)]$, where the COOH group of $[Co(H_2O)(Hedtra)]$ is replaced by the $CO_2Co(NH_3)_5$ moiety in **3**. The reduction of $[Co(H_2O)(Hedtra)]$ by Cr^{2+} [reaction (16)] also proceeds by an attack at CO_2 group of the cobalt(III) complex:

$$[Co(H_2O)(Hedtra)] + Cr^{2+} \rightarrow Co^{2+} + [Cr(H_2O)_5(H_nedta)]^{(n-1)+} \qquad (16)$$
$$\text{(main products)}$$

By using a competition method the rate constant was determined to be $1.3 \times 10^5 \ M^{-1} \ s^{-1}$ at 20°C and $I = 1.0 \ M$ (NaClO$_4$),[84] which is two orders of magnitude larger than the reduction of **3**. Since this value seems to be too large, kinetic measurements were repeated using the stopped-flow method.[85]

The redetermined value for (16) is $(7.6 \pm 0.4) \times 10^3 \ M^{-1} \ s^{-1}$ at 25°C and $I = 1.0 \ M$ (NaClO$_4$), and is the same order of magnitude as that for reaction (15). This supports the belief that Cr^{2+} attacks a carbonyl oxygen in reactions (15) and (16). Also, the result suggests that care is required in using the competition method to determine rate constants for fast reactions.

B. Reduction by aminopolycarboxylatoferrate(II) complexes

Reduction of $[CoX(NH_3)_5]^{2+}$ (X$^-$ = halides, NCS$^-$, NO$_3{}^-$, N$_3{}^-$) by Fe(II)L$^{(n-2)-}$ (L^{n-} = edta^{4-}, pdta^{4-}) were studied.[86,87] The pattern of rate constants with respect to the ligand X$^-$ indicates inner-sphere electron transfer mechanism. Reduction rates of $[CoCl(NH_3)_5]^{2+}$ by iron(II) complexes of edta^{4-}, cydta^{4-}, and hedtra^{3-} were measured.[87] The reaction rate increases in the order of $[Fe(cydta)]^{2-} < [Fe(hedtra)]^- < [Fe(edta)]^{2-} < [Fe(Hedta)]^-$. There is an indication that a linear free energy relationship exists between logarithmic rate constants and reduction potentials for the Fe(II)/Fe(III) couples. Both $[Fe(edta)]^{2-}$ and related complexes are widely employed as reducing reagents in studies on electron transfer reactions of metalloproteins, as described in Section III,H.

C. Reduction by aminopolycarboxylatochromium(II) complexes

An outer-sphere electron transfer mechanism was postulated for the reduction of $[CrX(H_2O)_5]^{2+}$ (X$^-$ = F$^-$, Cl$^-$, Br$^-$), $[CoCl(NH_3)_5]^{2+}$, Cl$_2$, Br$_2$, I$_2$, and IrCl$_6{}^{2-}$ by $[Cr(H_2O)(edta)]^{2-}$ based on the observation that the $[CrX(edta)]^{2-}$, which was expected as an inner-sphere oxidation product, was not detected after 3 min of the reaction.[88] In contrast, an inner-sphere mechanism was proposed for the reduction of $[CrX(NH_3)_5]^{2+}$ by $[Cr(H_2O)(edta)]^{2-}$ judged from the kinetic pattern.[89] Later, ligand substitution reactions of the ligand X in $[CrX(edta)]^{2-}$ and related aminopolycarboxylato complexes were found to be very fast with regard to the ligand substitution reactivity of chromium(III) complexes. Because the ligand substitution of X is complete within seconds,[19-21] the identification of the reaction product after 3 min of reaction was not immediate enough to determine the mechanism for reduction by $[Cr(H_2O)(edta)]^{2-}$. Further unequivocal information was the identification of the immediate product of the reaction between $[CoX(NH_3)_5]^{2+}$ ($X^- = OAc^-$, Cl^-, Br^-) and $[Cr(H_2O)(hedtra)]^-$ was $[CrX(hedtra)]^-$ which underwent subsequent aquation of X^-.[90,91] The reduction of $[CrCl(NH_3)_5]^{2+}$ by chromium(II) complexes ligating $edta^{4-}$, $hedtra^{3-}$, $mida^{2-}$, and $eddadp^{4-}$ also proceeds by an inner-sphere mechanism. A linear free-energy relationship was observed between logarithmic rate constants and the reduction potentials of the Cr(II)/Cr(III) couples.[92]

D. Reduction by titanium(III) complexes

Reduction reactions of $[Co(edta)]^-$ and $[CoCl(Hedta)]^-$ by titanium(III) aqua ions, which generally proceed via $[Ti(OH)(H_2O)_5]^{2+}$, have been investigated.[93] The question about inner-sphere vs outer-sphere mechanisms was not settled,[93] but indirect evidence which supports an outer-sphere mechanism has been proposed by comparing rate constants with those for the reduction by Cr^{2+} ions.[85]

Several electron transfer reactions of aminopolycarboxylatotitanium(III) complexes have been investigated. The complex $[Ti(H_2O)(Hedta)]$ is a more effective reductant than $[Ti(OH)(H_2O)_5]^{2+}$ and the rate constants for the reduction of $[Co(NH_3)_6]^{3+}$ (0.165 M^{-1} s^{-1}) and $[CoCl(NH_3)_5]^{2+}$ (9.6 M^{-1} s^{-1}) by $[Ti(H_2O)(Hedta)]$ are about 10 times greater than the reduction by $[Ti(OH)(H_2O)_5]^{2+}$ (0.015 M^{-1} s^{-1} for $[Co(NH_3)_6]^{3+}$ and 0.95 M^{-1} s^{-1} for $[CoCl(NH_3)_5]^{2+}$).[63] There are indications that reduction of $[Co(OH)(NH_3)_5]^{2+}$ by $[Ti(H_2O)(Hedta)]$[63] and reduction of $[Ru(NCS)(NH_3)_5]^{2+}$ by $[Ti(H_2O)(hedtra)]$[94] utilize an inner-sphere pathway.

Isomeric structures of aminopolycarboxylato-metal complexes are known, for example the two isomers **5** and **6** of $[Ti(H_2O)(hedtra)]$. It is possible that each isomer shows different redox reactivity. This is exemplified

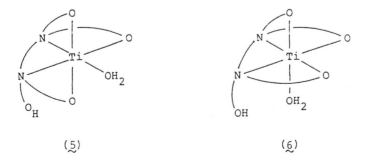

(5)   (6)

by the mechanism proposed for the reaction of $[V(IV)(O)(hedtra)]^-$ with $[Ti(H_2O)(hedtra)]$.[95] The rate-limiting isomerization of $[Ti(H_2O)(hedtra)]$ takes place prior to the electron transfer [reactions (17) and (18)],[95] although the isomers A and B are not definitely characterized.

$$[Ti(H_2O)(hedtra)] \rightleftharpoons [Ti(H_2O)(hedtra)] \tag{17}$$
$$\text{isomer A} \qquad\qquad \text{isomer B}$$

$$[Ti(H_2O)(hedtra)] + [V(IV)O(hedtra)]^- \rightarrow [Ti(IV)O(hedtra)]^- + [V(III)(H_2O)(hedtra)] \tag{18}$$
$$\text{isomer B}$$

E. Reduction by $[Co(edta)]^{2-}$

Reduction of $[Fe(CN)_6]^{3-}$ by $[Co(II)(edta)]^{2-}$ and related aminopolycarboxylatocobalt(II) complexes has been studied extensively presumably because of its interesting reaction sequence.[96–106] Formation of the cyanide bridged complex $[(edta)Co(III)–NC–Fe(II)(CN)_5]^{5-}$ is observed first, followed by slow formation of $[Co(III)(edta)]^-$ and $[Fe(II)(CN)_6]^{4-}$. It was first thought that the final products, $[Co(edta)]^-$ and $[Fe(CN)_6]^{4-}$, were obtained via the binuclear complex $[(edta)Co–NC–Fe(CN)_5]^{5-}$, which was the successor complex along the inner-sphere electron transfer.

$$[Co(edta)]^{2-} + [Fe(CN)_6]^{3-} \rightleftharpoons [(edta)Co(III)–NC–Fe(II)(CN)_5]^{5-} \tag{19}$$

$$[(edta)Co(III)–NC–Fe(II)(CN)_5]^{5-} \rightleftharpoons [Co(edta)]^- + [Fe(CN)_6]^{4-} \tag{20}$$

However, it has been shown that the $[(edta)Co(III)–NC–Fe(II)(CN)_5]^{5-}$ does not dissociate to give $[Co(edta)]^-$ and $[Fe(CN)_6]^{4-}$, and that an outer-sphere electron transfer between $[Co(edta)]^{2-}$ and $[Fe(CN)_6]^{3-}$ is the source of the final products.[99–104]

$$[Co(edta)]^{2-} + [Fe(CN)_6]^{3-} \rightleftharpoons [(edta)Co—NC—Fe(CN)_5]^{5-}$$
$$\quad\quad \rightarrow [Co(edta)]^- + [Fe(CN)_6]^{4-} \tag{21}$$

This type of reaction scheme is often referred to as a "dead-end" mechanism. The same reaction sequence as reaction (21) has been found in the reaction of $[Co(edta)]^{2-}$ with $[Ru(NH_3)_5(pz)]^{3+}$ (pz = pyrazine).[105] In a similar sense, the reaction between $[Ti(III)(H_2O)(hedtra)]$ and $[V(IV)O(hedtra)]^-$ forms a dead-end binuclear complex $[(hedtra)Ti(III)OV(IV)(hedtra)]^-$ in which no electron transfer takes place.[95] In contrast with the $[Co(edta)]^{2-} + [Fe(CN)_6]^{3-}$ and $[Co(edta)]^{2-} + [Ru(NH_3)_5(pz)]^{3+}$ reactions, the reduction of $[Ru(NH_3)_5(bpy)]^{3+}$ by $[Co(edta)]^{2-}$ proceeds predominantly by an outer-sphere mechanism, and no binuclear complex was observed.[107]

The use of Co(II)–edta as a reductant offers an approach to examine stereoselective outer-sphere electron transfer reactions, since the oxidized product, $[Co(III)(edta)]^-$, racemizes only very slowly. Thus, several percent of stereoselective electron transfer reactions were observed for the reduction of optically active $[Os(bpy)_3]^{3+}$, $[Ru(bpy)_3]^{3+}$, $[Co(bpy)_3]^{3+}$, $[Fe(bpy)_3]^{3+}$, $[Fe(phen)_3]^{3+}$, and $[Ni(IV)(S-Me_2L)]^{2+}$ $(S-Me_2L = (5S,12S)-4,7,10,13-$tetraaza-3,5,12,14-tetramethylhexadeca-3,13-diene-2,15-dionedioxime) by $[Co(edta)]^{2-}$.[108,109] The oxidation of $[Mo_2(V)O_4(pdta)]^{2-}$ with the optically active μ-superoxo complex $[(en)_2Co(\mu-NH_2, O_2)Co(en)_2]^{4+}$ to give molybdenum(VI) and the corresponding μ-amido-μ-peroxodicobalt(III) complex also shows stereoselectivity.[110]

It has been known that $[Co(edta)]^-$ and $[Co(pdta)]^-$ react easily with ethylenediamine to give $[Co(en)_3]^{3+}$ at room temperature.[111–113] The reactions exhibit stereoselectivity. Busch *et al.* proposed a mechanism based on the stepwise replacement of the $edta^{4-}$ or $pdta^{4-}$ from the cobalt(III) complex by ethylenediamine.[114] Doh *et al.*[4] studied the same reactions and found that several experimental findings contradictory with the mechanism proposed by Busch *et al.* However, they could not present an alternative mechanism. Recently, Geselowitz and Taube proposed a satisfactory mechanism which could account for the curious "substitution" of ethylenediamine onto $[Co(edta)]^-$.[108,115] This reaction is not substitution of $[Co(edta)]^-$ by ethylenediamine, but rather electron transfer reaction between $[Co(edta)]^-$ and $[Co(en)_3]^{2+}$. The latter species is derived from the reaction of ethylenediamine and cobalt(II) contained in $[Co(edta)]^-$ as an impurity. Therefore, the stereoselectivity observed for the reaction between $[Co(edta)]^-$ and ethylenediamine should be regarded as that of the electron transfer reaction between $[Co(edta)]^-$ and $[Co(en)_3]^{2+}$. There are reports that the reaction between Δ-$[Co(R-pdta)]^-$ and ethylenediamine gives almost 100% of Δ-$[Co(en)_3]^{3+}$.[114,116] From experiments, however, this does not seem to be true.[4]

F. Intramolecular electron transfer reactions

Cobalt(III) complexes containing a unidentate aminopolycarboxylate have been known, e.g., $[Co(NH_3)_5(H_n edta)]^{(n-1)+}$, $[Co(NH_3)_5(H_n nta)]^{n+}$, $[Co(NH_3)_5(H_n ida)]^{(n+1)+}$, and $[Co(NH_3)_5(H_n edda)]^{(n+1)+}$.[7,117,118] Reduction of these complexes by metal ions may proceed via complexation of the monodentate aminopolycarboxylato moiety toward the reductant prior to the electron transfer. The first example of this sort is shown in reaction (22).[117,119,120]

$$[Co(NH_3)_5(Hnta)]^+ + Fe^{2+} \rightleftharpoons [Co(NH_3)_5(nta)Fe(H_2O)_3]^{2+} + H^+$$
$$\downarrow k_{et}$$
$$Co(II) + Fe(III) \qquad (22)$$

The rate constant for the intramolecular electron transfer step was determined to be $0.115 \pm 0.004 \text{ s}^{-1}$ at $25°C$ and $I = 1.0 \text{ M}$ (NaClO$_4$).

In the Cr^{2+} reduction of $[Co(NH_3)_5(H_n\text{edta})]^{(n-1)+}$, the coordination of $H_n\text{edta}^{(4-n)-}$ moiety of the cobalt(III) complex to the Cr^{2+} center takes place prior to electron transfer, and the chromium(III) product is $[Cr(H_2O)(\text{edta})]^-$.[82] The chromium(III) products of the reaction of $[Co(NH_3)_5(H_n\text{ida})]^{(n+1)+}$ with Cr^{2+} are both $[Cr(H_2O)_5(H_n\text{ida})]^{(n+1)+}$ and $[Cr(H_2O)_3(\text{ida})]^+$ in which the $H_n\text{ida}^{(2-n)-}$ in the former acts as an unidentate ligand, and ida^{2-} in the latter acts as a terdentate ligand, respectively.[118] The product ratio of the above reaction depends on the pH of the solution.

Reduction of $[Co(NH_3)_5(H_2\text{nta})]^{2+}$ by Ti(III)$_{\text{aq}}$ proceeds by a mechanism analogous to reaction (22), but, in this case, the electron transfer step is strongly base catalyzed, and the hydroxo complex, $[Co(NH_3)_5$-$(\text{nta})Ti(OH)]^{2+}$, is suggested to be the active species for intramolecular electron transfer.[121]

The rates of the intramolecular electron transfer within the $[(NC)_5Fe(II)$–$CN\text{–}Co(III)L]^{4-}$ (L = hedtra^{3-}, bedtra^{3-}) complexes were studied by photoexcitation of the Co(III) $^1T_{1g} \leftarrow {}^1A_{1g}$ band which is followed by electron transfer from Fe(II) to Co(III).[122]

G. Reaction of aminopolycarboxylato complexes with peroxide and superoxide

The superoxide radical (O$_2^-$) is formed in normally functioning aerobic organisms. Superoxide and hydrogen peroxide in the presence of trace metal ions form hydroxyl radicals which are highly reactive and toxic to living systems. Therefore, it is important to scavenge superoxide for aerobic organisms. Disproportionation of superoxide to molecular oxygen and hydrogen peroxide (superoxide dismutation) is catalyzed by the metalloprotein superoxide dismutase. The catalysis appears to require the redox cycle of reactions (23) and (24).

$$M^{n+} + O_2^- \rightarrow M^{(n-1)+} + O_2 \tag{23}$$

$$M^{(n-1)+} + O_2^- + 2H^+ \rightarrow M^{n+} + H_2O_2 \tag{24}$$

Simple metal complexes have been examined to mimic the superoxide dismutases and to discuss mechanism of the dismutation. Again, edta complexes offer an advantage over simple metal ions due to the stability toward base hydrolysis at physiological pH. Among the metal complexes, Fe(II) and Fe(III) complexes in particular have been the subject of active research.

The $[Fe(III)(edta)]^-$ is reduced by O_2^- to produce $[Fe(II)(edta)]^{2-}$ and O_2. This reaction is strongly pH dependent. At pH 5.8–8.1, the rate constant is $(5–50) \times 10^5 \ M^{-1} \ s^{-1}$, but at around pH 9 there is a sharp decrease in the rate constant.[123] Thus the hydroxo species $[Fe(OH)(edta)]^{2-}$ is unreactive toward O_2^-.

In connection with the dismutation of superoxide, the reaction of H_2O_2 with metal complex is important. In 1956, the purple-colored species was found to form by the reaction between $[Fe(III)(edta)]^-$ and H_2O_2.[124] Since then efforts have been done to determine its structure and properties.[125–130] Recently obtained resonance Raman spectra indicates that the peroxo complex, $[Fe(III)(O_2)(edta)]^{3-}$, is responsible for the purple color.[131] To form a catalytic cycle of the superoxide dismutation it was suggested that the following three reactions should be combined.[132]

$$[Fe(III)(edta)]^- + O_2^- \rightarrow [Fe(II)(edta)]^{2-} + O_2 \qquad (25)$$

$$[Fe(II)(edta)]^{2-} + O_2^- \rightarrow [Fe(III)(O_2)(edta)]^{3-} \qquad (26)$$

$$[Fe(III)(edta)]^- + H_2O_2 \rightleftharpoons [Fe(III)(O_2)(edta)]^{3-} + 2H^+ \qquad (27)$$

There is dispute about the catalysis of dismutation of O_2^- by iron–edta complex. There are reports claiming that the iron–edta complex does not catalyze superoxide dismutation.[133,134] However, other reports agree with catalytic scavanging of superoxide by the iron–edta complex.[132,135,136]

Many studies have shown that the reaction between iron complex and H_2O_2 liberates oxygen. In contrast to these studies, the reaction of iron(II)–dtpa complex with H_2O_2 in phosphate buffer solution at pH 7.0 was found to result in consumption of oxygen rather than liberating oxygen.[137] It is thought that both free DTPA and the iron–dtpa complex participate in oxygen consumption by the following Fenton-type reaction, (28)–(30), where RH can be either dtpa or iron–dtpa complex:

$$[Fe(II)(dtpa)]^{3-} + H_2O_2 \rightarrow [Fe(III)(dtpa)]^{2-} + OH^- + \cdot OH \qquad (28)$$

$$\cdot OH + RH \rightarrow \cdot R + H_2O \qquad (29)$$

$$\cdot R + O_2 \rightarrow RO_2^- \qquad (30)$$

$[Mn(III)(edta)]^-$ and $[Mn(III)(cydta)]^-$ are rapidly reduced by O_2^- in dimethylsulfoxide to produce $[Mn(II)(edta)]^{2-}$ $(5 \times 10^4 \ M^{-1} \ s^{-1})$ and $[Mn(II)(cydta)]^{2-}$ $(\sim 1 \times 10^6 \ M^{-1} \ s^{-1})$ at 20°C, while $[Mn(II)(edta)]^{2-}$ reacts with superoxide in complicated manner. In an aqueous solution, on the other hand, both manganese(II)– and manganese(III)–edta complexes react with superoxide, but neither complex catalyzes the superoxide dismutation.[138,139] The $[Mn(cydta)]^-$ complex reacts with H_2O_2 to give $[Mn(cydta)]^{2-}$ and O_2. A mechanism which contains peroxo species, $[Mn(III)(O_2H)(cydta)]^{2-}$, is consistent with kinetic data.[140]

The manganese–edta system is not an effective model of superoxide dismutases. The basic difference between the iron–edta and manganese–edta systems is the metal(II)/metal(III) redox potentials. The standard redox potential of the $[Mn(edta)]^{2-}/[Mn(edta)]^-$ couple is 0.82 V, whereas the value is 0.12 V for the $[Fe(edta)]^{2-}/[Fe(edta)]^-$ couple. Both $[Fe(III)(edta)]^-$ and $[Mn(III)(edta)]^-$ can oxidize $O_2{}^-$. However, $[Mn(II)(edta)]^{2-}$ can be oxidized by $O_2{}^-$ only in acidic media, whereas $[Fe(II)(edta)]^{2-}$ can be oxidised by $O_2{}^-$ at any pH.

H. Electron transfer reactions of metal–aminopolycarboxylato complexes with metalloproteins

This section illustrates the use made of aminopolycarboxylatometal complexes in mechanistic studies into the reactions of metalloenzymes, especially the cytochromes and the single blue copper protein stellacyanin. Studies on iron–sulfur proteins are similarly of interest, but have been reviewed in a previous volume.[141]

1. Cytochromes

Electron transfer reactions of cytochrome c proceed in two steps involving (1) the approach of the inorganic complex to a region close to the heme edge which is partially exposed at the surface of the protein and (2) outer-sphere electron transfer. Electron transfer between an inorganic reagent and a large molecule such as a metalloprotein offers a good subject of investigation with regard to the Marcus theory.[142] The edta complexes of iron and, to a lesser extent of cobalt, have been employed as oxidant and/or reductant in examining such reactions.[143–150]

Kinetic studies of horse heart cytochrome c with several inorganic reducing and oxidizing reagents have been carried out. Based on observed electron exchange rate constants, the self-exchange rate constant for cytochrome c has been calculated with compensation for the electrostatic interaction.[144] Calculated self-exchange rate constants of the protein can vary over as much as three orders of magnitude depending on the identity of the inorganic reactant: $[Fe(CN)_6]^{3-} > [Co(phen)_3]^{3+} > [Ru(NH_3)_6]^{2+} > [Fe(edta)]^{2-}$. Such large variations in electrostatics-corrected self-exchange rate constants may indicate extra barriers for the electron transfer, most likely the barrier for association between the protein and the inorganic reagents. The following explanation has been offered for the observation that $[Fe(edta)]^{2-}$ is the least favorable reagent for the electron transfer with horse heart cytochrome c. The

$[Fe(edta)]^{2-}$ has a hydrophilic side comprising the carboxylates, and a hydrophobic side consisting of the methylene chain. In order to achieve efficient electron transfer, it has been suggested that π overlap between the carboxylate oxygens of $[Fe(edta)]^{2-}$ and the heme moiety of the protein has to be maximized. This requires that the carboxylates of $[Fe(edta)]^{2-}$ penetrate the hydrophobic region surrounding the heme edge, a process which could give an extra barrier to precursor complex formation prior to electron transfer.[144]

In some cases different reaction sites are available on the protein surface which the redox partner can select. Proton NMR measurement for the reaction of horse heart cytochrome c with $[Fe(H_2O)(edta)]^-$ and $[Cr(CN)_6]^{3-}$ have revealed two binding sites for the incoming complex are present near to the exposed heme edge.[149] Studies with cytochrome b_5 utilizing $[Co(edta)]^-$ as an oxidant have also been described.[150]

2. Stellacyanin

Aminopolycarboxylatocobalt(III) complexes have been used as a probe for the oxidant–protein interactions in the oxidation of stellacyanin from *Rhus vernicifera*.[151–154] The earlier report[151] of an association constant of 149 M^{-1} with $[Co(edta)]^-$ at 25°C and pH 7 now appears to be incorrect. Various inconsistencies have been addressed and no example is at present known in which prior association of stellacyanin is observed.[154]

References

[1] Fick, R.; Ulrich, H. German Patent 638071 (*Chem. Abstr.* **1937**, *31*, 1043⁹).

[2] Klemm, W. *Z. Anorg. Chem.* **1944**, *252*, 225.

[3] Schwarzenbach, G. *Helv. Chim. Acta* **1949**, *32*, 839.

[4] Doh, M.-K.; Ogino, H.; Fujita, J.; Saito, K.; Tanaka, N. *Chem. Lett.* **1974**, 1233.

[5] Thorneley, R. N. F.; Sykes, A. G. *J. Chem. Soc.* (*A*) **1969**, 742.

[6] Busch, D. H.; Bailar, J. C., Jr. *J. Am. Chem. Soc.* **1956**, *78*, 716.

[7] Ogino, H.; Tsukahara, K.; Tanaka, N. *Inorg. Chem.* **1977**, *16*, 1215.

[8] Takemoto, T.; Nomoto, K.; Fushiya, S.; Ouchi, R.; Kusano, G.; Hikino, H.; Takagi, S.; Matsuura, Y.; Kakudo, M. *Proc. Jpn. Acad.* **1978**, 54(*B*), 469.

[9] Fushiya, S.; Sato, Y.; Nozoe, S.; Nomoto, K.; Takemoto, T.; Takagi, S. *Tetrahedron Lett.* **1980**, *21*, 3071.

[10] Nomoto, K.; Yoshioka, H.; Arima, M.; Takemoto, T.; Fushiya, S.; Takagi, S. *Chimia* **1981**, *35*, 249.

[11] Budesinsky, M.; Budzikiewicz, H.; Prochazka, A.; Ripperger, H.; Romer, A.; Scholz, G.; Schreiber, K. *Phytochemistry* **1980**, *19*, 2295.

[12] Basolo, F.; Pearson, R. G. "Mechanisms of Inorganic Reactions"; Wiley: New York, 1967.

[13] Eigen, M.; Wilkins, R. G. *Adv. Chem. Ser.* **1965**, *49*, 55.

[14] Margerum, D. W.; Cayley, G. R.; Weatherburn, D. C.; Pagenkopf, G. K. *ACS Monogr.* **1978**, *174*, 1.
[15] Wiekliem, H. A.; Hoard, J. L. *J. Am. Chem. Soc.* **1959**, *81*, 549.
[16] Okamoto, K.; Tsukihara, T.; Hidaka, J.; Shimura, Y. *Chem. Lett.* **1973**, 145.
[17] Okazaki, H.; Tomioka, K.; Yoneda, H. *Inorg. Chim. Acta* **1983**, *74*, 169.
[18] Ogino, H.; Watanabe, T.; Tanaka, N. *Chem. Lett.* **1974**, *91*.
[19] Ogino, H.; Watanabe, T.; Tanaka, N. *Inorg. Chem.* **1975**, *14*, 2093.
[20] Sulfab, Y.; Taylor, R. S.; Sykes, A. G. *Inorg. Chem.* **1976**, *15*, 2388.
[21] Ogino, H.; Shimura, M.; Tanaka, N. *Inorg. Chem.* **1979**, *18*, 2497.
[22] Postmus, C.; King, E. L. *J. Phys. Chem.* **1955**, *59*, 1216.
[23] Guardalabene, J.; Gulnac, G.; Keder, N.; Shepherd, R. E. *Inorg. Chem.* **1979**, *18*, 22.
[24] Leupin, P.; Sykes, A. G.; Wieghardt, K. *Inorg. Chem.* **1983**, *22*, 1253.
[25] Schwarzenbach, G.; Heller, J. *Helv. Chim. Acta* **1951**, *34*, 576.
[26] Furlani, G.; Morpurgo, G.; Sartori, G. *Z. Anorg. Allg. Chem.* **1960**, *303*, 1.
[27] Hamm, R. E. *J. Am. Chem. Soc.* **1953**, *75*, 5670.
[28] Schwarzenbach, G.; Biedermann, W. *Helv. Chim. Acta* **1948**, *31*, 459.
[29] Thorneley, R. N. F.; Sykes, A. G.; Gans, P. *J. Chem. Soc. (A)* **1971**, 1494.
[30] Ogino, H.; Ito, S. Abstr. *Int. Conf. Coord. Chem. 22nd, Budapest*, **1982**, *TUC7*.
[31] Ogino, H.; Ito, S. Unpublished results.
[32] Peerce, P. J.; Gray, H. B.; Anson, F. C. *Inorg. Chem.* **1979**, *18*, 2593.
[33] Ogino, H.; Shimura, M.; Tanaka, N. *Inorg. Chem.* **1982**, *21*, 126.
[34] Schmidt, W.; Swinehart, J. H.; Taube, H. *J. Am. Chem. Soc.* **1971**, *93*, 1117.
[35] Chen, Z.; Cimolino, M.; Adamson, A. W. *Inorg. Chem.* **1983**, *22*, 3035.
[36] Brintzinger, H.; Thiele, H.; Muller, U. *Z. Anorg. Chem.* **1943**, *251*, 285.
[37] Dwyer, F. P.; Garvan, F. L. *J. Am. Chem. Soc.* **1960**, *82*, 4823.
[38] Gerdom, L. E.; Baenziger, N. A.; Goff, H. M. *Inorg. Chem.* **1981**, *20*, 1606.
[39] Sawyer, D. T.; McKinnie, J. M. *J. Am. Chem. Soc.* **1960**, *82*, 4191.
[40] Hoard, J. L.; Smith, G. S.; Lind, M. "Advances in the Chemistry of the Coordination Compounds"; Kirschner, S., Ed.; 1961, pp. 296–302.
[41] Hoard, J. L.; Kennard, C. H. L.; Smith, G. S. *Inorg. Chem.* **1963**, *2*, 1316.
[42] Kushi, Y.; Morimasa, K.; Yoneda, H. *Annu. Meet. Chem. Soc. Jpn., 49th, Tokyo*, **1984**, *1N31*.
[43] Jørgensen, C. K. *Acta Chem. Scand.* **1955**, *9*, 1362.
[44] Matsubara, T.; Creutz, C. *Inorg. Chem.* **1979**, *18*, 1956.
[45] Legg, J. I.; Bianchini, R. J.; Green, C. A.; Kaizaki, S.; Koine, N.; Wheeler, W. D. *Abstr. Int. Conf. Coord. Chem. 23rd, Boulder*, **1984**, *TUp 29-1*.
[46] Wheeler, W. D.; Legg, J. I. *Inorg. Chem.* **1984**, *23*, 3798.
[47] Lind, M. D.; Hamor, M. J.; Hamor, T. A.; Hoard, J. L. *Inorg. Chem.* **1964**, *3*, 34.
[48] Nesterova, Ya. M.; Polynova, T. N.; Martynenko, L. I.; Pechurova, N. I. *Zh. Strukt. Khim.* **1971**, *12*, 1110.
[49] Richards, S.; Pedersen, B.; Silverton, J. V.; Hoard, J. L. *Inorg. Chem.* **1964**, *3*, 27.
[50] Hoard, J. L.; Pedersen, B.; Richards, S.; Silverton, J. V. *J. Am. Chem. Soc.* **1961**, *83*, 3533.
[51] Polynova, T. M.; Anan'eva, N. N.; Porai-Koshits, M. A.; Martynenko, L. I.; Pechurova, N. I. *Zh. Strukt. Khim.* **1971**, *12*, 335.
[52] Novozhilova, N. V.; Polynova, T. N.; Porai-Koshits, M. A.; Pechurova, N. I.; Martynenko, L. I.; Khadi, A. *Zh. Strukt. Khim.* **1973**, *14*, 745.
[53] Ezerskaya, N. A.; Solovykh, T. S. *Russ J. Inorg. Chem. (Engl. Transl.)* **1966**, *11*, 991.
[54] Yoshino, Y.; Uehiro, T.; Saito, M. *Bull. Chem. Soc. Jpn.* **1979**, *52*, 1060.
[55] Yoshino, Y.; Uehiro, T.; Saito, M. *Chem. Lett.* **1978**, 487.
[56] Broomhead, J. A.; Basolo, F.; Pearson, R. G. *Inorg. Chem.* **1964**, *3*, 826.

[57] Ford, P. C. *Coord. Chem. Rev.* **1970**, *5*, 75.
[58] Kallen, T. W.; Earley, J. E. *Inorg. Chem.* **1971**, *10*, 1149.
[59] Edwards, J. O.; Monacelli, F.; Ortaggi, G. *Inorg. Chim. Acta* **1974**, *11*, 47.
[60] Oyama, N.; Anson, F. C. *J. Electroanal. Chem.* **1978**, *88*, 289.
[61] Oyama, N.; Anson, F. C. *J. Am, Chem. Soc.* **1979**, *101*, 739.
[62] Matsubara, T.; Creutz, C. *J. Am. Chem. Soc.* **1978**, *100*, 6255.
[63] Thompson, G. A. K.; Sykes, A. G. *Inorg. Chem.* **1979**, *18*, 2025.
[64] Chmelnick, A. M.; Fiat, D. *J. Chem. Phys.* **1969**, *51*, 4238.
[65] Adegite, A.; Earley, J. E.; Ojo, J. F. *Inorg. Chem.* **1979**, *18*, 1535.
[66] Swaddle, T. W.; Merbach, A. E. *Inorg. Chem.* **1981**, *20*, 4212.
[67] Swaddle, T. W. *Adv. Inorg. Bioinorg. Mech.* **1983**, *2*, 95.
[68] Hamm, R. E.; Suwyn, M. A. *Inorg. Chem.* **1967**, *6*, 139.
[69] Suwyn, M. A.; Hamm, R. E. *Inorg. Chem.* **1967**, *6*, 2150.
[70] Poh, B. L.; Stewart, R. *Can. J. Chem.* **1972**, *50*, 3432.
[71] Hamm, R. E.; Templeton, J. C. *Inorg. Chem.* **1973**, *12*, 755.
[72] Jones, T. E.; Hamm, R. E. *Inorg. Chem.* **1975**, *14*, 1027.
[73] Shimi, I. A. W.; Higginson, W. C. E. *J. Chem. Soc.* **1958**, 260.
[74] Dyke, R.; Higginson, W. C. E. *J. Chem. Soc.* **1960**, 1998.
[75] Wilkins, R. G.; Yelin, R. E. *J. Am. Chem. Soc.* **1970**, *92*, 1191.
[76] Woodruff, W. H.; Margerum, D. W.; Milano, M. J.; Pardue, H. L.; Santini, R. E. *Inorg. Chem.* **1973**, *12*, 1490.
[77] Evans, M. H.; Grossman, B.; Wilkins, R. G. *Inorg. Chim. Acta* **1975**, *14*, 59.
[78] Wood, P. B.; Higginson, W. C. E. *J. Chem. Soc.* **1965**, 2116.
[79] Pidcock, A.; Higginson, W. C. E. *J. Chem. Soc.* **1963**, 2798.
[80] Kurimura, Y.; Meguro, I.; Ohashi, K. *Bull. Chem. Soc. Jpn.* **1971**, *44*, 3367.
[81] Haim, A.; Sutin, N. *J. Am. Chem. Soc.* **1966**, *88*, 5343.
[82] Ogino, H.; Tsukahara, K.; Tanaka, N. *Inorg. Chem.* **1979**, *18*, 1271.
[83] Ogino, H.; Tsukahara, K.; Tanaka, N. *Inorg. Chem.* **1979**, *18*, 3290.
[84] Wood, P. B.; Higginson, W. C. E. *J. Chem. Soc.* **1966**, 1645.
[85] Ogino, H.; Kikkawa, E.; Shimura, M.; Tanaka, N. *J. Chem. Soc., Dalton Trans.* **1981**, 894.
[86] Grossman, B.; Wilkins, R. G. *J. Am. Chem. Soc.* **1967**, *89*, 4230.
[87] Kurimura, Y. *Bull. Chem. Soc. Jpn.* **1973**, *46*, 2093.
[88] Thorneley, R. N. F.; Kipling, B.; Sykes, A. G. *J. Chem. Soc. (A)* **1968**, 2847.
[89] Ogino, H.; Tanaka, N. *Bull. Chem. Soc. Jpn.* **1968**, *41*, 1622.
[90] Ogino, H.; Shimura, M.; Watanabe, T.; Tanaka, N. *Inorg. Nucl. Chem. Lett.* **1976**, *12*, 911.
[91] Ogino, H.; Shimura, M.; Tanaka, N. *Bull. Chem. Soc. Jpn.* **1978**, *51*, 1380.
[92] Davies, K. M.; Earley, J. E. *Inorg. Chem.* **1976**, *15*, 1074.
[93] Marčec, R.; Orhanović, M. *Inorg. Chem.* **1978**, *17*, 3672.
[94] Lee, R. A.; Earley, J. E. *Inorg. Chem.* **1981**, *20*, 1739.
[95] Kristine, F. J.; Shepherd, R. E. *Inorg. Chem.* **1981**, *20*, 215.
[96] Adamson, A. W.; Gonick, E. *Inorg. Chem.* **1963**, *2*, 129.
[97] Huchital, D. H.; Wilkins, R. G. *Inorg. Chem.* **1967**, *6*, 1022.
[98] Huchital, D. H.; Hodges, R. J. *Inorg. Chem.* **1973**, *12*, 998.
[99] Huchital, D. H.; Hodges, R. J. *Inorg. Chem.* **1973**, *12*, 1004.
[100] Rosenheim, L.; Speiser, D.; Haim, A. *Inorg. Chem.* **1974**, *13*, 1571.
[101] Huchital, D. H.; Lepore, J. *Inorg. Chem.* **1978**, *17*, 1134.
[102] Huchital, D. H.; Lepore, J. *J. Inorg. Nucl. Chem.* **1978**, *40*, 2073.
[103] Huchital, D. H.; Lepore, J. *Inorg. Chim. Acta* **1980**, *38*, 131.
[104] Reagor, B. T.; Huchital, D. H. *Inorg. Chem.* **1982**, *21*, 703.
[105] Seaman, G. C.; Haim, A. *J. Am. Chem. Soc.* **1984**, *106*, 1319.

[106] Ogino, H.; Takahashi, M.; Tanaka, N. *Bull. Chem. Soc. Jpn.* **1974**, *47*, 1426.
[107] Phillips, J.; Haim, A. *Inorg. Chem.* **1980**, *19*, 76.
[108] Geselowitz, D. A.; Taube, H. *J. Am. Chem. Soc.* **1980**, *102*, 4525.
[109] Lappin, A. G.; Laranjeira, M. C. M.; Peacock, R. D. *Inorg. Chem.* **1983**, *22*, 786.
[110] Kondo, S.; Sasaki, Y.; Saito, K. *Inorg. Chem.* **1981**, *20*, 429.
[111] Dwyer, F. P.; Gyarfas, E. C.; Mellor, D. P. *J. Phys. Chem.* **1955**, *59*, 296.
[112] Dwyer, F. P.; Garvan, F. L. *J. Am. Chem. Soc.* **1958**, *80*, 4480.
[113] Kirschner, S.; Wei, Y. K.; Bailar, J. C. Jr. *J. Am. Chem. Soc.* **1957**, *79*, 5877.
[114] Busch, D. H.; Swaminathan, K.; Cooke, D. W. *Inorg. Chem.* **1962**, *1*, 260.
[115] Geselowitz, D. A.; Taube, H. *Inorg. Chem.* **1981**, *20*, 4036.
[116] Irving, H.; Gillard, R. D. *J. Chem. Soc.* **1961**, 2249.
[117] Cannon, R. D.; Gardiner, J. *J. Am. Chem. Soc.* **1970**, *92*, 3800.
[118] Ogino, H.; Tsukahara, K.; Tanaka, N. *Inorg. Chem.* **1980**, *19*, 255.
[119] Cannon, R. D.; Gardiner, J. *Inorg. Chem.* **1974**, *13*, 390.
[120] Cannon, R. D.; Gardiner, J. *J. Chem. Soc., Dalton Trans.* **1976**, 622.
[121] Marčec, R.; Orhanović, M.; Wray, J. A.; Cannon, R. D. *J. Chem. Soc., Dalton Trans.* **1984**, 663.
[122] Reagor, B. T.; Kelley, D. F.; Huchital, D. H.; Rentzepis, P. M. *J. Am. Chem. Soc.* **1982**, *104*, 7400.
[123] Ilan, Y. A.; Czapski, G. *Biochim. Biophys. Acta* **1977**, *498*, 386.
[124] Cheng, K. L.; Lott, P. F. *Anal. Chem.* **1956**, *28*, 462.
[125] Ringbom, A.; Siitonen, S.; Saxen, B. *Anal. Chim. Acta* **1957**, *16*, 541.
[126] Mader, P. M. *J. Am. Chem. Soc.* **1960**, *82*, 2956.
[127] Orhanović, M.; Wilkins, R. G. *Croat. Chem. Acta* **1967**, *39*, 149.
[128] Kochanny, G. L., Jr.; Timnick, A. *J. Am. Chem. Soc.* **1961**, *83*, 2777.
[129] Walling, C.; Kurz, M.; Schuger, H. J. *Inorg. Chem.* **1970**, *9*, 931.
[130] Rizkalla, E. N.; El-Shafey, O. H.; Guindy, N. M. *Inorg. Chim. Acta* **1982**, *57*, 199.
[131] Hester, R. E.; Nour, E. M. *J. Raman Spectrosc.* **1981**, *11*, 35.
[132] McClune, G. J.; Fee, J. A.; McClusky, G. A.; Groves, J. T. *J. Am. Chem. Soc.* **1977**, *99*, 5220.
[133] Diguiseppi, J.; Fridovich, I. *Arch. Biochem. Biophys.* **1980**, *203*, 145.
[134] Butler, J.; Halliwell, B. *Arch. Biochem. Biophys.* **1982**, *218*, 174.
[135] Bull, C.; Fee, J. A.; O'Neill, P.; Fielden, E. M. *Arch. Biochem. Biophys.* **1982**, *215*, 551.
[136] Bull, C.; McClune, G. J.; Fee, J. A. *J. Am. Chem. Soc.* **1983**, *105*, 5290.
[137] Cohen, G.; Lewis, D.; Sinet, P. M. *J. Inorg. Biochem.* **1981**, *15*, 143.
[138] Stein, J.; Fackler, J. P. Jr.; McClune, G. J.; Fee, J. A.; Chan, L. T. *Inorg. Chem* **1979**, *18*, 3511.
[139] Bradić, Z.; Wilkins, R. G. *J. Am. Chem. Soc.* **1984**, *106*, 2236.
[140] Jones, T. E.; Hamm, R. E. *Inorg. Chem.* **1974**, *13*, 1940.
[141] Armstrong, F. *Adv. Inorg. Bioinorg. Mech.* **1982**, *1*, 65.
[142] Marcus, R. A. *Annu. Rev. Phys. Chem.* **1964**, *15*, 155.
[143] McArdle, J. V.; Gray, H. B.; Creutz, C.; Sutin, N. *J. Am. Chem. Soc.* **1974**, *96*, 5737.
[144] Wherland, S.; Gray, H. B. *Proc. Nat. Acad. Sci. U.S.A.* **1976**, *73*, 2950.
[145] Coyle, C. L.; Gray, H. B. *Biochem. Biophys. Res. Commun.* **1976**, *73*, 1122.
[146] Holwerda, R. A.; Read, R. A.; Scott, R. A.; Wherland, S.; Gray, H. B.; Millett, F. *J. Am. Chem. Soc.* **1978**, *100*, 5028.
[147] Mauk, A. G.; Scott, R. A.; Gray, H. B. *J. Am. Chem. Soc.* **1980**, *102*, 4360.
[148] Wei, J.-F.; Ryan, M. D. *J. Inorg. Biochem.* **1982**, *17*, 237.
[149] Williams, G.; Eley, C. G. S.; Moore, G. R.; Robinson, M. N.; Williams, R. J. P. *FEBS Lett.* **1982**, *150*, 293.
[150] Chapman, S. K.; Davies, D. M.; Vuik, C. P. J.; Sykes, A. G. *J. Am. Chem. Soc.* **1984**, *106*, 2692.

[151] Yoneda, G. S.; Holwerda, R. A. *Bioinorg. Chem.* **1978,** *8,* 139.

[152] Yoneda, G. S.; Mitchel, G. L.; Blackmer, G. L.; Holwerda, R. A. *Bioinorg. Chem.* **1978,** *8,* 369.

[153] Holwerda, R. A.; Clemmer, J. D. *J. Inorg. Biochem.* **1979,** *11,* 7.

[154] Sisley, M. J.; Segal, M. G.; Stanley, C. S.; Adzamli, I. K.; Sykes, A. G. *J. Am. Chem. Soc.* **1983,** *105,* 225.

ADVANCES IN INORGANIC AND BIOINORGANIC MECHANISMS, VOL. 4

Ligand Substitution and Redox Reactions of Gold(III) Complexes

L. H. Skibsted

Chemistry Department
Royal Veterinary and Agricultural University
Copenhagen, Denmark

I. INTRODUCTION

Gold(III) is a low-spin d^8 ion which forms square-planar complexes. The now well-established two-term rate law for square-planar ligand substitution reactions was in fact first observed for exchange of free chloride with tetrachloroaurate(III).[1] The remark by Rich and Taube in their 1954 paper about a possible generality of their observation was very provident, because other square-planar complexes, in particular those of isoelectronic Pd(II) and Pt(II), have since been shown to substitute by the same general mechanism. However, good nucleophiles toward these metal centers are in general also good reductants, and since gold(III) complexes are significantly more oxidizing than their Pd(II) and Pt(II) analogs, redox reactions often take place during or after ligand substitution reactions. Thus, the high oxidizing power of gold(III) complexes limits the number of nucleophiles for which ligand substitution can be investigated without complications from redox reactions, as noted in some of the earliest kinetic investigations on ligand substitutions of Au(III) complexes.[2] On the other hand, in studies of gold(III) complexes as oxidants, it has often been difficult to identify the kinetically important gold(III) species, since ligands are substituted with rates comparable to those for the redox reactions.

A number of recent developments in the chemistry of gold compounds have arisen from the use of gold compounds in the treatment of rheumatoid arthritis,[3] as a heavy-metal label in macromolecules,[4] or as a sensitizer in photographic emulsions.[5] However, recent efforts in gold chemistry have also yielded new knowledge of the more fundamental thermodynamic and kinetic properties of gold compounds in solution, and the following treatment attempts to systematize the mechanisms of reactions of Werner-type gold(III) complexes on the basis of such results.

II. THERMODYNAMIC CONSIDERATIONS

A. Gold aqua ions

The characteristic coordination numbers of gold(I) and gold(III) complexes are two and four, respectively. However, neither of the aqua ions $Au(H_2O)_2^+$ and $Au(H_2O)_4^{3+}$ have been characterized, either in solution or in the solid state. For the gold(I) state, the hypothetical diaquagold(I) ion is unstable toward disproportionation, in contrast to what is found for the solvento species in certain aprotic solvents such as acetonitrile[6] and dimethyl-sulfoxide.[7] For the gold(III) state, the tetraaquagold(III) ion would be expected to be a very strong acid in water, as clearly evidenced by the aqueous solution acidities of the mixed ligand complexes $Au(NH_3)_3(H_2O)^{3+}$ ($pK_a = -0.7^{[8]}$), $AuCl_3(H_2O)$ ($pK_a = 0.6^{[9]}$), and $AuBr_3(H_2O)$ ($pK_a = 2.1^{[10]}$).

The standard potentials of the aqua ions of "the noblest of the elements" are consequently not accessible by direct measurement, but must be derived from indirect calculations.[11,16]

B.   Gold(I) complexes

The standard reduction potentials of several gold(I) and gold(III) complexes other than the aqua ions are known from measurements with gold metal electrodes. Equilibrium between metallic gold, gold(I), and gold(III) must, however, be established before reliable results can be obtained, a fact which has not always been recognized.[19] The potentials for the series of gold(I) complexes in aqueous solutions quoted in Table 1, which have been based on a critical examination of literature data,[16] are characteristic of a soft acceptor, since the more reduceable ligands form the more stable complexes. For the reaction

$$AuCl_2^- + 2Br^- \rightleftharpoons AuBr_2^- + 2Cl^- \tag{1}$$

values of $\Delta H^\circ = -26 \text{ kJ mol}^{-1}$ and $\Delta S^\circ = -25 \text{ J mol}^{-1} \text{ K}^{-1}$ were found,[20] and the discrimination between ligands is clearly an enthalpy effect.

The stability constants for gold(I) complexes relative to the chloro complex $AuCl_2^-$, as calculated from the reduction potentials, have been shown[16] to correlate with the standard reduction potentials and the Brønsted basicity of the ligand according to Edwards' four-parameter equation

$$\tfrac{1}{2} \log \beta_2' = \alpha E_n + \beta H \tag{2}$$

In this form of Edwards' equation,[17] the mean stability constant for the complexes $\sqrt{\beta_2'}$ relative to a reference complex [i.e., $AuCl_2^-$; cf. Eq. (1)] is related to the ligand parameters $E_n = E^\circ$ (ref.) $- E^\circ(L_2^{(2v-2)-}, 2L^{v-})$ and $H = pK_a(HL^{(v+1)-}) - pK_a(\text{ref.})$. The metal ion parameters α and β determined from such correlations (cf. Fig. 1) have been shown to constitute a measure of the metal ion softness and metal ion hardness, respectively.[16] In addition, the stability of as yet unknown complexes can be predicted by Eq. (2) from combinations of ligand parameters and metal ion parameters. The value of $E_{1,0}^\circ$ for $Au(H_2O)_2^+$ in Table 1 has been calculated by this method, and in agreement with previous suggestions,[16] $E_{1,0}^\circ(Au(H_2O)_2^+/Au) = +1.83 \text{ V}$ forms the basis for the "absolute" stability constant β_2 of Table 1 (footnote b).

In aprotic solvents such as dimethyl sulfoxide[7] and acetonitrile[6,21] gold(I) forms even stronger complexes; however, the discrimination between the ligands is less pronounced than in water, as can be seen from the acetonitrile data in Table 2.

TABLE 1

Thermodynamic characterization of Au(I) and Au(III) complex formation at 25°C in aqueous solution

L^{v-}	Ligand parameters[a]		Au(I) complexes[b]			Au(III) complexes[c]			Au(I)/Au(III) equilibria[d,e]	
	$E°(L_2^{(2v-2)-}, 2L^{v-})$ (V)	$pK_a(HL^{(v+1)-})$	$E°_{1,0}{}^e$ (V)	$\log \beta'_2$	$\log \beta_2$	$E°_{3,0}{}^e$ (V)	$\log \beta'_4$	$\log \beta_4$	$\log K_{1,3}$	$\log K_{0a}$
H_2O	+2.60	−1.7	(+1.83)[f]	0	0	(+1.52)[f]		0	(15.7)	
Cl^-	+1.36	−6.1	+1.154	0	11	+1.002	0	26	7.71	14.7
Br^-	+1.09	−7.2	+0.959	3.29	15	+0.854	7.50	34	5.32	9.0
OH^-	+1.0	+15.7				+0.48	26.4	53		
SCN^-	+0.77	−0.8	+0.662	8.31	20	+0.636	18.55	45	1.32	5.0
NH_3	+0.76	+9.2	+0.563	10.0	21	+0.325	34.3	60	12.07	
I^-	+0.54	−8.7	+0.578	9.73	21	+0.56	22.4	49	0.9	0.4
$(NH_2)_2CS$	+0.42	−0.9	+0.380	13.1	24					
$S_2O_3^{2-}$	+0.08	+1.7	+0.153	16.9	28					
CN^-	−0.2	+9.4	−0.48	27.6	39	−0.10	56	82	−19.3	−9.8

[a] Standard reduction potentials from Refs. [11 and 17]. pK_a values taken from Refs. [16 and 17].
[b] Standard reduction potentials for $AuL_2^{(2v-1)-}$. β'_2 is the stability constant relative to $AuCl_2^-$, and β_2 is the stability constant relative to the hypothetical $Au(H_2O)_2^+$.
[c] Standard reduction potentials for $AuL_4^{(4v-3)-}$. β'_4 is the stability constant relative to $AuCl_4^-$, and β_4 is the stability constant relative to the hypothetical $Au(H_2O)_4^{3+}$.
[d] $K_{1,3}$ is the equilibrium constant for the disproportionation reaction $3AuL_2^{(2v-1)-} \rightleftharpoons AuL_4^{(4v-3)-} + 2Au + 2L^{v-}$.
[e] K_{0a} is the equilibrium constant for the oxidative addition $AuL_2^{(2v-1)-} + L_2^{(2v-2)-} \rightleftharpoons AuL_4^{(4v-3)-}$. Gold potentials taken from Ref. [16].
[f] Calculated from Edwards' equation.

TABLE 2

Stability constants for gold(I) complexes in acetonitrile at
24°C and ionic strength 0.1 M

L^{v-}	$\log K_1$	$\log K_2$	$\log \beta_2$	Ref.
Cl^-	12.1	7.8	19.9	[6]
Br^-	12.0	8.4	20.4	[6]
SCN^-			20.2	[6]
CN^-			33.7	[6]
NH_3	10.1	8.0	18.2	[6]
PPh_3	$(>2)^a$			[21]
$AsPh_3$	$(1.1)^a$			[21]
$SbPh_3$	$(-0.8)^a$			[21]

^{a} Equilibrium constant for $AuCl_2^- + L \rightleftharpoons AuClL + Cl^-$
(L = triphenylphosphine, -arsine, or -stibine) at 20°C.

C. Gold(III) complexes

Thermodynamic data for gold(III) complexes in solution are scarce, but as
for the gold(I) complexes, the stability data for gold(III) complexes in aqueous
solution.[24,27] In the solid state, penta- and hexacoordination is confirmed for
Fig. 1. The sequence of the stabilities for the ligands in Table 1 is the same for
the two oxidation states, but it should be noted that α(Au(III)) > α(Au(I)). This
indicates that gold(III) is an extremely soft acid and, in keeping herewith, the
discriminating power of gold(III) is significantly higher than that of gold(I)
and of other square-planar d^8 metal centers.[22,23] Over the years there have
been several reports of formation in solution of gold(III) halide or pseudo-
halide complexes with coordination numbers higher than four, and such
species have consequently been considered to be of kinetic importance.
However, critical examination of the experimental basis for these claims
provides no real evidence for the existence of such higher complexes in
solution.[24,27] In the solid state, penta- and hexacoordination is confirmed for
several gold(III) complexes,[18] but only with chelate ligands. In all cases one
or more bonds have been found to be longer than is normal for Au(III)
complexes.[25,26]

The consecutive equilibrium constants for the overall process

$$AuCl_4^- + 4Br^- \rightleftharpoons AuBr_4^- + 4Cl^- \tag{3}$$

have been determined in aqueous solution by both spectrophotometric and
potentiometric methods.[22,28–30] As might be expected, the equilibrium is
relatively insensitive to ionic strength changes, and reasonable agreement has

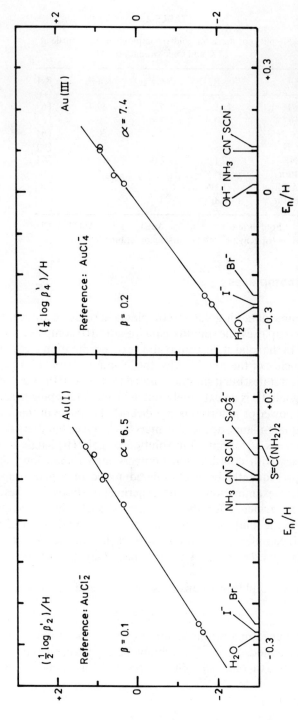

Fig. 1. Stability constants for gold(I) and gold(III) complexes relative to dichloroaurate(I) and tetrachloroaurate(III), respectively, in aqueous solution, plotted according to Edwards' equation[16]: $(\frac{1}{N}\log\beta'_N)/H = \alpha E_n/H + \beta$.

been found between the results obtained by different methods. However, potentiometric measurements tend to overestimate the value of the first consecutive stability constant K'_1. The consecutive equilibrium constants, including that for the cis/trans equilibrium determined spectrophotometrically by Elding and Gröning,[22] are combined in Table 3 with the calorimetric results of Cudey et al.[31] cis-$AuCl_2Br_2^-$ is statistically favored over its trans counterpart, and the observation that $\Delta H_1^\circ = \Delta H_2^\circ$ (mainly cis) is more negative than ΔH_3° (mainly cis) $= \Delta H_4^\circ$ suggests that bromide has some trans influence[1] relative to chloride. However, this is not in agreement with the results (also included in Table 3) obtained for the process[23]

$$trans\text{-}Au(NH_3)_2Cl_2^+ + 2Br^- \rightleftharpoons trans\text{-}Au(NH_3)_2Br_2^+ + 2Cl^- \qquad (4)$$

which indicate that the difference between K'_1 and K'_2 is statistically determined, since the ΔH_n° values are identical and the difference between the reaction entropies has a value close to that expected from statistical arguments: $R \ln 4 = 12 \text{ J mol}^{-1} \text{ K}^{-1}$. Solvation of chloride is stronger than that of bromide, and the statistically corrected reaction entropies for the chloride/bromide exchange (certain reasonable assumptions were introduced during the calculations; see footnote to Table 3) are all negative and not very different from $S_{Cl^-}^\circ - S_{Br^-}^\circ = -26.1 \text{ J mol}^{-1} \text{ K}^{-1}$.[23] For the process

$$Au(dien)Cl^{2+} + Br^- \rightleftharpoons Au(dien)Br^{2+} + Cl^- \qquad (5)$$

comparable results were obtained,[32,33] and the reaction enthalpies for the chloride/bromide exchange at all these Au(III) centers are not very different from each other. In other words, the halide–gold(III) bond strength is fairly independent of whether the cis or trans ligand is a chloride, a bromide, or an aliphatic amine, and none of these ligands have any significant trans or cis influence relative to each other.

The force constant for gold(III)–halide bonds decreases with increasing atomic weight of the halogen atoms,[34] and thermochemical calculations suggest that hydration effects are solely responsible for the higher aqueous solution stability of $AuBr_4^-$ relative to $AuCl_4^-$.[35] However, the results for mixed cyanohalogold(III) complexes do not fit into this simple scheme. Ligand-to-metal charge transfer spectra of AuX_4^- in comparison with those of $trans\text{-}Au(CN)_2X_2^-$ (X = Cl, Br) show that the σ-π mixing in $trans$-$Au(CN)_2X_2^-$ involves the halide orbitals and the more stable cyanide orbitals which serve to destabilize both the σ and the π halide orbitals.[37] Thermodynamic results seem to confirm that both σ and π effects are important when comparing halide bonding in the mixed ligand complexes with halide

---

[1] Trans and cis influence refers to thermodynamic phenomena, whereas trans and cis effects refer to kinetic phenomena.

TABLE 3

Thermodynamic parameters for the exchange of chloride with bromide in gold(III) complexes in aqueous solution at 25°C

Equilibrium	K'_n	$K'_{n,\text{corr.}}$[a]	ΔH_n° (kJ mol^{-1})	ΔS_n° (J mol^{-1} K^{-1})	$\Delta S_{n,\text{corr.}}^\circ$[a] (J mol^{-1} K^{-1})	Medium	Ref.
$AuCl_4^- + Br^- \rightleftharpoons AuCl_3Br^- + Cl^-$	243±40	61±10	−15.94±0.05	−8±2	−19±2	K'_n: 1.0 M NaClO$_4$	[22]
$AuCl_3Br^- + Br^- \rightleftharpoons cis\text{-}AuCl_2Br_2^- + Cl^-$	64±13	64±13	−15.94±0.05[b]	(−19±2)[c]	(−19±2)[c]		
$AuCl_3Br^- + Br^- \rightleftharpoons trans\text{-}AuCl_2Br_2^- + Cl^-$	34±7	68±14		(−24±2)[c]	(−18±2)[c]	ΔH_n°: 0.10–0.15 M[d]	[31]
$cis\text{-}AuCl_2Br_2^- + Br^- \rightleftharpoons AuClBr_3^- + Cl^-$	73±15	73±15	−13.66±0.05[b]	(−10±2)[c]	(−10±2)[c]		
$trans\text{-}AuCl_2Br_2^- + Br^- \rightleftharpoons AuClBr_3^- + Cl^-$	153±15	77±8		(−4±1)[c]	(−10±1)[c]		
$AuClBr_3^- + Br^- \rightleftharpoons AuBr_4^- + Cl^-$	17±5	68±20	−13.66±0.05	−22±3	−11±3		
$trans\text{-}Au(NH_3)_2Cl_2^+ + Br^- \rightleftharpoons trans\text{-}Au(NH_3)_2ClBr^+ + Cl^-$	122±2	61±1	−12.1±1.2	−1±4	−7±4	1.0 M H(Cl, Br, ClO$_4$)	[23]
$trans\text{-}Au(NH_3)_2ClBr^+ + Br^- \rightleftharpoons trans\text{-}Au(NH_3)_2Br_2^+ + Cl^-$	29.9±0.5	60±1	−13.5±0.6	−17±2	−12±2		
$Au(dien)Cl^{2+} + Br^- \rightleftharpoons Au(dien)Br^{2+} + Cl^-$ [e]	39±1	39±1	−12±2	−10±7	−10±7	1.0 M Na(Cl, ClO$_4$); pH=2.3	[33]
$trans\text{-}Au(CN)_2Cl_2^+ + Br^- \rightleftharpoons trans\text{-}Au(CN)_2ClBr^+ + Cl^-$	1100	550				0.51 M NaClO$_4$	[36]
$trans\text{-}Au(CN)_2ClBr^+ + Br^- \rightleftharpoons trans\text{-}Au(CN)_2Br_2^+ + Cl^-$	120	240					

[a] Statistically corrected; cf. Ref. [23].

[b] Experimental results are for the cis/trans equilibrium mixtures.

[c] ΔS_2° and ΔS_3° are calculated with the assumption that ΔH_2° and ΔH_3° are independent of stereochemistry.

[d] The reaction enthalpies are considered independent of ionic strength and adapted for the calculation of ΔS_n° at unit ionic strength.

[e] dien = 1,4,7-triazaheptane.

bonding in the tetrahalo complexes. Two cyanide ligands thus increase the discrimination between bromide and chloride (last two entries of Table 3), and an even more significant effect was found for the discrimination between chloride and iodide,[38,39] as can be seen from the mean stability constants:

$$AuCl_4^- + 4I^- \rightleftharpoons AuI_4^- + 4Cl^-, \qquad \sqrt[4]{\beta_4'} = 10^6 \qquad (6)$$

$$\textit{trans-}Au(CN)_2Cl_2^- + 2I^- \rightleftharpoons \textit{trans-}Au(CN)_2I_2^- + 2Cl^-, \qquad \sqrt{\beta_2'} = 10^8 \qquad (7)$$

$$Au(CN)_3Cl^- + I^- \rightleftharpoons Au(CN)_3I^- + Cl^-, \qquad \beta_2' = 10^4 \qquad (8)$$

The degree of hydrolysis of gold(III) in aqueous solution decreases along the series $Au(NH_3)_4^{3+} > AuCl_4^- > AuBr_4^-$, and when statistically corrected, the hydrolysis constants for the first halides are the same as for halide hydrolysis in $Au(dien)Cl^{2+}$ and for $Au(dien)Br^{2+}$ ($K_h = \sim 10^{-7}$ and $\sim 10^{-8}$ M, respectively[9,32,40]). Such processes are markedly endothermic: $\Delta H°$ for the hydrolysis of the first ligand is ~ 20 kJ mol^{-1} for $Au(NH_3)_4^{3+}$[8] and ~ 30 kJ mol^{-1} for $AuCl_4^-$.[40]

D. Equilibrium between gold(I) and gold(III)

The equilibrium constants in Table 1 for the disproportionation reactions

$$3AuL_2^{(2v-1)-} \rightleftharpoons AuL_4^{(4v-3)-} + 2Au + 2L^{v-}, \qquad K_{1,3} \qquad (9)$$

indicate that the softer ligands favor the gold(I) state relative to the gold(III) state. This does not conflict with higher discriminating power of gold(III) relative to gold(I), but is simply a result of the fact that the number of Au(I)–ligand bonds is higher than the number of Au(III)–ligand bonds in the equilibrium of Eq. (9). The values $\Delta H° = -78$ kJ mol^{-1}, $\Delta S° = -120$ J mol^{-1} K^{-1} for $L^{v-} = Cl^-$, and $\Delta H° = -61$ kJ mol^{-1}, $\Delta S° = -110$ J mol^{-1} K^{-1} for $L^{v-} = Br^-$ show that the low oxidation state is entropy favored.[20] Amine ligands in particular stabilize gold(III), and gold(III) coordinated to porphyrin ligands could not be reduced electrochemically.[27] The combination of Edwards' equation for the two oxidation states (cf. Fig. 1) permits the estimation of $K_{1,3}$ from the ligand parameters $E°$ and pK_a for disproportionation reactions for which experimental values are not yet accessible

$$\log K_{1,3} = -3.85 + 9.4\,E° + 0.2\,pK_a \qquad (10)$$

Gold(I) complexes add halogens or pseudo-halogens

$$AuL_2^{(2v-1)-} + L_2^{(2v-2)-} \rightleftharpoons AuL_4^{(4v-3)-} \qquad (11)$$

and the equilibrium constants for the oxidative addition reactions and for the disproportionation reactions depend in the same qualitative manner on the ligand properties (cf. Table 1). Equilibrium data for the addition of halogens to $Au(CN)_2^-$ confirm, when compared to that for addition to dihaloaurate(I), the strong cis influence of cyanide.[38]

III. LIGAND SUBSTITUTION IN GOLD(III) COMPLEXES

Models involving transition states of increased coordination number have been successful in accounting for the experimental observations on ligand substitution in gold(III) complexes[1,2,41] and in d^8 low-spin square-planar complexes in general. In early studies of ligand substitutions in gold(III) complexes it was found that the rates of ligand substitutions are very sensitive to the nature of both the entering and the leaving ligand, and that the discrimination between the nucleophiles depends on the leaving group. On the basis of these observations, Cattalini and Tobe concluded that bond formation and bond breaking are nearly synchronous processes.[41] Subsequent studies have fully confirmed this proposal of concerted substitution involving a pentacoordinated transition state rather than a pentacoordinated intermediate, and have also confirmed early suggestions by Baddley and Basolo of the importance of charge neutralization in this transition state.[2]

A.   Reactions involving unidentate ligands

The substrates of choice for the study of substitution reactions in gold(III) complexes have been the easily accessible tetrachloro- and tetrabromo-aurate(III) ions.[1,10,22,41–48] However, even for the reactions of these simple complex ions with unidentate nucleophiles such as halides or thiocyanate, the reaction stoichiometries were not fully established until recently.[22,45,48] The exact nature of the reaction intermediates in such substitution reactions has also been the subject of some discussion[45–48] (see Section III,D).

Results from several comprehensive studies of simple unidentate complex systems now permit patterns of reactivity in gold(III) complexes to be established with certainty.[22,47–53] Elding and Gröning[22] have thus studied the kinetics of substitution for the reversible reaction of Eq. (3), and established the stoichiometric mechanism depicted in Fig. 2. Figure 2 also includes the reaction scheme adapted for the anation of tetraamminegold(III) by chloride or by bromide

$$Au(NH_3)_4{}^{3+} + 4H^+ + 4X^- \rightarrow AuX_4{}^- + 4NH_4{}^+ \tag{12}$$

which has been studied under similar aqueous solution conditions.[50-52] For all of the bromide/chloride exchange reactions of Fig. 2, thermodynamic parameters are available (cf. Table 3).

1. Rate expression

The ligand substitution reactions of Fig. 2 were investigated in acidic aqueous solution of unit ionic strength (with the exception of the hydrolysis reactions) and with excess of the entering ligand. For all of these reactions, the substitution rate was found to be proportional to the concentration of the entering ligand, and the first-order term in the usual two-term rate law for ligand substitution in square-planar complexes

$$\text{Rate} = (k_1 + k_2[Y])[\text{complex}] \tag{13}$$

was in no case significant. The first-order k_1 term corresponds to a solvent path and the k_2 term to a direct replacement by the entering ligand, Y, corresponding to the reaction scheme of Fig. 3. The full rate expression simplifies to Eq. (13) if (1) pseudo-first-order conditions are employed, (2) the solvento intermediate (i.e., AuL_3H_2O in Fig. 3) is present in a steady-state concentration, (3) reverse reactions are suppressed, and, furthermore, if (4) $k_{-1}[X] \ll k_3[Y]$.[22,54]

Water is relatively ineffective as an entering ligand in gold(III) complexes, and consequently, the k_1 term is often of little importance or vanishes completely for substitution reactions in aqueous solutions.[22] k_1 has been determined directly for several complexes (cf. Table 4), and for tetra-chloroaurate(III) consistency was found between the directly determined value for k_1 and that determined from chloride exchange according to Eq. (13).[55]

The marked acidity of aquagold(III) complexes (cf. Section II,A) can, if hydrolysis is not suppressed, complicate the observed kinetics, and it should also be noted that in some cases it is a matter of taste (or confidence) as to whether one decides that a small intercept in a plot of k_{obs} versus [Y] is significant and corresponds to a k_1 path. For example, for the exchange of bromide with chloride in *trans*-$Au(CN)_2Br_2{}^-$ a value for a k_1 term was determined corresponding to $k_2/k_1 = 800$ liter mol^{-1},[49] whereas for the corresponding reaction in $AuBr_4{}^-$ the k_1 term was found to be insignificant (k_1 is in this case independently known and $k_2/k_1 = 45$ liter mol^{-1}; Table 4).[22]

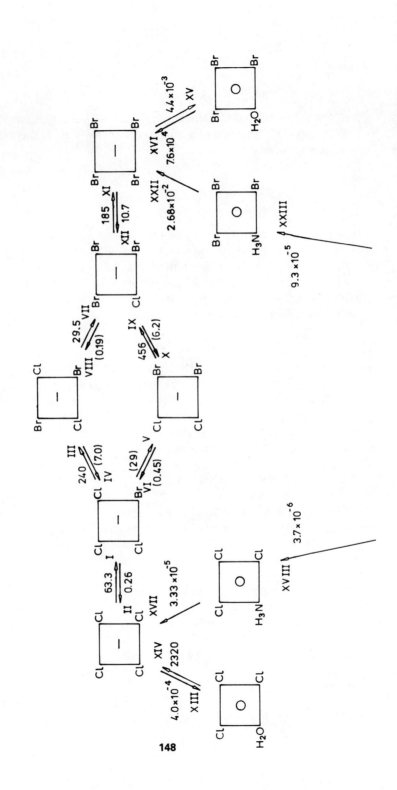

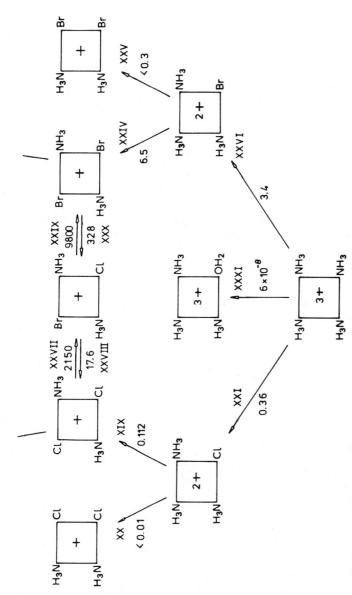

Fig. 2. Ligand substitutions by chloride, bromide, or water in gold(III) complexes with unidentate ligands in aqueous solution at 25°C. The symbol in the center of the square shows the charge of the complex. Rate constants are in liters mol^{-1} s^{-1}; hydrolysis reactions corrected to second order. Reactions I–XII: data taken from Ref. [22]; ionic strength, 1.0 M; rate constants in brackets are determined indirectly. Reactions XIII–XIV: data from Refs. [44 and 9]; ionic strength, 0.8 M. Reactions XV–XVI: data from Ref. [10]; ionic strength, 0.1 M. Reactions XVII–XXXI: data from Refs. [8 and 50–53]; ionic strength, 1.0 M.

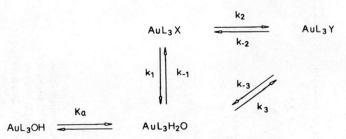

Fig. 3. Reaction scheme for the ligand substitution reaction $AuL_3X + Y \rightarrow AuL_3Y + X$.

2. Effect of entering ligand

The relative substitution efficiency for series of nucleophiles has been established for several gold(III) substrates. For $Au(dien-H)Cl^+$ (dien-H = deprotonated 1,4,7-triazaheptane) the efficiency increases in the order $OH^- \sim H_2O < N_3^- < Br^- < SCN^- < I^-$ in aqueous solution,[2] similar results being found in methanol.[56] For the substrate $AuCl_4^-$ in methanol, the series was established as $CH_3OH < NO_2^- < Br^- < N_3^- <$ pyridine $< SCN^- < I^-$.[41] For $Au(dien-H)Cl^+$ in aqueous solution, I^- reacts 10^5 times faster than H_2O, and for $AuCl_4^-$ in methanol, I^- reacts 10^7 times faster than CH_3OH. (The possibility of a rate-determining direct reduction of the latter substrate by iodide is discussed in Section IV,A.) The span of reactivity is very large and larger than for other square-planar complexes for the same nucleophiles.[41] From rate data for the chloride/bromide exchanges and hydrolysis reactions of Fig. 2 and the similar reactions in platinum(II) complexes, Elding[22,54] calculated that bromide reacts $\sim 2 \times 10^6$, chloride $\sim 10^5$, and water $\sim 10^3$ times faster with gold(III) complexes than with the corresponding platinum(II) complexes. This shows not only that gold(III) reacts faster than platinum(II), but also, since the influence of the leaving group on these rates is similar for gold(III) and platinum(II) complexes, the greater discriminating ability of gold(III) than of platinum(II) complexes. This conclusion is certainly not new,[2,41] but the Au(III)/Pt(II) rate ratios calculated by Elding result from comparisons of several reactions for both metals studied under identical conditions and thus provide a quantitative measure of the difference in discriminating ability between the two metal centers.

With the exception of hydroxide, the above-mentioned order of reactivity toward gold(III) complexes resembles closely the order of stability for the same ligands in such complexes. For the substitution of one ligand by various nucleophiles in the three substrates $AuCl_4^-$, $AuBr_4^-$, and $Au(NH_3)_4^{3+}$ in aqueous solution, second-order rate constants have been compiled in Table 4,

TABLE 4

Second-order rate constants for the substitution of one ligand in tetrachloroaurate(III), tetrabromoaurate(III), and tetraaminegold(III) in aqueous solution at 25°C

Entering ligand	$AuCl_4^-$			$AuBr_4^-$			$Au(NH_3)_4^{3+}$		
	k_2 (liter mol^{-1} s^{-1})	medium	Ref.	k_2 (liter mol^{-1} s^{-1})	medium	Ref.	k_2 (liter mol^{-1} s^{-1})	medium	Ref.
H_2O	$4.0 \times 10^{-4\,a}$	0.8 M NaClO$_4$	[44]	$4.4 \times 10^{-3\,a}$	0.1 M NaClO$_4$	[10]	$6 \times 10^{-8\,a,c}$	1.0 M NaClO$_4$	[8]
Cl^-	2.6^b	0.09 M Cl$^-$	[1]	11	1.0 M NaClO$_4$	[22]	0.36	1.0 M H(Cl, ClO$_4$)	[52]
Br^-	63	1.0 M NaClO$_4$	[22]				3.4	1.0 M H(Br, ClO$_4$)	[50]
I^-	$<9 \times 10^{4\,d}$	1.0 M NaClO$_4$	[47]	$<1 \times 10^{7\,d}$	1.0 M NaClO$_4$	[47]	7.5×10^3	0.03 M NaClO$_4$	[33]
SCN^-	1.3×10^4	1.0 M NaClO$_4$	[48]	8.7×10^4	1.0 M NaClO$_4$	[48]			

[a] Corrected to second order by dividing by the concentration of water.
[b] Extrapolated to 25°C from 0–10°C.
[c] Extrapolated to 25°C from 80–50°C.
[d] Limits are rate constants for reduction. Substitution was found to be slower than reduction.

and from the equilibrium data in Table 1, reaction free energies can be estimated for the same reactions from

$$\Delta G^\circ = -\tfrac{1}{4}RT\,2.30[\log \beta_4(\mathrm{AuY_4}^{(4v-3)-}) - \log \beta_4(\mathrm{AuL_4}^{(4v-3)-})] \quad (14)$$

Although this calculation explicitly assumes the distribution between the consecutive stability constants to be the same for the reacting complex, $\mathrm{AuL_4}^{(4v-3)-}$, and for $\mathrm{AuY_4}^{(4v-3)-}$ (i.e., the gold(III) complexes with four of the nucleophiles coordinated as ligands), reasonable values for the reaction free energy are expected to be obtained from this equation. For the $\mathrm{AuCl_4}^-/\mathrm{Cl}^-$ reaction this is inherently true, and for both the $\mathrm{AuCl_4}^-/\mathrm{Br}^-$ reaction and the $\mathrm{AuBr_4}^-/\mathrm{Cl}^-$ reaction, the consecutive stability constants of Table 3 verify this assumption directly. Figure 4 establishes linear relationships between the free energies of activation and the free energies of reaction for reactions of the type

$$\mathrm{AuL_4}^{(4v-3)-} + \mathrm{Y}^{v-} \rightarrow \mathrm{AuL_3Y}^{(4v-3)-} + \mathrm{L}^{v-} \quad (15)$$

The available rate and equilibrium data are very limited. However, they cover energy differences corresponding to many orders of magnitude in both rate

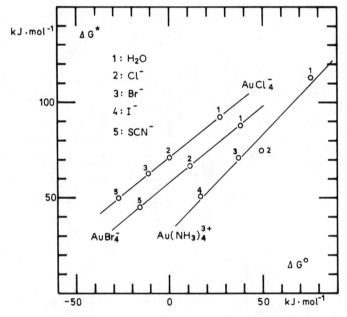

Fig. 4. Relationship between free energy of reaction and free energy of activation for ligand substitution reactions in tetrachloroaurate(III), tetrabromoaurate(III), and tetraamminegold(III) in aqueous solutions at 25°C.

and stability constants and reflect marked differences in donor properties of the nucleophiles.

For a series of reactions of a common substrate with various nucleophiles, a slope of unity in a linear free energy relationship is indicative of associative activation, whereas at the other extreme, a zero slope indicates a transition state of decreased coordination number in cases where long-lived intermediates in the reactions can be excluded.[57,58] For ligand substitution reactions in gold(III) complexes there is no evidence of such persistent intermediates of either increased or decreased coordination number, as discussed below in Section III,D.

The linear free energy relationships for $AuCl_4^-$ and $AuBr_4^-$ have a common slope of 0.78 ± 0.02, whereas the slope for $Au(NH_3)_4^{3+}$ is 1.0 ± 0.1, showing that the discriminating abilities of these complexes are equally $(Au(NH_3)_4^{3+})$, or almost equally $(AuCl_4^-$ and $AuBr_4^-)$, as great in a kinetic as in a thermodynamic sense; in other words, the rate of substitution is very dependent on the entering ligand. For $AuCl_4^-$ and $AuBr_4^-$, the 0.78 slope indicates bond breaking to be of some importance in the transition state. It should be noted that the common slope indicates that bond breaking is equally important in these two complexes, or, as also shown by Elding and Gröning,[22] that chloride and bromide are equivalent as leaving groups. For the weak acid $Au(NH_3)_4^{3+}$ ($pK_a = 7.5$[8]), some ambiguity exists owing to the possibility of proton transfer in the transition state. However, a resulting amido complex is not expected to have enhanced reactivity, as judged from the similar reaction rates observed for chloride/bromide exchange in $Au(dien)Cl^{2+}$ and in its conjugate base $Au(dien-H)Cl^+$.[26]

Linear relationships between rate and nonkinetic parameters have also been established for series of nucleophiles in which only the σ donor properties are varied. In methanol, the second-order rate constant for the substitution of a chloride in tetrachloroaurate(III) by a heterocyclic amine

$$AuCl_4^- + am \underset{k_{-2}}{\overset{k_2}{\rightleftharpoons}} AuCl_3am + Cl^- \tag{16}$$

depends on the basicity and the steric properties of the amine.[42] For the reverse reaction, the second-order rate constant depends only on the amine basicity.[41] For series of amines with similar steric properties, linear relationships are found between $\log k_2$ and pK_a for the conjugate acid of the entering amine. Similar results were obtained for the reaction of cationic gold(III) substrates with heterocyclic amines in acetone, and the slopes of such plots show that the discriminating power of these gold(III) substrates toward σ donor nucleophiles depends on the effective positive charge on the gold atom.[59] For $AuCl_4^-$ the slope of 0.15 indicates the minor role of basicity in the discrimination between nucleophiles, whereas the surprisingly large slope of 0.89 for the 5-nitro-1,10-phenanthroline complex $Au(5-NO_2-phen)Cl_2^+$

TABLE 5

Second-order rate constants and activation parameters for substitution of one ligand in tetrachloroaurate(III) in methanol and in tetraaminegold(III) in aqueous solution at 25°C

Entering ligand	$AuCl_4^-$ [a]			Entering ligand	$Au(NH_3)_4^{3+}$ [b]		
	k_2 (liter mol^{-1} s^{-1})	$\Delta H^\ddagger$ (kJ mol^{-1})	$\Delta S^\ddagger$ (J mol^{-1} K^{-1})		k_2 (liter mol^{-1} s^{-1})	$\Delta H^\ddagger$ (kJ mol^{-1})	$\Delta S^\ddagger$ (J mol^{-1} K^{-1})
NO_2^-	0.028	35	−138	H_2O	6×10^{-8}	~109	~ −13
Br^-	0.140	51	−92	Cl^-	0.36	81	20
N_3^-	0.35	27	−163	Br^-	3.4	73	8
Pyridine	1.6	44	−92	$S_2O_3^{2-}$	195	48	−42
SCN^-	7.4	50	−59				
I^-	84	50	−38				

[a] Ionic strength, 0.1 M; from Ref. [41].
[b] Ionic strength, 1.0 M; from Refs. [8, 51, 52, and 61].

demonstrates considerable discrimination and, consequently, a greater importance of bond making for the cationic complex.

The thermodynamic discrimination of gold(III) complexes can be concluded to be an ethalpy effect (see Section II,C). Activation parameter data for gold(III) substitution reactions are scarce and do not allow any clear conclusions to be drawn. Thus, for the reaction of tetrachloroaurate(III) with various nucleophiles in methanol it is apparent that small enthalpies of activation are eclipsed by very unfavorable entropies of activation.[41] Furthermore, there is no simple correlation between reaction rate and enthalpy of activation, as can be seen from Table 5. Their values must be regarded with some reservation since more recent studies[22,47,48] with bromide, iodide, and thiocyanate in aqueous solution have shown that the reaction scheme originally invoked is an oversimplification. However, these and other available data for reactions in different solvents[1,2,49–53,62–63] indicate that negative entropies of activation are characteristic of ligand substitutions in gold(III) complexes.

For the reactions of tetraamminegold(III), the faster the reaction the smaller the enthalpy of activation. This qualitative correlation seems very reasonable, since the kinetic discrimination of this substrate strictly parallels the thermodynamic discrimination, the latter being a result of differences in reaction enthalpies. This tripositive substrate is, however, atypical since the entropies of activation are very close to zero, showing that the decrease in entropy of the transition state due to a net increase in bonding is, in this particular case, counteracted by the increase in entropy due to the partial charge neutralization in the transition state.

3. Effect of leaving ligand

For the displacement of heterocyclic amines in $AuCl_3am$ by thioethers in acetone, the rate of reaction decreases with increasing basicity of the amine, and for a given thioether linear relationships were found between log k_2 and pK_a for the conjugate ammonium ion. The slopes of such plots depend on the nature of the entering thioether, and the influence of the leaving ligand amine decreases with increasing rate of reaction of the thioethers.[60,64] Although the difference in discrimination between the substrates by different thioethers is not overwhelming, these observations show that an increase in reactivity of the entering ligand parallels a decrease in influence of the leaving ligand. In other words, bond breaking plays a greater role for the less efficient nucleophiles.

Bromide and, in particular, chloride are examples of such less efficient nucleophiles, and the role of the leaving ligand on the rate of substitution is expected to be more significant. For the displacement of heterocyclic amines in $AuCl_3am$ by chloride [Eq. (16)] or bromide in methanol, the rate of reaction also decreases with increasing basicity of the amine, corresponding to increasing strength of the gold(III)–nitrogen bond.[41,65] For chloride and bromide as entering ligand in $AuCl_3X^{v-}$ and $AuBr_3X^{v-}$, respectively, in aqueous solution

$$AuCl_3X^{v-} + Cl^- \rightarrow AuCl_4^- + X^{v-} \tag{17}$$

$$AuBr_3X^{v-} + Br^- \rightarrow AuBr_4^- + X^{v-} \tag{18}$$

the ratios between the second-order rate constants $k_{NH_3} : k_{Br^-} : k_{Cl^-} : k_{H_2O} = 1 : 8 \times 10^3 : 5 \times 10^3 : 7 \times 10^7$ for $AuCl_3X^{v-}$ (data from Refs. [1, 9, 22, and 52]; k_{Cl^-} is statistically corrected), and the second-order rate constants $k_{NH_3} : k_{SCN^-} : k_{Cl^-} : k_{H_2O} = 1 : 5 \times 10^2 : 9 \times 10^3 : 3 \times 10^6$ for $AuBr_3X^{v-}$ [10,22,48,51] indicate not only that the rate of substitution decreases with increasing stability of the gold(III) leaving-ligand bond (cf. stability data in Table 1), but also that the effect is more important for the less efficient nucleophile chloride. Chloride and bromide are equivalent as leaving ligands, in agreement with the common slope for tetrachloroaurate(III) and tetrabromoaurate(III) in the linear free energy relationships of Fig. 4, and in agreement with the proposal[35] of near-equal bond strength of gold(III)–chloride and –bromide bonds discussed in Section II,C. The gold(III)–nitrogen bond is remarkably robust compared to gold(III)–halide bonds, and it is very surprising that heterocyclic amines were found to be better leaving ligands than chloride in the reaction of $AuCl_3am$ with bromide in methanol.[65] This latter result might deserve a closer examination, especially since a change of solvent, such as from water to methanol, is not likely to cause any reversal of ligand lability.[66–68]

Another puzzling result is that the rate of substitution of X^- by chloride in $Au(dien-H)X^+$ in methanol increases as $N_3^- < Br^- < SCN^-$, which, at least for Br^- and SCN^-, corresponds to an increase in gold(III) leaving-ligand bond strength.[56]

4. Effect of nonreacting ligands

The reaction of tetrachloroaurate(III) with bromide to give tetrabromoaurate proceeds $\sim 90\%$ via the trans isomer of the dibromodichloroaurate(III) intermediate, whereas the reverse reaction takes place mainly via the cis isomer, as can be seen from the rate constants of Fig. 2.[22] Consequently, among the nonreacting ligands the trans ligand rather than the cis ligand is important in determining the relative rate and the course of reaction for these ligand substitutions. This is certainly not surprising, given the well-established importance of the trans effect in platinum(II) chemistry, but until recently hardly any quantitative data have been available for a discussion of trans and cis effects in gold(III) complexes, as also noted by Peshchevitskii *et al.*[62] In fact, some results could be,[65] and some results have indeed been,[46,69–74] interpreted within models which take no account of trans effects.

For the consecutive replacements of the chloride ligands in tetrachloro-aurate(III) by bromide and for the reverse reaction, Elding and Gröning[22] used the empirical relation

$$k/m = q \times T \times C_1 \times C_2 \qquad (19)$$

to rationalize the experimentally determined rate constants k normalized by m, the number of equivalent leaving ligands. T is a relative trans effect factor and C_1 and C_2 are likewise relative cis effect factors, and the constant q is characteristic of the particular substitution at fixed conditions. For either bromide or chloride as entering ligand, the rate of substitution is accelerated by a factor of 14 ± 3 by a *trans*-bromide relative to a *trans*-chloride, whereas a 1.1 ± 0.1 relative Br^-/Cl^- cis effect is hardly significant. For the consecutive replacements of bromide by chloride in *trans*-Au(CN)$_2$Br$_2^-$

$$trans\text{-}Au(CN)_2Br_2^- + Cl^- \rightarrow trans\text{-}Au(CN)_2BrCl^- + Br^- \qquad (20)$$

$$trans\text{-}Au(CN)_2BrCl^- + Cl^- \rightarrow trans\text{-}Au(CN)_2Cl_2^- + Br^- \qquad (21)$$

Mason[36] found second-order rate constants of 1670 and 143 liter mol^{-1} s^{-1}, respectively, at 25°C in aqueous solution of ionic strength 0.51 M. This corresponds to a relative Br^-/Cl^- trans effect of 5.8, whereas a value of 9.2 ± 0.3 emerges from a comparison of reaction XXVII–XXX of Fig. 2. For these reactions, the relative Br^-/Cl^- trans effect is independent of whether the entering or the leaving ligand is chloride or bromide, and for thiocyanate as an entering ligand in AuCl$_4^-$ and AuBr$_4^-$ a comparable Br^-/Cl^- trans effect of ~ 8 was found.[48] Like the Br^-/Cl^- cis effect, the SCN^-/Br^- cis effect is very small, and a factor of ~ 1.9 is estimated from a comparison of the reactions of thiocyanate with AuBr$_4^-$ and *trans*-Au(SCN)$_2$Br$_2^-$.[48]

This is in contrast to the behavior of cyanide, which manifests a cis effect relative to chloride of 23 [from comparison of reaction II of Fig. 2 and the reaction of Eq. (21)], and a cis effect relative to bromide of 18 [from comparison of reaction XII of Fig. 2 and the reaction of Eq. (20)]. However, a comparison of the rates for the displacement of heterocyclic amines in *trans*-Au(CN)$_2$Clam by bromide, azide, and nitrite in methanol with the rates for the corresponding reactions of AuCl$_3$am indicates much smaller and less consistent cis effects of cyanide relative to chloride, and again leaves us either with some ambiguities concerning the role of solvent or with some doubt about reaction stoichiometries.[65]

The chloride/bromide interchange reactions are faster in *trans*-diamminedihalogold(III) complexes than in tetrahaloaurate(III) complexes, suggesting a significant NH_3/halide cis effect.[53] From a study of the chloride and bromide anation reactions of tetraamminegold(III) it is possible to obtain further information concerning both the cis and the trans effect of ammonia, although for these reactions it is necessary to compare rates for complexes with different charges.[50–52] Two distinct stages with rates differing by five orders of magnitude are observed in the reaction sequence in which the four ammonia ligands in tetraamminegold(III) are replaced by either chloride or bromide. The first stage giving *trans*-diamminedihalogold(III)[53] as product consists of two consecutive reactions of comparable rates (cf. Fig. 2 and Table 6), whereas in the second and slower stage, the two consecutive reactions are quite well separated.

The occurrence of two separate stages in these anation reactions shows that cis-coordinated ammonia strongly accelerates the substitution of coordinated ammonia by halides. The relative NH_3/Cl$^-$ cis effect and the relative NH_3/Br$^-$ cis effect correspond to ~ 20 kJ mol^{-1} and ~ 15 kJ mol^{-1}, respectively, if it is assumed that charge neutralization in the transition state contributes solely to the activation entropy, and also that cis and trans effects contribute only to the magnitude of the activation enthalpy. This crude approximation receives at least some support from the observation that only the activation enthalpies and not the activation entropies depend on the nature of the trans ligand, both in *trans*-diamminedihalogold(III) and in *trans*-dicyanodihaloaurate(III), and that the Br$^-$/Cl$^-$ trans effect discussed above corresponds in both of the latter substrates to ~ 5 kJ mol^{-1}.[36,53]

Coordinated chloride and, in particular, coordinated bromide labilize trans ligands relative to ammonia, and only *trans*-diamminedihalogold(III) was observed as product from the reactions of triamminechlorogold(III) and triamminebromogold(III) with chloride and bromide, respectively.[53] However, the relative Cl$^-$/NH_3 trans effect and the relative Br$^-$/NH_3 trans effect amount to differences in activation enthalpies of only a few kilojoules per mole (cf. Table 6), and for these anation sequences the halide/ammonia trans effect is counteracted by the effect of charge neutralization in the transition state and by statistical factors (number of equivalent leaving ligands), which taken together render the activation entropies more favorable. As a result, in the anation reactions of tetraamminegold(III) by halides, the relative rates of the first two ammonia/chloride exchange reactions are reversed compared to the analogous ammonia/bromide exchange reactions. This unique substitution behavior for a planar complex stresses the importance of charge neutralization in the transition state.

cis-Diamminedihalogold(III) complexes have not yet been characterized, but due to a combination of NH_3/Cl$^-$ cis effect and Cl$^-$/NH_3 trans effect,

TABLE 6

Rate constants and activation parameters for the reaction of tetraamminegold(III) ions with chloride and bromide in acidic aqueous solution of unit ionic strength at 25°C

Reaction	m^c	k_n (liter mol^{-1} s^{-1})	$\Delta H_n^{\ddagger}$ (kJ mol^{-1})	$\Delta S_n^{\ddagger}$ (J mol^{-1} K^{-1})	$\Delta S_{n,\text{corr.}}^{\ddagger}$[d] (J mol^{-1} K^{-1})
		Chloride anation[a]			
$Au(NH_3)_4{}^{3+} + Cl^- \xrightarrow{k_1} Au(NH_3)_3Cl^{2+} + NH_3$	4	0.36 ± 0.02	81 ± 2	20 ± 4	9 ± 4
$Au(NH_3)_3Cl^{2+} + Cl^- \xrightarrow{k_2} trans\text{-}Au(NH_3)_2Cl_2{}^+ + NH_3$	1	0.112 ± 0.002	78 ± 1	-1 ± 3	-1 ± 3
$trans\text{-}Au(NH_3)_2Cl_2{}^+ + Cl^- \xrightarrow{k_3} AuNH_3Cl_3 + NH_3$	2	$(3.7 \pm 0.5) \times 10^{-6}$	105 ± 4	2 ± 10	-4 ± 10
$AuNH_3Cl_3 + Cl^- \xrightarrow{k_4} AuCl_4{}^- + NH_3$	1	$(3.33 \pm 0.02) \times 10^{-5}$	97 ± 2	-7 ± 4	-7 ± 4
		Bromide anation[b]			
$Au(NH_3)_4{}^{3+} + Br^- \xrightarrow{k_1} Au(NH_3)_3Br^{2+} + NH_3$	4	3.40 ± 0.08	73 ± 3	8 ± 3	-3 ± 3
$Au(NH_3)_3Br^{2+} + Br^- \xrightarrow{k_2} trans\text{-}Au(NH_3)_2Br_2{}^+ + NH_3$	1	6.5 ± 0.4	69 ± 3	2 ± 10	2 ± 10
$trans\text{-}Au(NH_3)_2Br_2{}^+ + Br^- \xrightarrow{k_3} AuNH_3Br_3 + NH_3$	2	$(9.3 \pm 0.3) \times 10^{-5}$	88 ± 3	-26 ± 8	-32 ± 8
$AuNH_3Br_3 + Br^- \xrightarrow{k_4} AuBr_4{}^- + NH_3$	1	$(2.68 \pm 0.09) \times 10^{-2}$	84 ± 4	7 ± 9	7 ± 9

[a] From Ref. [52].
[b] From Refs. [50 and 51].
[c] m is the number of equivalent leaving ligands.
[d] Corrected for the number of equivalent leaving ligands.

anation is expected to be much faster than for their trans counterparts, as is also indirectly confirmed by a comparison of the rate of reaction XVIII of Fig. 2 with that for the anation reaction of the trimethylenediamine complex $Au(tn)Cl_2^+$ with chloride,

$$Au(tn)Cl_2^+ + Cl^- + H^+ \rightarrow Au(tnH)Cl_3^+ \tag{22}$$

for which a second-order rate constant of 1.52×10^{-3} liter $mol^{-1} s^{-1}$ (aqueous solution, ionic strength 1.7 M, at 25°C) was found.[75]

The chloride/bromide exchange in $Au(dien)Cl^{2+}$ was originally found to be faster than in the conjugate base $Au(dien-H)Cl^+$,[2] but a reinvestigation[26] has shown that the reaction rates are comparable (rate constants of 164 and 162 liter $mol^{-1} s^{-1}$, respectively, in aqueous solution of ionic strength 0.2 M, at 25°C). These reactions are slow compared to reaction XXVII of Fig. 2, confirming that chloride has not only a stronger trans effect than an amine, but also, which is somewhat surprising, stronger than an amide ligand. Other types of ligands are expected to exhibit significant trans and/or cis effects, but only qualitative information is available. Thus, in the highly distorted trichloro(triphenylphosphine)gold(III), the trans chloride is substituted orders of magnitude faster than in the planar tetrachloroaurate(III) ion.[75] Likewise, the *trans*-halide in $AuCl_3SCN^-$ and in $AuBr_3SCN^-$ was found to be very labile.[48]

In conclusion, trans effects appear to be fully as important in ligand substitutions of gold(III) complexes as is the case with other square-planar complexes. On the basis of the limited number of quantitative studies which have been made, the following reasonably certain, but admittedly very fragmentary sequence of trans effectiveness can be assembled:

$$NH_3 < Cl^- < Br^- < SCN^-, \text{phosphines} \tag{23}$$

However, as more data become available, cis effects may prove to be equally or even more important. The sequence

$$Cl^- \leq Br^- < NH_3 < CN^- \tag{24}$$

for acceleration of cis replacement, which is at least valid for chloride/bromide interchanges as discussed above, parallels the bonding properties of these ligands (cf. Table 1). Cyanide is a π acceptor, and relative to the π donors chloride and bromide, cyanide increases the effective positive charge on the metal center and thus facilitates bond formation with the incoming nucleophile. The cis effect of ammonia relative to chloride is small in platinum(II) complexes but much greater in palladium(II) complexes.[76] The available data suggest this cis effect of ammonia to be even more pronounced in gold(III) complexes.

B. Reactions involving polydentate ligands

Tetrachloroaurate(III) reacts readily with ammonia and with amines, even in moderately acidic solution,[8,77] and for polydentate amines it is generally found that the entering of the first nitrogen donor is the rate-determining step, and that subsequent ring-closing reactions are fast.[78–82] In acidic aqueous solution, both $AuCl_4^-$ and $AuCl_3OH^-$ were found to react with the mono-protonated form of ethylenediamine and with the diprotonated form of diethylenetriamine ($dienH_2^{2+}$). The complex $AuCl_3OH^-$ reacts faster than $AuCl_4^-$ despite the fact that chloride is expected to be a better leaving ligand than hydroxide, and Louw and Robb[79,80] have proposed that an initial proton transfer step leads to the formation of coordinated water, which is a good leaving group relative to chloride and, in particular, relative to hydroxide. The reactions of $dienH_2^{2+}$ with the two substrates have very similar enthalpies of activation (58 and 54 kJ mol^{-1}, respectively), whereas the entropy of activation is more positive for the reaction with $AuCl_3OH^-$ than for the reaction with $AuCl_4^-$ (8 and -86 J mol^{-1} K^{-1}, respectively), in keeping with the proton transfer model which involves charge neutralization.

For *N*-methyl-substituted ethylenediamines and tetra- and penta-methyl-substituted diethylenetriamines, rate expressions for the binding of the amines show that chloride participates in the rate-determining step, in contrast to what is found for the unsubstituted amines.[81,82] There was no spectral indication of coordination of a fifth chloride to gold(III) or of formation of any other higher complexes of gold(III) (cf. Section II,C), whereas conductivity measurements gave evidence for ion pairing between the protonated form of especially the more heavily substituted (more lipophilic) amines and chloride. This supports the very convincing idea of a reaction between the gold(III) complex and such a chloride/protonated amine ion pair, rather than the protonated amine itself.

In the rate laws for the detachment of coordinated 2,2-bipyridine from $Au(bipy)Cl_2^+$ in aqueous methanol,

$$Au(bipy)Cl_2^+ + H^+ + 2Cl^- \rightarrow AuCl_4^- + Hbipy^+ \qquad (25)$$

a similar higher order term with respect to chloride concentration was found, and a competing reaction via a distorted octahedral hexacoordinated gold(III) complex $Au(bipy)Cl_4^-$ was suggested by Annibale *et al.*[83] Hexacoordinated gold(III) was likewise invoked in the mechanism for displacement of 5-nitro-1,10-phenanthroline from $Au(5\text{-}NO_2\text{-}phen)Cl_2^+$ by chloride.[84] However, the "extra" chloride term can equally well be accounted for by postulating ion pair formation during the detachment of the chelate ligand, and it has been argued by Louw and de Wall[82] that the displacement of the second chelate nitrogen is facilitated when the hydrogen-bonding capacity of the initially displaced and protonated nitrogen is "neutralized" by ion pair formation with, e.g., a chloride ion. Again this mechanism seems very convincing, given the lack of spectroscopic evidence for hexacoordination in these gold(III) complexes and also given the fact that, in the anation of $Au(NH_3)_4^{3+}$ by chloride and bromide, strictly first-order dependences on the entering halide concentration were found.[50–52] Also, for the displacement of ethylenediamine or trimethylenediamine, for which the corresponding ammonium ions have less tendency to form ion pairs, no higher chloride concentration dependent term was observed.[85]

For the gold(III) complexes of the aliphatic amines, ring opening and subsequent slower displacement of the singly bonded amine are kinetically separated processes and each proceeds via the normal associative nucleophilic attack known for substitution of unidentate ligands:

$$
\begin{array}{ccc}
\underset{Cl}{\overset{Cl}{\boxed{+}}}\!\!\!\!\!\!\!\!\!\!\overset{N}{\underset{N}{\Big)}}
& \rightleftharpoons &
\underset{Cl}{\overset{Cl}{\boxed{O}}}\!\!\!\!\!\!\!\!\!\!\overset{N}{\underset{Cl}{\Big\backslash}}_{N}
& \rightleftharpoons &
\underset{Cl}{\overset{Cl}{\boxed{O}}}\!\!\!\!\!\!\!\!\!\!\overset{N}{\underset{Cl}{\Big\backslash}}_{NH^+}
\end{array}
\tag{26}
$$

$$
\underset{Cl}{\overset{Cl}{\boxed{-}}}\!\!\!\!\!\!\!\!\overset{Cl}{\underset{Cl}{}}
$$

The same mechanism is also operative for the displacement of N-substituted ethylenediamines from gold(III) complexes, with ring opening being faster for N-substituted ethylenediamines than for ethylenediamine itself.[86] For ethylenediamine, the closing of the five-membered ring is faster, by a factor of 10, than the closing of the six-membered ring for trimethylenediamine, whereas for the ring-opening process, the rates are very similar.[85] The replacement of the tridentate amine diethylenetriamine (and its 3-methyl analog) from gold(III) was also investigated by Annibale *et al.*[87] The experimental conditions employed were such that the predominant gold(III) reactant had partially detached amines, and the experimentally determined rate equation was completely analogous to that found for the bidentate amines. An observation widely quoted as providing "direct evidence" for hexacoordination in gold(III) solution chemistry is the difference between the UV spectra of $Au(dien)Cl^{2+}$ dissolved in a 0.5 M perchlorate and in a 0.5 M chloride medium, respectively, both at pH 2.5.[32] However, the conditions prevailing in the chloride medium are very similar to those used by Annibale *et al.*[87] in their kinetic study, and the observed spectral effect of chloride is thus more reasonably explained on the basis of partial detachment of the tridentate amine. In the sterically hindered chlorogold(III) complex of N-substituted diethylenetriamines, substitution of chloride by certain other unidentate ligands has likewise been postulated to proceed via a ring-opening mechanism.[88,89]

C.   Reactivity of coordinated ligands

The very stable bonds formed between gold(III) and more polarizable ligands exert a strong influence on the reactivity of ligands coordinated to gold(III). The acidity of amines is thus enhanced by more than 30 orders of magnitude upon coordination to gold(III) (pK_a of ammonia as an acid in water has been estimated to 41,[90,91] compared to 7.5 when coordinated in $Au(NH_3)_4^{3+}$[8,73]), and the formation of potentially nucleophilic coordinated amide functions in gold(III) complexes is extensive even at neutral pH, a fact which has been utilized for the synthesis of macrocyclic complexes by condensation with β-diketones.[92,93]

The catalytic influence of gold(III) complexes on a number of reactions is similarly a result of increased reactivity of the reactants when coordinated to gold(III). Satchell *et al.* have found that various gold(III) complexes are very powerful promoters of the hydrolysis of N-cyclohexylthiobenzamide[69,72,73] and N-cyclohexyl methylthiobenzimidate[74] in aqueous solution to yield N-cyclohexylbenzamide as product, whereas the kinetic stability of the carboxamide is apparently unaffected by the presence of gold(III) complexes. The course of the reaction suggests, as do spectroscopic results, coordination of the organic substrate through sulfur to gold(III) prior to a rate-determining hydrolysis. Gold(III) complexes were also found to promote hydrolysis of thiol esters of benzoic acid to give gold mercaptides and benzoic acid as products.[70,71] The suggested mechanism for the accelerated thiol ester hydrolysis involves a slow, rate-determining substitution of one of the ligands coordinated to gold(III) by the ester. For the thioamides, hexacoordinated intermediates, e.g. $(dien)Cl_2AuS{=}CR'R''^+$, and pentacoordinated intermediates, e.g. $(phen)Cl_2AuS{=}CR'R''^+$, were proposed. A rate-determining attack by water on $(dien)Cl_2AuS{=}CR'R''^+$ to give the gold(III) sulfide complex $Au(dien)S^+$ plus chloride and carboxamide was postulated, whereas for $(phen)Cl_2AuS{=}CR'R''^+$, a *slow* dissociation of a chloride ligand was invoked as rate determining, followed by fast attack by water.[72] With our present knowledge concerning ligand substitutions in gold(III) complexes these mechanisms seem rather implausible, and it is not reasonable that two so closely related gold(III) complexes should react by such different mechanisms.

Indeed, these reactions are complicated, and the role of redox processes is uncertain since iodide was found to accentuate the effect of tetrachloro-aurate(III)[74] under conditions in which iodide is known to reduce gold(III) complexes.[47]

D. Intimate mechanism for ligand substitutions

In the preceding discussion of kinetic data for ligand substitutions in gold(III) complexes a general mechanism with associative activation has been implied:

$$Y + \qquad \longrightarrow \qquad Y \longrightarrow \qquad \longrightarrow \qquad + X \tag{27}$$

For ligand substitution reactions in square-planar complexes in general, elaborate experiments have been designed and much speculation has been devoted to rationalizing kinetic results in terms of an intimate mechanism which includes a detailed description of the relative roles of bond formation and bond breaking and of geometries of the involved transition states. For gold(III) in particular, recent progress in that respect appears to have clarified the situation, in the sense that a relatively simple picture is now emerging.

1. The substrate

The kinetic data available can, as discussed in previous sections, all be accommodated within models in which the substrates are the "bare" tetracoordinated gold(III) complexes. Certainly, such species are solvated to different degrees depending on the nature of the complex and on the nature of the solvent. However, an investigation of the reaction of Eq. (16) in acetone[59] and various alcohols[66] showed that the solvent had little, if any, effect on the ability of the substrate to differentiate between the various amines [forward reaction of Eq. (16)], indicating that the solvent molecules are not, in a kinetic sense, integrated parts of the substrate. Also, when substitution reactions are investigated in solutions of high and constant ionic strength, it is not necessary to invoke ion pairing between anionic nucleophiles and cationic substrates such as $Au(NH_3)_4^{3+}$ to account for the observed kinetics,[50–52] a result again supporting the concept of a "bare" substrate.

2. The transition state

A "typical" ligand substitution in a gold(III) complex has a small enthalpy of activation and a negative entropy of activation. The small enthalpy of activation verifies the ease with which a fifth ligand binds to form the transition state, and the negative entropy of activation is in agreement with the resulting net increase in bonding in such a transition state. Both the entering and the leaving ligand are of importance for the substitution rates, but data are not available to determine in what manner the bond formation and the bond-breaking parts of the processes influence the activation parameters. However, it is obvious from the data of Table 6 that charge neutralization during bond formation makes the entropies of activation significantly more positive.

Bond formation is more important than bond breaking in the transition state. This is evident from the slopes of the linear free energy relationships of Fig. 4, which, however, also show that the two parts of the substitution process are not separated. Preliminary experiments show that an increase in pressure accelerates the reactions XXVII–XXX of Fig. 2 only moderately,[94] findings which are also in agreement with the concept of a transition state in which both association of the entering ligand and dissociation of the leaving ligand are significant, although clearly the first part dominates since rates increase with increasing pressure.

Double-humped energy profiles with each hump corresponding to separate transition states for bond formation and bond breaking, respectively, have been used in the discussion of the two aspects of the substitution process in square-planar complexes. However, for gold(III) complexes, the kinetic data appear to be accommodated within a model with one transition state, corresponding to concerted bond formation and bond breaking.[18,41] Some attention was therefore drawn when spectroscopic evidence for persistent pentacoordinate intermediates was reported for the reaction of tetrachloroaurate(III) with bromide or thiocyanate,[45,46] corresponding on an energy profile to the presence of gaps between two well-separated energy barriers. Rate constants for the formation and subsequent decay of $AuCl_4Br^{2-}$ and $AuCl_4SCN^{2-}$ were also given, but more detailed studies have now conclusively shown that the observed processes are due to stepwise substitution in agreement with Fig. 2.[22] The pentacoordinated trigonal bipyramidal species depicted in Eq. (27) thus corresponds to a transition state rather than to a reaction intermediate.

The structure of the transition states in the substitution processes can only be deduced indirectly, but X-ray structures of gold(III) complexes of different reactivities may provide hints in that respect. Tetrabromoaurate[95] is strictly planar and has ligand–metal–ligand angles close to 90°, whereas the angles in *trans*-diamminedibromogold(III), which has marginally more labile bromide ligands (cf. Fig. 2), are distorted from 90°, although the complex is still planar.[53] A tetragonal distortion was observed in the amido complex $Au(dien-H)Cl^+$, but this distortion is not accompanied by any significant rate acceleration for chloride/bromide exchange relative to exchange in the almost planar $Au(dien)Cl^{2+}$.[26] Trichloro(triphenylphosphine)gold(III), in which chloride exchange is very fast compared to exchange in the planar tetra-chloroaurate(III), is highly distorted from planarity.[75] X-Ray structures clearly describe the ground state and provide no direct information about the transition state, but it is noteworthy that only distortions which tend toward trigonal bipyramidal structures, such as for the former triphenylphosphine complex, are accompanied by major enhancement of reactivity.

E. Comparison with other square-planar complexes

Gold(III) complexes react faster than palladium(II) and, in particular, platinum(II) complexes with the exception of some hydrolysis reactions.[44,96] The NMR study of Pesek and Mason[63] on the cyanide exchange kinetics for tetracyano complexes provides us with a rare example in which the same reaction has been studied for nickel(II), palladium(II), platinum(II), and gold(III) under identical conditions. The cyanide exchange rates are in the order $Ni(II) \gg Au(III) > Pd(II) > Pt(II)$, as can be seen in Table 7, and a

TABLE 7

Rate constants (24°C) and activation parameters for cyanide exchange in tetracyano complexes in deuterium oxide[a]

Reaction	k_2 (liter mol^{-1} s^{-1})	$\Delta H^{\ddagger}$ (kJ mol^{-1})	$\Delta S^{\ddagger}$ (J mol^{-1} K^{-1})
$Ni(CN)_4{}^{2-} + CN^- \rightarrow$	$> 5 \times 10^{5}$[b]		
$Pd(CN)_4{}^{2-} + CN^- \rightarrow$	120	17 ± 2	-178 ± 7
$Pt(CN)_4{}^{2-} + CN^- \rightarrow$	26	26 ± 3	-143 ± 8
$Au(CN)_4{}^- + CN^- \rightarrow$	3900	28 ± 1	-100 ± 3

[a] From Pesek and Mason.[63]

[b] Only a lower limit could be determined.

similar reactivity order has been observed for other square-planar sub-stitutions as the metal ion is changed.[2] The exceptionally high reactivity of $Ni(CN)_4^{2-}$ was explained by the stability of five-coordinate $Ni(CN)_5^{3-}$, of which salts have been isolated and which may be similar to the transition state. The complex $Au(CN)_4^-$ reacts faster than $Pd(CN)_4^{2-}$ and $Pt(CN)_4^{2-}$ as a result of a more favorable entropy of activation, since the enthalpy of activation is larger than for either $Pd(CN)_4^{2-}$ or $Pt(CN)_4^{2-}$.

IV. REDUCTION OF GOLD(III) COMPLEXES

The αE term dominates over the βH term in Edwards' equation for gold(III) (Fig. 1), corresponding to the fact that more reduceable ligands form the more stable complexes. Thermodynamically, gold(III) complexes such as $AuCl_4^-$ and $AuBr_4^-$ can be reduced by good nucleophiles (e.g., thiocyanate, iodide, thiourea, thiosulfate, and cyanide), as can be seen from the standard potentials of Table 1. This implies, however, that there is competition between substitution and reduction when gold(III) complexes react with good nucleophiles. The kinetics of such competing substitution reactions have been investigated in detail for only a few systems.

A. Reduction by halides

From the stability constants for AuI_4^- relative to $AuCl_4^-$ and $AuBr_4^-$ ($\beta'_4 = 10^{22}$ and $\beta'_4 = 10^{15}$, respectively; Table 1), the rate of substitution of a chloride or a bromide ligand by an iodide in the latter two substrates can be estimated from the linear free energy relationships of Fig. 4, giving the values 4×10^4 and 2×10^5 liter mol^{-1} s^{-1}, respectively. These estimated constants are, however, smaller than the experimental rate constants deter-mined using stopped-flow spectrophotometry to follow the reaction of iodide with $AuCl_4^-$ and $AuBr_4^-$ in aqueous solution of unit ionic strength at 25°C: $(9 \pm 1) \times 10^4$ and $(1.0 \pm 0.3) \times 10^7$ liter mol^{-1} s^{-1}, respectively.[47] In agree-ment with the above arguments, the observed reactions were identified by Elding and Olsson[47] as direct reductions by iodide. With an excess of iodide relative to gold(III) complex one single reaction was observed, whereas with an excess of gold(III) complex relative to iodide, reduction took place in two consecutive steps, owing to the fact that rapid equilibria involving ICl_2^- and I_2Cl^- influence the kinetics. The mechanism is intermolecular, involving a direct attack on one of the halide ligands of the complex by an outer-sphere iodide in the rate-determining step. However, it was not possible to differentiate between the two mechanisms of Eq. (28) (X = Br, Cl):

$$
\begin{array}{c}
\left[X\!-\!\!\!\begin{array}{c}X\\|\\Au\\|\\X\end{array}\!\!\!-\!X\right]^{-} \xrightarrow{+I^{-}} \left[X\!-\!\!\!\begin{array}{c}X\\|\\Au\\|\\X\end{array}\!\!\!-\!X\cdots I\right]^{2-} \longrightarrow
\end{array}
$$

$$
X^{-} + \left[\begin{array}{c}X\\|\\Au\\|\\X\end{array}\right]^{-} + XI \qquad (28a)
$$

$$
\left[X\!-\!\!\!\begin{array}{c}X\\|\\Au\\|\\X\end{array}\!\!\!-\!X\right]^{-} \xrightarrow{+I^{-}} \left[X\!-\!\!\!\begin{array}{c}X\\|\\Au\\|\\X\end{array}\!\!\!-\!X\cdots I\right]^{2-} \xrightarrow{+I^{-}}
$$

$$
\left[X\!-\!\!\!\begin{array}{c}X\\|\\Au\\|\\X\end{array}\!\!\!-\!X\!-\!I\cdots I\right]^{3-} \longrightarrow X^{-} + \left[\begin{array}{c}X\\|\\Au\\|\\X\end{array}\right]^{-} + I_2X^{-} \qquad (28b)
$$

In the mechanism of Eq. (28a), the reaction proceeds by a direct two-electron transfer, whereas in the mechanism of Eq. (28b) a formally gold(II) intermediate is involved. However, neither mechanism invokes initial substitution before reduction, such as has been suggested in earlier studies of the reaction between tetrachloroaurate(III) and iodide in aqueous solution[46,74] or in methanol.[41] For the latter reaction medium, a rate constant of 84 liter mol^{-1} s^{-1} was found for the reaction between tetrachloroaurate(III) and iodide at

25°C (cf. Table 5). Methanol has a smaller dielectric constant than water, resulting in a lower reaction rate. With the information available, it is not possible to establish with certainty whether the observed reaction is a chloride/iodide exchange or a direct reduction, although the observation of only one single reaction (which is common for the reaction in water and in methanol when iodide is in excess) could indicate a common mechanism, i.e., direct reduction without any preceding substitution reactions.

Tetrabromoaurate(III) was found by Ettore[97] to oxidize uridine (1-β-D-ribofuranosyl-1H,3H-pyrimidine-2,4-dione) in water to give 5-bromo-6-hydroxy-5,6-dihydrouridine. The same product is expected from direct bromination, and a mechanism involving reductive elimination of bromine from tetrabromoaurate(III) was suggested, in agreement with the observed kinetics. Two mechanisms for this reductive elimination were considered:

$$(29)$$

$$(30)$$

The unimolecular mechanism for reductive elimination [Eq. (29)] for square-planar complexes requires, however, rather extreme distortion of the square-planar ligand arrangement, as pointed out by Elding and Olsson,[47] and the mechanism of Eq. (30) is probably generally operative for reductive elimination of halogen from tetrahaloaurates.

The opposite reaction, the oxidative addition of halogens to gold(I) complexes, has, in the case of bis(μ-dibutyldithiocarbamato-S,S')digold(I), been proposed to proceed via a charge transfer complex.[98] For dicyanoaurate(I), the addition of iodine was found to be catalyzed by iodide,[38] and it was suggested that the triiodide ion reacts faster than iodine in a single-step mechanism giving *trans*-$Au(CN)_2I_2^-$ via the following transition states:

$$
\begin{array}{cc}
\text{(structure with CN, Au, I, CN, charge } - \text{)} & \text{(structure with CN, Au, I, CN, I, charge } 2- \text{)}
\end{array}
\qquad (31)
$$

However, at least for the reaction between $Au(CN)_2^-$ and iodine, trans addition is symmetry forbidden,[99] and another mechanism has been suggested for oxidative addition to gold(I) complexes (charges omitted)[61]:

$$
\qquad (32)
$$

L is either a solvent molecule (uncatalyzed reaction) or an iodide. The uncatalyzed reaction would, according to this latter mechanism, require a fast substitution of the water molecule by an iodide since no intermediate $Au(CN)_2I(H_2O)$ species was experimentally detectable.[38] However, the intermediate A of Eq. (32) resembles the transition state for ligand substitution [cf. Eq. (27)] so closely that an enhanced substitution rate is understandable.

B. Reduction by main Group VI donors

When gold(III) complexes are allowed to react with thiocyanate,[48,100,101] thiosulfate,[61] thiourea,[102–106] or various sulfur-containing organic compounds,[107–113] two-step reactions are observed. The first step is accompanied by a considerable increase in absorbance in the visible region and has been identified as substitution. The kinetics of such reactions have been discussed in Section III. During the second step the solutions become decolorized since colorless gold(I) complexes are formed by reduction of gold(III) complexes. In several cases, the intermediate gold(III) complex with the reductants coordinated as ligands has been isolated,[48,102,111] and for the ethylenediammonium salt of the sulfite complex $Au(en)(SO_3)_2^-$, an X-ray structure analysis showed that the strongly reducing ligands cause no significant distortion of the square-planar arrangement.[112]

Certain carboxylates[101,114-123] and ascorbate[124] have been found to be oxidized by tetrachloroaurate(III) in similar two-step reactions, although detection of the substitution intermediates has been hampered by the spectral similarities between mixed carboxylato halo gold(III) complexes and hydrolyzed tetrahaloaurate(III) species.

1. Thiocyanate, thiosulfate, and thiourea

On the time scale of the redox reaction between thiocyanate and tetrachloroaurate(III) or tetrabromoaurate(III) the ligand substitutions can be regarded as rapid equilibrium processes. Elding et al.[48] have investigated these reactions in aqueous solution of unit ionic strength at 25°C and found that the reductions take place as reductive eliminations from the mixed complexes $AuCl_{4-i}(SCN)_i^-$ and $AuBr_{4-i}(SCN)_i^-$, following a direct attack on a chloride, bromide, or a thiocyanate ligand of the complex ions by a thiocyanate ($X = Cl$, Br, or SCN):

$$(33)$$

Subsequent reactions of the product X-SCN give the final oxidation products, sulfate and cyanide, the latter of which was shown to inhibit the reaction at high gold concentrations. For $AuBr_4^-$ and $Au(SCN)_4^-$, the rate constants for reduction by SCN^- had the values $(5 \pm 2) \times 10^4$ and $(2.4 \pm 0.2) \times 10^3$ liter $mol^{-1} s^{-1}$, respectively. The 20-fold difference in rate was explained on the basis of better bridging properties of bromide than of thiocyanate,[48] but correlates also with the difference in reduction potential of the gold(III) substrates (cf. Table 1).

Tetraamminegold(III) and partially hydrolyzed tetrachloroaurate(III) were also found to react with thiosulfate in aqueous solution via two-step processes giving the final products tetrathionate and dithiosulfatoaurate(I).[61] A mechanism alternative to that of Eq. (33), involving a direct attack on the metal center and invoking a transition state resembling intermediate A of Eq. (32), was proposed for the reductive elimination part of this reaction. The rate of reduction and the reaction order with respect to thiosulfate were found to be pH dependent, but the reaction was in all cases first order in complex. The explanation offered was that the reactant is the weak acid $Au(S_2O_3)_3(H_2O)^{3-}$ ($pK_a = 10.0$) which reacts by an intermolecular mechanism [cf. Eq. (32)], whereas the conjugate base $Au(S_2O_3)_3(OH)^{4-}$ reacts mainly as a result of attack by water molecules analogous to the solvent path for ligand substitution reactions.[61]

Thiourea and N-substituted thioureas are oxidized by gold(III) complexes giving disulfides, and the reactions have been investigated in aqueous propanol and in chloroform.[102–106] A slow intramolecular electron transfer was proposed as the second reaction step,[105] but a reaction mechanism analogous to that of Eq. (33) or that discussed above for thiosulfate reduction could equally well be operative.

2. Sulfides and disulfides

Tetrachloroaurate(III) oxidizes dialkyl sulfides to sulfoxides, and the kinetics of this type of reaction have been investigated in aqueous methanol by Annibale *et al.*[110] As for the reaction with thiocyanate, it was found that an initial substitution equilibrium was established prior to the redox reaction. The substitution equilibrium position depends strongly on the bulkiness of the entering sulfide, while electronic effects seem to be important in the subsequent redox reaction since the more basic sulfides react faster. This led to the conclusion that nucleophilic attack at the metal center is not rate determining in the redox reaction, but that a mechanism similar to that of Eq. (33) is operative ($X = Cl$ or SR_2, charges omitted):

$$
\underset{X}{\overset{X}{Cl-Au-Cl}} \xrightarrow{+SR_2} \underset{X}{\overset{X}{Cl-Au-Cl\cdots SR_2}} \longrightarrow \underset{X}{\overset{X}{Au}} + [ClSR_2]Cl \tag{34}
$$

The primary product, a $[ClSR_2]Cl$ salt, is thus formed by transfer of a chlorine atom from gold(III) to sulfide, and is hydrolyzed in a subsequent fast reaction. For a given sulfide, the sequence of reactivity is $AuCl_4^- \ll AuCl_3(SR_2) < AuCl_2(SR_2)_2^+$, paralleling the increasing positive charge on the complex, and added chloride is consequently inhibiting the reaction by mass action law retardation. Under the experimental conditions employed, a higher degree of substitution than two was not detected. However, a change of the experimental conditions such that gold(III) complexes with four sulfides coordinated are thermodynamically favored would be valuable since such complexes must react by a mechanism which is more or less modified compared to that of Eq. (34). Thiomorpholin-3-one is also oxidized by tetrachloroaurate(III) yielding the corresponding sulfoxide,[107] and the reaction is inhibited by chloride ions. On the basis of this inhibition, it was originally suggested that $AuCl_3OH^-$ was the reactive species and that a hydroxyl group was transferred to the sulfide.[107] However, as pointed out by Annibale *et al.*,[110] the observed kinetics can equally well be explained by the same mechanism as that for the simple thioethers.

(S)-Methionine and (R)-methionine are both stereospecifically oxidized by tetrachloroaurate(III) to give (S)-methionine (S)-sulfoxide and (R)-methionine (R)-sulfoxide, respectively,[108,109] in a two-step reaction as for the simple thioethers. It was found that two methionine molecules were required for the reduction of each gold(III) and that coordination via the amino group was involved.[109] If a mechanism similar to that of Eq. (34) is operative, it would require that the asymmetry of the initially coordinated methionine is controlling the approach of the second methionine, and that the subsequent hydrolysis of the $[ClSR_2]Cl$ intermediate is proceeding stereospecifically, either with or without inversion.

Tetrachloroaurate(III) has successfully been used as catalyst for the oxidation of sulfides to sulfoxides by nitric acid under phase-transfer conditions, as outlined in Fig. 5.[113] The oxidation of the thioether is selective and can be carried out in the presence of other oxidizable groups which are poor nucleophiles toward gold(III), e.g., vinyl, tertiary amines, hydroxy, and diol groups. Moreover, asymmetric disulfides are oxidized regiospecifically, leading to the formation of a single monosulfoxide.

Cystine and other disulfides are oxidatively cleaved by tetrabromoaurate(III) to form sulfinic and ultimately sulfonic acid derivatives and elementary gold.[125,126] No gold(III) intermediates were observed, in agreement with the expected poor coordinating abilities of the disulfides. The latter property of disulfides also explains the formation of elementary gold rather than a gold(I) complex. Oxidation of 1-β-D-thioglucose disulfide by tetrabromoaurate(III) and by bromine gave the same products, i.e., glucose and sulfate, thus suggesting a mechanism for these reactions involving an initial reductive elimination of bromine from tetrabromoaurate(III), such as found for oxidation of the likewise weakly coordinating uridine [cf. Eq. (30)].

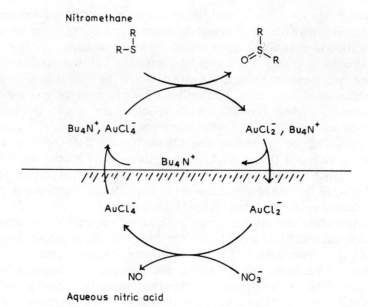

Fig. 5. Gold(III)-catalyzed oxidation of sulfide to sulfoxides by nitric acid under phase-transfer conditions.

3. Carboxylates

In acidic aqueous solution carboxylates are oxidized by tetrachloro-aurate(III) to give carbon dioxide and, eventually, elementary gold, and the kinetics and mechanism of such reactions have been investigated in detail by Maritz and van Eldik for a series of mono- and dicarboxylic acids.[116–122] At pH values close to the pK_a value of the carboxylic acids in question, formic acid reacts faster than the other monocarboxylic acids investigated, and of the dicarboxylic acids examined, oxalic acid, malonic acid, and substituted malonic acids react at comparable rates, which are higher than those for dicarboxylic acids with longer carbon chains.[120] All these redox reactions are accelerated considerably by irradiation at 253 nm,[116,121] but the Au(III) excited states responsible for the photochemical reactions have not been identified.

The reactions were followed spectrophotometrically at 313 nm using an excess of reductant and were in most cases found to be first order in reductant. For the oxidation of [^{14}C]oxalic acid[117] and [^{14}C]formic acid,[119] a linear relationship between the decrease in absorbance at 313 nm and the count rate obtained for released $^{14}CO_2$ was found. These observations were taken as evidence for a rate-determining substitution in tetrachloroaurate(III) or in a hydrolysis product of it. In the case of oxalic acid, the substitution reaction is followed by a fast ring closure and a subsequent and fast intramolecular two-electron transfer:

$$
\left. \begin{array}{c} AuCl_4^- \\ \Updownarrow \\ AuCl_3H_2O \\ \Updownarrow \\ AuCl_3OH^- \end{array} \right\} + \left\{ \begin{array}{c} HOOCCOOH \\ \Updownarrow \\ {}^-OOCCOOH \\ \Updownarrow \\ {}^-OOCCOO^- \end{array} \right\} \longrightarrow
$$

$$
\left\{ \begin{array}{c} AuCl_3OOCCOOH^- \\ \Updownarrow \\ AuCl_3OOCCOO^{2-} \end{array} \right\} \longrightarrow AuCl_2(OOC)_2^- \longrightarrow AuCl_2^- + 2\,CO_2
$$

(35)

The rate increases with increasing pH (≤ 3.5) and decreasing chloride ion concentration ($10^{-3} \leq [Cl^-] \leq 0.8\ M$), and in agreement with the mechanism of Eq. (35) it was concluded that the efficiency of the various oxalate species as nucleophiles increases with increasing negative charge: oxalic acid < hydrogen oxalate < oxalate, and that $AuCl_3H_2O$ and $AuCl_3OH^-$ both react faster than $AuCl_4^-$. Similar results were found for formic acid,[119] implying that hydroxide is a better leaving group than chloride, as also found for the reaction of polydentate amines with the same substrate (see Section III,B).

However, the proton transfer mechanism proposed for these latter reactions is not applicable to the reaction with carboxylic acid/carboxylate pairs, since only the carboxylic acid itself is capable of donating the proton necessary for the transformation of the poor leaving group, hydroxide, to the better leaving group, water. Moreover, a serious objection against a proton transfer mechanism for these reactions arises from the observation of generally higher reactivity of the carboxylate compared to the free acid for all the carboxylic acid/carboxylate pairs investigated[117–119,122]; e.g., $HCOO^-$ reacts nine times faster than HCOOH with $AuCl_3OH^-$.[119] It is therefore tempting to suggest that the inhibiting effect of chloride on these reactions is due to mass action law retardation, as discussed for the redox reactions between sulfides and tetrachloroaurate(III), and that initial substitution equilibria precede rate-determining redox reactions. Also, the second-order rate constants measured at 50°C for the reactions of monocarboxylic acids (aqueous 0.8 M chloride and unit ionic strength at pH close to the pK_a value for the carboxylic acid in question), which range from 8×10^{-4} (formic acid) to 4×10^{-5} liter mol^{-1} s^{-1} (acetic acid), seem suspiciously small for substitution reactions when compared with those of Table 4.

The rate of reaction for malonate, methylmalonate, and chloromalonate increases with increasing chloride concentration,[118,122] in marked contrast to the decrease observed for the carboxylates mentioned above, including dimethylmalonate.[120] The explanation offered was that the activated hydrogens of malonate and monosubstituted malonates interact with a chloride ion creating a more reactive adduct,[118,122] in close analogy to the ion pair mechanism discussed in Section III,B for reactions involving polydentate amines. In a mechanism including a substitution equilibrium, these observations can alternatively be accounted for by proposing variation in the order of redox reactivity of the mixed carboxylatochloroaurate(III) species on changing from a given carboxylate to another. In this context it should also be noted that no linear relationship between decrease in absorbance at 313 nm and $^{14}CO_2$ count rate was found for the reaction between tetrachloroaurate(III) and malonic acid which had been ^{14}C labeled at the carboxylic carbons,[118] but that an induction period of about one "spectrophotometric" half-life preceded the onset of significant $^{14}CO_2$ formation. Thus, also in this respect malonic acid behaves differently from both oxalic and formic acid (*vide supra*), supporting the proposal of differences in redox reactivities mentioned above.

C. Reduction by main Group V donors

In acetonitrile solution, triphenylphosphine forms gold(I) complexes with stability constants more than one order of magnitude larger than triphenylarsine and more than three orders of magnitude larger than triphenylstibine, as can be seen from Table 2. As discussed in Section II,C, a similar trend is expected for gold(III) complexes and, for a given substrate, the nucleophilicity is thus expected to decrease with increasing atomic weight of the donor atom: PPh$_3$ > AsPh$_3$ > SbPh$_3$ (cf. Fig. 4). However, AsPh$_3$ and SbPh$_3$ are both stronger reductants than PPh$_3$,[111] and the results of the investigation by Roulet *et al.*[211] of the reaction between tetrachloroaurate(III) and the triphenyl derivatives of main Group V elements in aprotic solvents thus permit a very valuable comparison of reactivities for members of a unique series of reductants in which nucleophilicity and oxidizability do not parallel each other.

Tetrachloroaurate(III) oxidizes the triphenyl derivatives to give the corresponding dichloro compounds Cl$_2$PPh$_3$, Cl$_2$AsPh$_3$, and Cl$_2$SbPh$_3$, respectively. $AuCl_4^-$ is reduced to $AuCl_2^-$, and the latter rapidly establishes equilibrium with species in which reductant acts as a ligand. The three reactions were all first order in tetrachloroaurate(III) and in reductant, although a path independent of $[PPh_3]$ was found for dichloromethane as solvent. For the reaction with triphenylphosphine, two molecules were consumed in the rate-determining step, consistent with the formation of the substitution product trichloro(triphenylphosphine)gold(III) prior to a fast reduction by the second phosphine. It is noteworthy that analogous substitution products were not identified as intermediates in the reaction with the arsine or the stibine derivative. The activation entropies (Table 8) are all negative and, with the exception of the $[PPh_3]$-independent term in dichloromethane, are close to $-130 \ kJ \ mol^{-1}$. The sign and magnitude suggest an associatve activation for all the reductions in the three solvents investigated, and the lack of variation in $\Delta S^{\ddagger}$ shows that specific solvation is of little importance, in agreement with the conclusions of Section III,D.

The reaction mechanism suggested by Roulet *et al.*[211] includes a nucleophilic attack by the reductant perpendicular to the plane of $AuCl_4^-$ as the rate-determining step, followed by a rapid intramolecular redox process. However, the observed kinetics can be equally well accounted for by a mechanism which involves a direct attack by the reductant on a coordinated chloride as the rate-determining step. Such a mechanism is also more compatible with what is known for the reduction of tetrachloroaurate(III) by halides and sulfides, and, moreover, provides us with a simple explanation of the difference in behavior of triphenylphosphine relative to triphenylarsine and triphenylstibine. For the arsine and the stibine, which are both good reductants but poor nucleophiles, reduction is faster than substitution and the mechanism resembles that for reduction by iodide [Eqs. (28a) and (28b)]. For triphenylphosphine, substitution becomes faster than reduction when the substrate is $AuCl_4^-$, but the sequence is reversed when the substrate is $AuCl_3PPh_3$, and the mechanism is thus similar to that of Eq. (34) for reduction of $AuCl_4^-$ by sulfides.

In acidic aqueous chloride solution, hydrazinium[127] and hydroxylammonium[128] ions reduce tetrachloroaurate(III) directly without any detectable initial substitution. These reactions were found to initiate polymerization of acrylamide, which was taken as evidence for participation of free radicals generated by tetrachloroaurate(III) acting as a one-electron donor toward these substrates, and a reaction mechanism similar to that for reduction by iodide seems therefore likely [cf. Eq. (28)]. In Table 9, the activation parameters are compared with those determined for the reduction of tetrachloroaurate(III) by hypophosphite[129] under similar conditions. Charge neutralization in the transition state seems to have little effect on the rates of these reductions, since the negative entropies of activation found for

TABLE 8

Activation parameters for the rate-determining step in the reduction of tetrachloroaurate(III) by triphenyl derivatives of main Group V elements in aprotic solvents[a]

		Dichloromethane		Acetone		Acetonitrile	
Reductant[b]		$\Delta H^{\ddagger}$ (kJ mol^{-1})	$\Delta S^{\ddagger}$ (J mol^{-1} K^{-1})	$\Delta H^{\ddagger}$ (kJ mol^{-1})	$\Delta S^{\ddagger}$ (J mol^{-1} K^{-1})	$\Delta H^{\ddagger}$ (kJ mol^{-1})	$\Delta S^{\ddagger}$ (J mol^{-1} K^{-1})
PPh$_3$	k_1	52.7 ± 0.4	-50 ± 4				
	k_2	10 ± 3	-125 ± 13	13 ± 7	-142 ± 25	7 ± 3	-130 ± 13
AsPh$_3$	k_2	~ 25	~ -130	32 ± 4	-125 ± 13		
SbPh$_3$	k_2	29 ± 3	-117 ± 13	31 ± 2	-121 ± 8		

[a] From Ref. [21].
[b] Rate = $(k_1 + k_2[\text{reductant}])[\text{AuCl}_4^-]$. The k_1 path is only detectable for PPh$_3$ and only in dichloromethane.

TABLE 9

Activation parameters for the reduction of
tetrachloroaurate(III) in acidic aqueous solution

Reductant	$\Delta H^{\ddagger}$ (kJ mol^{-1})	$\Delta S^{\ddagger}$ (J mol^{-1} K^{-1})	Ref.
$H_2NNH_3^+$	50 ± 2	-61 ± 3^a	[127]
$HONH_3^+$	76 ± 2	-60 ± 3^a	[128]
$H_2PO_2^-$	128 ± 2	136 ± 7	[129]

^{a} In the original papers, these two values are quoted as
having positive signs, presumably owing to a typograph-
ical error.

both of the cationic reductants (Table 9) are replaced by a positive entropy of
activation for the negative hypophosphite ion. This is in marked contrast to
what is found for ligand substitutions (Section III,A,4) and provides support
for a mechanism involving no initial substitution before the reduction of
tetrachloroaurate(III) by these reductants.

D. Reduction by metal ions

Gold(III) is reduced by platinum(II) in aqueous chloride solution and the
reaction rate was found to be independent of the concentration of chloride
$(0.25 \leq [Cl^-] \leq 1.0 \ M)$ and to increase with increasing ionic strength,
suggesting that the reactants are $AuCl_4^-$ and $PtCl_4^{2-}$.[131] Steric effects
in uncharged platinum(II) complexes with bulky ligands were shown to
be important for the reaction rate in acetonitrile,[130] and the observed
rate law was different for *trans*- and *cis*- platinum(II) complexes. For the
trans geometry the rate law is dominated by a third-order term,
$k[Pt(II)][Au(III)][Cl^-]$, whereas for the cis geometry, equilibria between
chloro- and solventoplatinum(II) complexes complicate the observed rate
expressions.[132] The dependence on chloride concentration observed for this
medium was taken as evidence for a bridged transition state with a Pt–Cl–Au
bridge.[130]

The noncomplementary reaction between tetrachloroaurate(III) and
iron(II) is retarded by iron(III), which is indicative of a one-electron reaction with a gold(II) intermediate[131]:

$$AuCl_4^- + Fe^{2+} \rightleftharpoons AuCl_4^{2-} + Fe^{3+} \tag{36}$$

$$AuCl_4^{2-} + Fe^{2+} \rightarrow AuCl_2^- + Fe^{3+} + 2Cl^- \tag{37}$$

The rate of reduction increases with increasing chloride concentration, which suggests that chloro complex formation reduces Fe(III) retardation and/or increases the effectiveness of Fe(II) as reductant.[131] Au(II) is substitution labile and was found to catalyze chloride exchange in tetrachloroaurate(III).[133]

V. CONCLUSION

An increasing body of experimental results is providing evidence for a common, intermolecular mechanism for the reduction of gold(III) complexes by ligands, in which the rate-determining step is an attack by an outer-sphere ligand on a coordinated ligand. As pointed out by both Annibale *et al.*[110] and Elding *et al.*,[48] such a mechanism resembles that normally invoked to describe octahedral/square-planar redox reactions, of which those involving Pt(IV)/Pt(II) are the best understood. It also appears that hydrolyzed species play a less important role in the redox chemistry of gold(III) than originally believed.

For ligand substitution reactions, the initial step is a direct attack on the gold(III) center. Diverging interpretations of experimental results for ligand substitution reactions have led to some confusion, but the picture now emerging confirms that gold(III) complexes are "normal" in the sense that they behave qualitatively like platinum(II) and palladium(II) complexes in their ligand substitution reactions. However, significant differences are observed; e.g., gold(III) reacts faster than palladium(II) and, in particular, than platinum(II), and cis effects and charge neutralization are more important in determining the ligand substitution rate for gold(III) than for platinum(II) and palladium(II). Kinetically, gold(III) discriminates very efficiently between nucleophiles, and the stability constants for gold(III) complexes show that Au(III) probably ranks as the "softest" of all metal ions for which complex formation has been quantitatively characterized.

The initial step for the reduction reactions differs from the initial step for the ligand substitution reactions in the point of attack, and for gold(III)–substrate/ligand combinations it is often a close run whether a ligand substitution or a redox reaction is the result of the approach of a reducing ligand. Tetrachloroaurate(III) is, for example, reduced directly by iodide, triphenylarsine, and triphenylstibine, whereas with triphenylphosphine and sulfides substitution occurs initially. For tetrabromoaurate(III)/thiocyanate, reduction and substitution take place at almost equal rates, at least under certain conditions, and it would be interesting to see how variations in temperature, medium, ionic strength, etc. effect the balance between these two types of reaction.

Acknowledgment

The author is grateful to Dr. Lone Melchior Larsen for help with the literature search, to Dr. Martin P. Hancock for many helpful discussions, and to Rigmor Jensen and Per Jensen for their assistance in the preparation of the article.

References

[1] Rich, R. L.; Taube, H. *J. Phys. Chem.* **1954**, *58*, 1.
[2] Baddley, W. H.; Basolo, F. *Inorg. Chem.* **1964**, *3*, 1087.
[3] Lippard, S. J., Ed.; "Platinum, Gold and Other Metal Chemotherapeutic Agents"; *Am. Chem. Soc. Symp. Ser.* **1983**, *209*.
[4] Sadler, P. J. *Struct. Bond.* **1976**, *29*, 171.
[5] Hartung, J.; Schröter, C.; Reinhold, J.; Zwanziger, H.; Dietzsch, W.; Hoyer, E. *J. Signal AM* **1980**, *8*, 95.
[6] Goolsby, A. D.; Sawyer, D. T. *Anal. Chem.* **1968**, *40*, 1978.
[7] Foll, A.; Le Démézet, M.; Courtot-Coupez, J. *Bull. Soc. Chim. France* **1972**, 408.
[8] Skibsted, L. H.; Bjerrum, J. *Acta Chem. Scand.* **1974**, *A28*, 740.
[9] van Z. Bekker, P.; Robb, W. *Inorg. Nucl. Chem. Lett.* **1972**, *8*, 849.
[10] Louw, W. J.; Robb, W. *Inorg. Chim. Acta* **1974**, *9*, 33.
[11] Latimer, W. M. "The Oxidation Potentials"; 2nd ed., Prentice-Hall: New York, 1952.
[12] Erenburg, A. M.; Peshchevitskii, B. I. *Russ. J. Inorg. Chem.* **1969**, *14*, 1429.
[13] Hawkins, C. J.; Mønsted, O.; Bjerrum, J. *Acta Chem. Scand.* **1970**, *24*, 1059.
[14] Hancock, R. D.; Finkelstein, N. P. *Inorg. Nucl. Chem. Lett.* **1971**, *7*, 477.
[15] Hancock, R. D.; Finkelstein, N. P.; Evers, A. *J. Inorg. Nucl. Chem.* **1974**, *36*, 2539.
[16] Skibsted, L. H.; Bjerrum, J. *Acta Chem. Scand.* **1977**, *A31*, 155.
   Skibsted, L. H.; Bjerrum, J. *J. Indian Chem. Soc.* **1977**, *54*, 102.
[17] Edwards, J. O. *J. Am. Chem. Soc.* **1954**, *76*, 1540.
[18] Puddephatt, R. J. "The Chemistry of Gold"; Elsevier: Amsterdam, 1978.
[19] Johnson, P. R.; Pratt, J. M.; Tilley, R. I. *J. Chem. Soc. Chem. Commun.* **1978**, 606.
[20] Nikolaeva, N. M.; Erenburg, A. M.; Antipina, V. A. *Isv. Sib. Otd. Akad. Nauk. SSSR, Ser. Khim. Nauk.* **1972**, 126.
[21] Roulet, R.; Lan, N. Q.; Mason, W. R.; Fenske, G. P. *Helv. Chim. Acta* **1973**, *56*, 2405.
[22] Elding, L.-I.; Gröning, A.-B. *Acta Chem. Scand.* **1978**, *A32*, 867.
[23] Mønsted, O.; Skibsted, L. H. *Acta Chem. Scand.* 1983, *A38*, 23.
[24] Ryan, J. L. *Inorg. Chem.* **1969**, *8*, 2059.
[25] Hollis, L. S.; Lippard, S. J. *J. Am. Chem. Soc.* **1983**, *105*, 4293.
[26] Nardin, G.; Randiaccio, L.; Annibale, G.; Natile, G.; Pitteri, B. *J. Chem. Soc., Dalton Trans.* **1980**, 220.
[27] Jamin, M. F.; Iwamoto, R. T. *Inorg. Chim. Acta.* **1978**, *27*, 135.
[28] Pouradier, J.; Coquard, M. J. *Chim. Phys.* **1966**, *63*, 1072.
[29] Peshchevitskii, B. I.; Belevantsev, V. I. *Russ. J. Inorg. Chem.* **1967**, *12*, 161.
[30] Almgren, L. *Acta Chem. Scand.* **1971**, *25*, 3713.
[31] Cudey, G.; Schuffenecker, L.; Bourdet, J. B.; Lozar, J. *Thermochim. Acta* **1983**, *67*, 1.
[32] Baddley, W. H.; Basolo, F.; Gray, H. B.; Nölting, C.; Poë, A. J. *Inorg. Chem.* **1963**, *2*, 921.
[33] Stevns, M.; Skibsted, L. H. To be published.
[34] Hendra, P. J. *J. Chem. Soc. A* **1967**, 1298.
[35] Peschevitskii, B. I.; Belevantsev, V. I. *Russ. J. Inorg. Chem.* **1969**, *14*, 1256.
[36] Mason, W. R. *Inorg. Chem.* **1970**, *9*, 2688.

[37] Isci, H.; Mason, W. R. *Inorg. Chem.* **1983**, *22*, 2266.

[38] Ford-Smith, M. H.; Habeeb, J. J.; Rawsthorne, J. H. *J. Chem. Soc., Dalton Trans.* **1972**, 2116.

[39] Dubinskii, V. I.; Demidova, G. V. *Russ. J. Inorg. Chem.* **1971**, *16*, 134.

[40] Belevantsev, V. I.; Kolonin, G. R.; Ryakhovskaya, S. K. *Russ. J. Inorg. Chem.* **1972**, *17*, 1303.

[41] Cattalini, L.; Tobe, M. L. *Inorg. Chem.* **1966**, *5*, 1145.
Cattalini, L.; Orio, A.; Tobe, M. L. *J. Am. Chem. Soc.* **1967**, *89*, 3130.

[42] Cattalini, L.; Nicolini, M.; Orio, A. *Inorg. Chem.* **1966**, *5*, 1674.

[43] Shamovskaia, G. I.; Peshchevitskii, B. I. *Izv. Sib. Otd. Akad. Nauk. SSR, Ser. Khim. Nauk.* **1972**, 53.

[44] Robb, W. *Inorg. Chem.* **1967**, *6*, 382.

[45] Hall, A. L.; Satchell, D. P. N. *J. Chem. Soc. Chem. Commun.* **1976**, 163.

[46] Hall, A. L.; Satchell, D. P. N. *J. Chem. Soc., Dalton Trans.* **1977**, 1403.

[47] Elding, L. I.; Olsson, L. F. *Inorg. Chem.* **1982**, *21*, 779.

[48] Elding, L. I.; Gröning, A.-B.; Gröning, Ö. *J. Chem. Soc. Dalton Trans.* **1981**, 1093.

[49] Mason, W. R. *Inorg. Chem.* **1970**, *9*, 2688.

[50] Skibsted, L. H. *Acta Chem. Scand.* **1979**, *A33*, 113.

[51] Skibsted, L. H. *Acta Chem. Scand.* **1983**, *A37*, 613.

[52] Ventegodt, J.; Øby, B.; Skibsted, L. H. *Acta Chem. Scand.* **1985**, *A39*, in press.

[53] Kaas, K.; Skibsted, L. H. *Acta Chem. Scand.* **1985**, *A39*, 1.

[54] Elding, L. I. *In* "Inorganic Reaction Mechanisms", Vol. 7; Sykes, A. G., Ed.; The Royal Society of Chemistry: London, 1981, p. 133.

[55] Fry, F. H.; Hamilton, G. A.; Turkevich, J. *Inorg. Chem.* **1966**, 5, 1943.

[56] Peshchevitskii, B. I.; Shamovskaya, G. I. *Russ. J. Inorg. Chem.* **1972**, *17*, 1386.

[57] Swaddle, T. W. *Coordin. Chem. Rev.* **1974**, *14*, 217.

[58] Swaddle, T. W. *Adv. Inorg. Bioinorg. Mech.* **1983**, *2*, 93.

[59] Cattalini, L.; Doni, A.; Orio, A. *Inorg. Chem.* **1967**, *6*, 280.

[60] Cattalini, L.; Martelli, M.; Marangoni, G. *Inorg. Chem.* **1968**, *7*, 1492.

[61] Nord, G.; Skibsted, L. H.; Halonin, A. S. *Acta Chem. Scand.* **1975**, *A29*, 505.

[62] Peshchevitskii, B. I.; Shamovskaya, G. I.; Mal'chikov, C. D. *Russ. J. Inorg. Chem.* **1971**, *16*, 181.

[63] Pesek, J. J.; Mason, W. R. *Inorg. Chem.* **1983**, *22*, 2958.

[64] Cattalini, L.; Martelli, M.; Marangoni, G. *Inorg. Chem.* **1968**, *7*, 1492.

[65] Cattalini, L.; Orio, A.; Tobe, M. L. *Inorg. Chem.* **1967**, *6*, 75.

[66] Cattalini, L.; Ricevuto, V.; Orio, A.; Tobe, M. L. *Inorg. Chem.* **1968**, *7*, 51.

[67] Blandamer, M. J.; Burgess, J.; Hamshere, S. J.; Wellings, P. *Transition Met. Chem. (Weinheim)* **1979**, *4*, 161.

[68] Alexander, R. D.; Holper, P. N. *Transition Met. Chem. (Weinheim)* **1980**, *5*, 108.

[69] Hall, A. J.; Satchell, D. P. N. *J. Chem. Soc. Perkin Trans. II* **1975**, 1351.

[70] Patel, G.; Satchell, R. S.; Satchell, D. P. N. *J. Chem. Soc. Perkin Trans. II* **1981**, 1406.

[71] Patel, G.; Satchell, R. S.; Satchell, D. P. N. *Inorg. Chim. Acta* **1981**, *54*, L 97.

[72] Micallef, J. V.; Satchell, D. P. N. *Inorg. Chim. Acta* **1982**, *64*, L 187.
Micallef, J. V.; Satchell, D. P. N. *J. Chem. Soc. Perkin Trans. II* **1982**, 971.

[73] Micallef, J. V.; Satchell, D. P. N. *J. Chem. Soc. Perkin Trans. II* **1982**, 1379.

[74] Hall, A. J.; Satchell, D. P. N. *J. Chem. Soc. Perkin Trans. II* **1976**, 1278.

[75] Bandoli, G.; Clemente, D. A.; Marangoni, G.; Cattalini, L. *J. Chem. Soc., Dalton Trans.* **1973**, 886.

[76] Poë, A. J.; Vaughan, D. H. *Inorg. Chim. Acta* **1967**, *1*, 255.

[77] Beran, P.; Vleck, A. A. *Coll. Czech. Chem. Commun.* **1959**, *24*, 3572.

[78] Louw, W. J.; Robb, W. *Inorg. Chim. Acta* **1969**, *3*, 29.

[79] Louw, W. J.; Robb, W. *Inorg. Chim. Acta* **1969**, *3*, 303.

[80] van Z. Bekker, P.; Louw, W. J.; Robb, W. *Inorg. Chim. Acta* **1972**, *6*, 564.

[81] Louw, W. J.; Robb, W. *Inorg. Chim. Acta* **1974**, *8*, 253.
[82] Louw, W. J.; de Wall, D. J. A. *Inorg. Chim. Acta* **1978**, *28*, 35.
[83] Annibale, G.; Cattalini, L.; El-Awady, A.; Natile, G. *J. Chem. Soc., Dalton Trans.* **1974**, 802.
[84] Annibale, G.; Natile, G.; Cattalini, L. *J. Chem. Soc., Dalton Trans.* **1976**, 285.
[85] Annibale, G.; Natile, G.; Pitteri, B.; Cattalini, L. *J. Chem. Soc., Dalton Trans.* **1978**, 728.
[86] Annibale, G.; Cattalini, L.; Natile, G. *J. Chem. Soc., Dalton Trans.* **1975**, 188.
[87] Annibale, G.; Natile, G.; Cattalini, L. *J. Chem. Soc., Dalton Trans.* **1976**, 1547.
[88] Weick, C. F.; Basolo, F. *Inorg. Chem.* **1966**, *5*, 576.
[89] Font, D. L.; Weick, C. F. *Inorg. Chem.* **1973**, *12*, 1864.
[90] Coulter, L. V.; Sinclair, J. R.; Cole, A. G.; Roper, G. C. *J. Am. Chem. Soc.* **1959**, *81*, 2986.
[91] Jolly, W. L. *J. Am. Chem. Soc.* **1954**, *58*, 250.
[92] Brawner, S. A.; Lin, I. J. B.; Kim, J.-H.; Everett, G. W., Jr. *Inorg. Chem.* **1978**, *17*, 1304.
[93] Kim, J.-H.; Everett, G. W., Jr. *Inorg. Chem.* **1979**, *18*, 3145.
[94] van Eldik, R.; Skibsted, L. H. Work in progress.
[95] Strähle, J.; Gelinek, J.; Kölmel, M. *Z. Anorg. Allg. Chem. (Leipzig)* **1979**, *456*, 241.
[96] Elding, L. I. *Inorg. Chim. Acta* **1972**, *6*, 683.
[97] Ettore, R. *J. Chem. Soc., Dalton Trans.* **1983**, 2329.
[98] Kita, H.; Itoh, K.; Tanaka, K.; Tanaka, T. *Bull. Chem. Soc. Jpn.* **1978**, *51*, 3530.
[99] Pearson, R. G. "Symmetry Rules for Chemical Reactions"; Wiley: New York, 1976, p. 286.
[100] Bjerrum, N.; Kirschner, Aa. *Kgl. Dan. Vidensk. Selsk. Math. Afd.* **1918**, *8*, 5.
[101] Kazakov, V. P.; Konovalova, M. V. *Russ. J. Inorg. Chem.* **1968**, *13*, 231.
[102] Shulmann, V. M.; Saveleva, Z. A.; Novoselov, R. I. *Russ. J. Inorg. Chem.* **1973**, *8*, 376.
[103] Makotchenko, E. V.; Peschevitskii, B. I.; Novoselov, R. I. *Izv. Sib. Otd. Akad. Nauk. SSR, Ser. Khim. Nauk.* **1978**, 44.
[104] Makotchenko, E. V.; Peschevitskii, B. I.; Novoselov, R. I. *Izv. Sib. Otd. Akad. Nauk. SSR, Ser. Khim. Nauk.* **1981**, 47.
[105] Makotchenko, E. V.; Peschevitskii, B. I.; Novoselov, R. I. *Izv. Sib. Otd. Akad. Nauk. SSR, Ser. Khim. Nauk.* **1981**, 52.
[106] Makotchenko, E. V.; Peschevitskii, B. I.; Novoselov, R. I. *Izv. Sib. Otd. Akad. Nauk. SSR, Ser. Khim. Nauk.* **1981**, 56.
[107] De Filippo, D.; Devillanova, F.; Preti, C. *Inorg. Chim. Acta* **1971**, *5*, 103.
[108] Bordignon, E.; Cattalini, L.; Natile, G.; Seatturin, A. *J. Chem. Soc. Chem. Commun.* **1973**, 878.
[109] Natile G.; Bordignon, E.; Cattalini, L. *Inorg. Chem.* **1976**, *15*, 246.
[110] Annibale, G.; Canovese, L.; Cattalini, L.; Natile, G. *J. Chem. Soc. Dalton Trans.* **1980**, 1017.
[111] Brown, D. H.; McKinley, G. C.; Smith, W. E. *J. Chem. Soc., Dalton Trans.* **1978**, 199.
[112] Dunand, A.; Gerdil, R. *Acta Crystallogr.* **1975**, *B31*, 370.
[113] Gasparrini, F.; Giovannoli, M.; Misti, D.; Natile, G.; Palmieri, G. *Tetrahedron* **1983**, *39*, 3181.
[114] Kazakov, V. P.; Konovalova, M. V. *Russ. J. Inorg. Chem.* **1968**, *13*, 1226.
[115] Kazakov, V. P.; Metveeva, A. I.; Erenburg, A. M.; Peshchevitskii, B. I. *Russ. J. Inorg. Chem.* **1965**, *10*, 563.
[116] Maritz, B. S.; van Eldik, R.; van den Berg, J. A. *J. South Afr. Chem. Inst.* **1975**, *28*, 14.
[117] Maritz, B. S.; van Eldik, R. *Inorg. Chim. Acta* **1976**, *17*, 21.
[118] Maritz, B. S.; van Eldik, R. *Inorg. Chim. Acta* **1976**, *20*, 43.
[119] Maritz, B. S.; van Eldik, R. *J. Inorg. Nucl. Chem.* **1976**, *38*, 1545.
[120] Maritz, B. S; van Eldik, R. *J. Inorg. Nucl. Chem.* **1976**, *38*, 1749.
[121] Maritz, B. S.; van Eldik, R. *J. Inorg. Nucl. Chem.* **1976**, *38*, 2124.
[122] Maritz, B. S.; van Eldik, R. *J. Inorg. Nucl. Chem.* **1977**, *39*, 1935.
[123] Hassan, S. A.; Shahine, A.; Ba-Isa, A. *Egypt. J. Chem.* **1980**, *23*, 157.
[124] Ripan, R.; Pop, G.; Pop, I.; Nascu, C. *Rev. Roum. Chim.* **1977**, *22*, 361.

[125] Shaw, C. F., III; Cancro, M. P.; Witkiewicz, P. L.; Eldridge, J. E. *Inorg. Chem.* **1980,** *19,* 3198.
[126] Witkiewicz, P. L.; Shaw, C. F., III, *J. Chem. Soc. Chem. Commun.* **1981,** 1111.
[127] Sen Gupta, K. K.; Basu, B. *Transition Met. Chem. (Weinheim)* **1983,** *8,* 3.
[128] Sen Gupta, K. K.; Basu, B. *Transition Met. Chem. (Weinheim)* **1983,** *8,* 6.
[129] Sen Gupta, K. K.; Basu, B.; Sen Gupta, S.; Naudi, S. *Polyhedron* **1983,** *2,* 983.
[130] Peloso, A. *Coord. Chem. Rev.* **1975,** *16,* 95.
[131] Moodley, K.; Nicol, M. J. *J. Chem. Soc., Dalton Trans.* **1977,** 993.
[132] Peloso, A. *J. Chem. Soc., Dalton Trans.* **1983,** 1285.
[133] Rich, R. L.; Taube, H. *J. Phys. Chem.* **1954,** *58,* 6.

Kinetics and Mechanisms of Actinide Redox and Complexation Reactions

K. L. Nash

U.S. Geological Survey
Denver Federal Center
Denver, Colorado, USA

J. C. Sullivan

Chemistry Division
Argonne National Laboratory
Argonne, Illinois, USA

I. INTRODUCTION

Actinide elements are those following actinium (element No. 89) in the periodic table. Elements thorium through californium (No. 98) are considered in this review. The aqueous solution chemistry of the 5f transition elements, in contrast to the 4f series, is concerned with the III, IV, V, VI, and VII oxidation states as has been previously documented.[1,2] The available oxidation states are not common to all members of the 5f series; Th exhibits only the IV state, while both Np and Pu solutions have been prepared in five oxidation states. Beyond Am the trivalent state is the phenomenologically stable state.

The trivalent and tetravalent actinides are aquo ions with coordination numbers between 6 and 10. The pentavalent and hexavalent states are linear dioxo cations with formal charges of $1+$ and $2+$, respectively, with the coordination sphere satisfied by 6 molecules of water in the equatorial plane. Interactions of the cations with water (or other molecules) in the primary coordination spheres is predominantly, if not solely, electrostatic. This is a reflection of the shielded nature of the 5f electrons.

Thermodynamic values for the actinide ions in acidic aqueous media are available.[3] From this compilation, we note that U(III) is a powerful reductant (U(III)/(IV) = -0.607 V) and Np(VII) a powerful oxidant (Np(VI)/(VII) = $+2.04$ V). In addition to such thermodynamic considerations, it is important to note that the radioactive nature of the actinides (except Th and U) provides an *in situ* source of H_2O_2 which must be accommodated within the particular experiment.

The kinetics of redox reactions of the lighter 5f transition elements (U, Np, Pu, and Am) have been reviewed by Newton[4] in an earlier publication. A summary of the solution chemistry of the transplutonium elements (Am, Cm, Bk, Cf, and Es)[5] includes some aspects of oxidation–reduction kinetics of these elements. This work is intended, in part, to selectively update and expand Newton's original work.

We have, in addition, collected the data available on the kinetics of complex ion formation and dissociation reactions of actinides in aqueous solutions. Although there are no previous surveys in this area, there are two papers[6,7] summarizing the rates of ligand and solvent exchange (studied by NMR techniques) on U(VI) in nonaqueous solvents.

Research in actinide solution chemistry is motivated, among other considerations, by the intellectual stimulus of attempting to understand the chemistry of these man-made elements. A pragmatic motivation for such studies is the fact that aqueous solution chemistry of the lighter actinides is ubiquitous in the entire nuclear fuel cycle. Actinide ions are here referred to collectively by the abbreviation An.

II. KINETICS OF URANIUM OXIDATION–REDUCTION REACTIONS

The reactions for which redox kinetics results for uranium are available since 1974 fall into four groups: oxidation of uranium(III), oxidation of uranium(IV), disproportionation of uranium(V), and reduction of U(VI).

A. Uranium(III) oxidation

By far the most studied system is the oxidation of U(III) by a variety of oxidants. The oxidants whose reaction kinetics have been studied can be divided into three classes: metal ion oxidants, nonmetal inorganic oxidants, and organic oxidants. There is some overlap between the first and the last category with several studies of the effect of coordinated and noncoordinated Co(III)–organic complexes on the rate of U(III) oxidation. The recurring point in these papers is whether the reaction occurs by an inner- or outer-sphere mechanism, and whether the Marcus[8] cross-relation can be applied to these systems.

An extensive list of the rate parameters for reactions involving trivalent transition metal ions and complexes is given in Table 1. The smallest second-order rate constant for a reaction with no $[H^+]$ dependence is for the Cr(III)–U(III) reaction while the fastest reactions are with the iodo- and bromopentammine ruthenium(III) complexes and the Co(III) aquo ion. The reaction rate constants in this group span 12 orders of magnitude.

The reactions of U(III) with a series of ruthenium(III) amine complexes have been investigated in a 1 M trifluoromethanesulfonate medium,[9] and in 1 M perchlorate.[10] The empirical form of the rate law in the two studies are identical but the rate constants determined in perchloric acid are consistently 10–20 times higher than those in the former medium. The discrepancy may be attributed to oxidation of U(III) and/or the product Ru(II) by ClO_4^- producing an artificially high rate constant.

The rates of reduction of $Co(NH_3)_5^{3+}$ carboxylate and amine complexes[11,12] by U(III) have been reported. The log k values for outer-sphere reactions are linearly related to corresponding log k values for reduction of the same derivatives by Cr(II), V(II), Eu(II), and $Ru(NH_3)_6^{2+}$. The rate of reaction with carboxylato-bridged dicobalt(III) complexes[13] is highly dependent on the presence of a "pendant" carbonyl group. Those bridging ligands with active carbonyl groups show a 10^2–10^7 acceleration in the rates, implying electron transfer from a remote point of attachment of the reductant.

The rates of the oxidation of U(III) by several trivalent metal ions[14–18] have been reported. The rate law for the U(III) hexaaquo Co(III) reaction in

TABLE 1

Summary of kinetic results for uranium(III) oxidation[a]

Oxidant	Rate expression	Parameters	Ref.	T (°C)	I (M)
$Ru(NH_3)_5H_2O^{3+}$	$-d[U(III)]/dt = k[U(III)][Ru(III)]$	$k = 1.11(\pm 0.33) \times 10^4$ $\Delta H^* = -2.5(\pm 5.9)$ $\Delta S^* = -159(\pm 21)$	[9]	25	1.0
$Ru(NH_3)_5OH^{2+}$	$-d[U(III)]/dt = k[U(III)][Ru(III)]$	$k = 1.73(\pm 0.12) \times 10^5$ $k = 9.5 \times 10^4$	[10] [10]	25 25	0.25 0.25
$Ru(NH_3)_6^{3+}$	$-d[U(III)]/dt = k[U(III)][Ru(III)]$ $-d[U(III)]/dt = k[U(III)][Ru(III)]$	$k = 1.10(\pm 0.46) \times 10^4$ $\Delta H^* = +4.2(\pm 2.9)$ $\Delta S^* = -159(\pm 8)$ $k = 1.00(\pm 0.15) \times 10^5$ $\Delta H^* = +12.7(\pm 1.7)$ $\Delta S^* = -107(\pm 11)$	[9] [10]	25 25	1.0 0.25
$Ru(NH_3)_5Cl^{2+}$	$-d[U(III)]/dt = k[U(III)][Ru(III)]$	$k = 7.1 \times 10^5$ $\Delta H^* = +2.4(\pm 1.9)$ $\Delta S^* = -125(\pm 4)$	[10]	25	0.25
$Ru(NH_3)_5Br^{2+}$	$-d[U(III)]/dt = k[U(III)][Ru(III)]$	$k = 1.2 \times 10^6$ $\Delta H^* = +3.47(\pm 2.09)$ $\Delta S^* = -118(\pm 3)$	[10]	25	0.25
$Ru(NH_3)_5I^{2+}$	$-d[U(III)]/dt = k[U(III)][Ru(III)]$	$k \geq 2 \times 10^6$	[10]	25	0.25
$Ru(en)_3^{3+}$	$-d[U(III)]/dt = k[U(III)][Ru(III)]$	$k = 1.62(\pm 0.27) \times 10^5$ $\Delta H^* = -5.4(\pm 5.4)$ $\Delta S^* = -163(\pm 17)$	[9]	25	0.1
$Co(en)_3^{3+}$	$-d[U(III)]/dt = k[U(III)][Co(III)]$	$k = 0.18$	[11]	25	0.2
$Co(NH_3)_6^{3+}$	$-d[U(III)]/dt = k[U(III)][Co(III)]$	$k = 1.38$	[11]	25	0.2
$Co(NH_3)_5L^{3+}$	$-d[U(III)]/dt = k[U(III)][Co(III)]$		[11]	25	0.2
L = imidazole		$k = 2.7$	[11]	25	0.2
L = pyrazole		$k = 14$	[11]	25	0.2

	k	Ref	T	
L = pyridine	$k = 26$	[11]	25	0.2
L = dimethylformamide	$k = 33$	[11]	25	0.2
L = N,N-dimethylnicotinamide	$k = 4.2 \times 10^4$	[11]	25	0.2
L = triethylacetate	$k = 6.3$	[11]	25	0.2
L = trimethylacetate	$k = 42$	[11]	25	0.2
L = cyclopentanecarboxylate	$k = 3.7 \times 10^3$	[11]	25	0.2
L = formate	$k = 8 \times 10^4$	[11]	25	0.2
L = lactate	$k = 7 \times 10^4$	[11]	25	0.2
L = propionate	$k = 5.3 \times 10^3$	[12]	22	0.2
L = $NH_3CH_2COO^-$	$k = 5.8 \times 10^2$	[12]	22	0.2
L = $(CH_3)_3CCH_2COO^-$	$k = 1.01 \times 10^3$	[12]	22	0.2
L = $(CH_3)_3NCH_2COO^-$	$k = 66$	[12]	22	0.2
L = $NH_3C(CH_3)_2COO^-$	$k = 5.1$	[12]	22	0.2
L = $C_6H_5CH_2COO^-$	$k = 3.1 \times 10^3$	[12]	22	0.2
L = 1-py-CH_2COO^-	$k = 3.1 \times 10^2$	[12]	22	0.2
L = o-CH_3-$C_6H_4COO^-$	$k = 3.4 \times 10^2$	[12]	22	0.2
L = o-NH_3-C_6H_4-COO^-	$k = 28$	[12]	22	0.2

$$(NH_3)_3\text{Co} \quad \text{Co(NH}_3)_3 = I - d[I]/dt = k[I][U(III)]$$

	k	Ref	T	
X = L-COO$^-$		[13]	25	0.2
X = formate	$k = 4.1$	[13]	25	0.2
X = acetate	$k = 1.26$	[13]	25	0.2
X = trifluoroacetate	$k = 11.8$	[13]	25	0.2
X = benzoate	$k = 2.8$	[13]	25	0.2
X = phthalate	$k = 7.3 \times 10^4$	[13]	21	0.2
X = isophthalate	$k = 4.5 \times 10^2$	[13]	21	0.2
X = pyruvate	$k = 1.5 \times 10^5$	[13]	21	0.2
X = phenylglyoxalate	$k = 1.2 \times 10^5$	[13]	21	0.2
X = 4-methoxyphenylglyoxylate	$k = 1.7 \times 10^5$	[13]	21	0.2
X = 2,4-dimethoxyphenyl-glyoxalate	$k = 1 \times 10^5$	[13]	21	0.2

(continued)

TABLE 1 (continued)

Oxidant	Rate expression	Parameters	Ref.	T (°C)	I (M)
X = 2,4,6-trimethoxyphenyl-glyoxalate		$k = 8 \times 10^4$	[13]	21	0.2
Co(III)	$-d[U(III)]/dt = k[U(III)][Co(III)]$	$k = 1.12(\pm0.1) \times 10^6$ $\Delta H^* = +8.37(\pm1.05)$ $\Delta S^* = -107(\pm3)$	[14]	20	2.0
Tl(III)	$-d[U(III)]/dt = k[U(III)][Tl(III)]$ $k = k_{-1}K/([H] + K)$	$k_{-1} = 3.6(\pm0.2) \times 10^5$ $\Delta H^* = +33(\pm2)$ $\Delta S^* = -17(\pm6)$	[15]	20	2.0
Cr(III)	$-d[U(III)]/dt = k[U(III)][Cr(III)]$ $k = k_0 + k_1/[H]^2$	$k_0 = 6.2(\pm0.3) \times 10^{-2}$ $\Delta H^* = +27.7(\pm4.0)$ $\Delta S^* = -172(\pm13)$ $k_1 = 3.8(\pm0.1) \times 10^{-3}$ $\Delta H^* = +63.2(\pm5.4)$ $\Delta S^* = -79(\pm21)$	[15]	20	2.0
Fe(III)	$-d[U(III)]/dt = k[Fe(III)][U(III)]$	$k = 3.92(\pm0.20) \times 10^5$ $\Delta H^* = +6.40(\pm3.72)$ $\Delta S^* = -116(\pm13)$	[16]	20	2.0
V(IV)	$-d[U(III)]/dt = k[V(IV)][U(III)]$	$k = 3.66(\pm0.20) \times 10^5$ $\Delta H^* = 25.3(\pm1.8)$ $\Delta S^* = -51.5(\pm6.3)$	[16]	20	2.0
V(III)	$-d[U(III)]/dt = k[V(III)][U(III)]$ $k = k_0 + k_1/[H]$	$k_0 = 54.2(\pm3.7)$ $\Delta H^* = +40.9(\pm2.6)$ $\Delta S^* = -71.6(\pm8.8)$	[16] [16]	20 20	2.0 2.0

Species	Rate law and constants	Ref.		
Eu(III)	$-d[\text{U(III)}]/dt = k[\text{Eu(III)}][\text{U(III)}]$ $k = k_0 + k_1[[\text{H}]$ $k_1 = 278(\pm2)$ $\Delta H^* = +39.7(\pm1.2)$ $\Delta S^* = -62.8(\pm4.1)$ $k_0 = 0.0134(\pm0.0001)$ $\Delta H^* = +33.1(\pm1.3)$ $\Delta S^* = -169(\pm5)$ $k_1 = 1.84(\pm0.01) \times 10^{-3}$ $\Delta H^* = +32.7(\pm1.0)$ $\Delta S^* = -187(\pm3)$	[18] [18]	25 25	2.0 2.0
N_3H	$-d[\text{U(III)}]/dt = k[\text{U(III)}][\text{N}_3\text{H}]$ $k = 6.5 \times 10^{-3}$ $\Delta H^* = +43.1$ $\Delta S^* = -142.3$	[19]		
CCl_3COOH	$d[\text{U(III)}]/dt = k[\text{RCl}_3][\text{U(III)}]$ $k = 1.9$ $\Delta H^* = +54.7$ $\Delta S^* = 50.4$	[23]	15	2.0
CCl_2HCOOH	$-d[\text{U(III)}]/dt = k[\text{RCl}_2][\text{U(III)}]$ $k = 0.23$ $\Delta H^* = +46.2$ $\Delta S^* = -94.6$	[24] [24]	15 15	2.0 2.0
CClH_2COOH	$-d[\text{U(III)}]/dt = k[\text{RCl}][\text{U(III)}]$ $k = 0.018$ $\Delta H^* = +55.4$ $\Delta S^* = 85.7$	[24]	15	2.0
Benzaldehyde	$-d[\text{U(III)}]/dt = k[\text{Bza}][\text{U(III)}]$ $k = 0.049$ $\Delta H^* = +14.2(\pm2.5)$ $\Delta S^* = -230(\pm5)$	[25]	15	0.5
Salicylaldehyde	$-d[\text{U(III)}]/dt = k[\text{Sa}][\text{U(III)}]$ $k = 0.034$ $\Delta H^* = 16.3(\pm2.5)$ $\Delta S^* = -226(\pm5)$	[25]	15	0.5

[a] Units for rate constants determined by the reaction order: for zero-order reactions (M s^{-1}), for first-order reactions (s^{-1}), for second-order reactions (M^{-1} s^{-1}), for third-order reactions (M^{-2} s^{-1}). Units for activation parameters: ΔH^* (kJ mol^{-1}), ΔS^* (J mol^{-1} K^{-1}).

$2\,M\,HClO_4-NaClO_4^{[14]}$ is reported to be independent of acidity $(0.1-1.0\,M$ acid):

$$-d[Co(III)]/dt = k_1[Co(III)][U(III)]$$

where $k_1 = 1.12(\pm 0.1) \times 10^6\,M^{-1}\,s^{-1}$ $(T = 20°C)$. The rate-controlling step postulated for the U(III)–Tl(III) reaction[15] may be expressed as

$$U^{3+} + TlOH^{2+} \rightarrow [U\cdot OH\cdot Tl]^{5+}$$

The rate of formation of this complex at $T = 20°C$, $[H^+] = 1.0\,M$, $I = 2.0\,M$ is $3.6 \times 10^5\,M^{-1}\,s^{-1}$. In the U(III)–Cr(III) system[15] the rate law has an acid-independent step and one dependent on $[H^+]^{-2}$, in contrast to the previous study[14] which has terms dependent on $[H^+]^{-1}$ and $[H^+]^{-2}$. The postulated reactions are

$$U^{3+} + Cr^{3+} \rightarrow [U\cdot Cr]^{6+}$$

$$U^{3+} + Cr(OH)_2^{\,+} \rightarrow [U\cdot Cr(OH)_2]^{4+}$$

The authors[15] concluded that the discrepancy in the rate laws is probably due to the higher ionic strength and wider range of acidity investigated.

Reactions of Fe(III) and V(IV) with U(III) exhibited a slight $[H^+]$ dependence[16] (attributed to a medium effect in terms of a Harned-type equation), while reaction of V(III) with U(III) exhibited an inverse $[H^+]$ dependence. Ionic strength dependency of the Fe(III) and V(IV) reactions showed that, in accord with predictions of the Debye–Hückel equation for cation–cation reactions, the rate increased with increasing ionic strength. The inverse acid-dependent step in the V(III) reaction was attributed to hydrolysis of V(III). As the hydrolysis constants of V(III) and Fe(III) are quite similar (pKa's in the range 2.4–2.9), it is difficult to explain an acid-dependent step in one system and not in the other.

In a follow-up study, the activation parameters for the Fe(III)–U(III) reaction were determined as a function of ionic strength.[17] The values of ΔS^* for these reactions were found to be independent of ionic strength, the changes in the rate being reflected exclusively in ΔH^*. A model equation based on coulombic interactions fails to reproduce the observed values for the activation parameters. Several plausible reasons for the discrepancy between theory and fact are presented but none is completely satisfactory.

The postulated mechanism for the U(III)–Eu(III) reaction[18] involves reaction between the two hydrated metal ions and a parallel path in which one of the two is hydrolyzed. Since the hydrolysis constants for both metal ions are comparable, it is impossible to say which is hydrolyzed.

Finally, the rates of U(III) oxidation by hydrazoic acid,[19] hydroxyl-amine,[20] a series of halogenated organic acids,[21–24] and benzaldehyde and salicylaldehyde[25] have been reported in Table 1.

B. Uranium(IV) oxidation

In the oxidation of U(IV), several interesting mechanistic possibilities develop (Table 2). While U(III) generally behaves as a one electron reductant, U(IV) has the potential to react by either a one- or two-electron transfer

TABLE 2

Summary of kinetics of uranium(IV) oxidation[a]

Oxidant	Rate expression	Parameters	Ref.	T (°C)	I (M)
HOCl	$-d[U(IV)]/dt =$ $k[U(IV)][HOCl]/[H]$	$k = 1.08\,(\pm 0.08)$ $\Delta H^* = +77.0\,(\pm 0.4)$ $\Delta S^* = +13\,(\pm 2)$	[27]	25	3.0
V(V)	$-d[V(V)]/dt =$ $k[U(IV)][V(V)]$		[30]	24	2.0
in LiClO$_4$	$k = k_1 + k_2/[H]$	$k_1 = 2.49\,(\pm 0.07) \times 10^4$ $\Delta H^* = +31.1\,(\pm 2.8)$ $\Delta S^* = -55.6\,(\pm 9.6)$ $k_2 = 2.47\,(\pm 0.7) \times 10^4$ $\Delta H^* = 79.5\,(\pm 0.4)$ $\Delta S^* = +107\,(\pm 5)$			
in NaClO$_4$		$k_1 = 2.31\,(\pm 0.07) \times 10^4$ $\Delta H^* = +35.7\,(\times 7.1)$ $\Delta S^* = -41.8\,(\pm 24.7)$ $k_2 = 2.72\,(\pm 0.04) \times 10^4$ $\Delta H^* = +84.9\,(\times 1.7)$ $\Delta S^* = +126\,(\pm 5)$			
Fe(CN)$_6{}^{3-}$	$-d[Fe(CN)_6{}^{3-}]/dt =$ $k[U(IV)][Fe(CN)_6{}^{3-}]$		[33]	25	1.0
	$k = \dfrac{(k_1/[H] + k_2/[H]^2)}{(1 + K/[H])}$	$k_1 = 2.7$ $\Delta H_1^* = +41\,(\pm 1)$ $\Delta S_1^* = +71.5\,(\pm 4.6)$ $k_2 = 0.87$			
IrCl$_6{}^{2-}$	$-d[U(IV)]/dt =$ $k[U(IV)][IrCl_6{}^{2-}]$		[34]	25	1.0
	$k = \dfrac{(k_1/[H] + k_2/[H]^2)}{(1 + K/[H])}$	$k_1 = 0.21$ $\Delta H^{1*} = +96.4\,(\pm 6.3)$ $\Delta S^{1*} = 63.6\,(\pm 21.5)$ $k^2 = 0.12$ $\Delta H^{2*} = +108.3\,(\pm 5.0)$ $\Delta S^{2*} = +99.1\,(\pm 17.0)$			

[a] Units for rate constants determined by the reaction order: for zero-order reactions (M s^{-1}), for first-order reactions (s^{-1}), for second-order reactions (M^{-1} s^{-1}), for third-order reactions (M^{-2} s^{-1}). Units for activation parameters: ΔH^* (kJ mol^{-1}), ΔS^* (J mol^{-1} K^{-1}).

(thermodynamically, the conversion of U(IV)–U(VI) is favorable, but when U(V) is formed as an intermediate, it either is further oxidized to U(VI) or disproportionates). The existence of U(V) (the dioxo cation UO_2^+) as an intermediate is a crucial point in deciding whether a particular reaction of U(IV) occurs by a one- or two-electron transfer. It has been shown[26] that the lability of the coordinated oxygens on U(V) is considerably greater than those on U(VI), a fact which can be of importance in deriving a reaction mechanism.

The question of one- vs two-electron transfer in the reduction of halogens by U(IV) was addressed in two investigations.[27,28] In one study,[27] the reaction of U(IV) with HClO in aqueous perchlorate was examined, while the second[28] investigated the oxidation of U(IV) by aqueous halogens. These are particularly interesting systems as both the oxidants and reductants are capable of multiple electron transfers.

For the former system, the rate law includes an inverse dependence on $[H^+]$ which the authors attribute to the loss of a proton by one of the reactants in a rapid preequilibrium. They propose a uranyl-like activated complex $[H_2UO_2Cl^{3+}]$ with one proton residing on each oxygen. In addition to other evidence, they cite the results of labeled oxygen study in which the concentration of U(V) was less than 2×10^{-6} M, implying that the reaction occurs by a single two-electron transfer rather than through a U(V) intermediate. The reaction of U(IV) with aqueous Cl_2 is closely related to the U(IV)–HOCl reaction since aqueous Cl_2 readily disproportionates to produce measurable amounts of HOCl in solution. The results obtained in the latter study[28] are suspect since the authors did not adequately correct for the HOCl produced.

The U(IV) ion is oxidized by I_2 with a rate law having a complex dependence on $[I^-]$ and $[H^+]^{-2}$.[29] With bromine, however, reaction occurs by a simple rate law directly dependent on bromine. Rate constants and activation parameters for the bimolecular reactions of U(IV) with the halogens show that rate constants and activation entropies decrease in the order Cl > Br > I while the activation energies are ~ 96 kJ mol^{-1} for Cl, Br and $+71$ kJ mol^{-1} for I. A plot of ΔG^* vs ΔG° for the three reactions is not obviously linear and the slope is much less than the 0.5 required by the simple Marcus theory.

The oxidation of U(IV) by V(V)[30] proceeds by steps involving an acid-independent path and one inversely dependent on the acidity (this despite the formation of a dioxo cation as the oxidation product). The author suggests that the lower than expected dependence on acidity and the unusually high rate of reaction are due to a configuration of the activated complex which allows the transfer of one of the "yl" oxygens of vanadium to uranium. Three possible intermediates are

A complex dependence on the medium was observed at high ionic strengths. In addition, a difference in acid dependence of the rates was observed at high ionic strengths when $LiClO_4$ was replaced by $NaClO_4$. These effects have not been satisfactorily explained.

The rate of the U(IV)–Ce(IV) reaction was reported previously[4] as being too fast to measure in aqueous perchlorate. The rate is measurable in sulfate[31] and fluoride[32] solutions, but the effect of sulfate and fluoride on rate and mechanism of oxidation is complex.

Two further studies in this category are the oxidation of U(IV) by two oxidants which are substitution inert and normally react predominantly by an outer-sphere mechanism: ferricyanide[33] and hexachloroiridate(IV).[34] The mechanism for both systems involves reaction of the mono- and dihydroxides of U(IV) with the oxidant. There is no apparent correlation between reduction potentials of the oxidants and reported rate constants.

C. Disproportionation of uranium(V)

The disproportion of U(V) has been discussed previously.[4] This system has been reinvestigated,[35] confirming and extending the original results. The rate increases with the concentration of U(VI) suggesting the existence of a precursor binuclear complex:

$$U(V) + U(VI) \rightleftharpoons U(VI) \cdot U(V) \qquad K_e$$

$$2U(V) \rightarrow U(IV) + U(VI) \qquad k_5$$

$$U(V) \cdot U(VI) + U(V) \rightarrow U(IV) + 2U(VI) \qquad k_6$$

Both k_5 and K_e are larger in D_2O than in water and have a marked dependence on ionic strength. An alternate mechanism for the disproportionation reaction of U(V) has been suggested.[6] This mechanism involves the protonation of UO^{2+} in a rapid preequilibrium followed by the reaction between the protonated intermediate and UO_2^+:

$$UO_2^+ + H^+ \rightleftharpoons UO_2H^{2+} \qquad K$$

$$UO_2H^{2+} + UO_2^+ \rightarrow products \qquad k_d$$

The influence of U(VI) on the reaction rate was not investigated in this study and the proposed U(V)–U(VI) complex was therefore not identified.

D. Uranium(VI) reductions

Studies of the reduction of U(VI) by the hydrated electron in perchloric acid solutions[36] as well as the U(V) + CO_3^- reaction[37] will be discussed later. A study of the reduction of a series of U(VI) complexes by the hydrated electron[38] demonstrated that the rate was not markedly influenced by changes in the composition of the equatorial coordination sphere of U(VI). On the other hand, the rate of the reaction between e_{aq}^- and U(VI) is drastically inhibited by negatively charged micelles.[39]

III. NEPTUNIUM OXIDATION–REDUCTION KINETICS

A. Np(III) oxidation

In contrast to U(III) studies the significant reactions reported for Np(III) are very sparse. An interesting study[40] of the reaction

$$Np(III) + Cr(III) \cdot Np(V) = 2\,Np(IV) + Cr(III)$$

demonstrated the empirical form of the rate law as

$$-d\text{Np(III)}/dt = k[\text{Np(III)}][\text{Cr(III)} \cdot \text{Np(V)}][\text{H}^+]^0$$

at 25°C, $I = 0.5\ M$, $k = 1.6\ (\pm 0.1) \times 10^3\ M^{-1}\ s^{-1}$, $\Delta H^* = 11\ (\pm 0.4)$ kJ mol^{-1} and $\Delta S^* = -148\ (\pm 2)$ J mol^{-1} K^{-1}. The increased rate of the Np as compared to the Pu system[4] is reflected in more favorable ΔS^* and ΔH^* for the Np system.

The kinetic and equilibrium parameters were determined for the reaction of neptunium(III) with tris(ethylenediamine)ruthenium(III)[41] in aqueous trifluoromethanesulfonate media. At 25°C, $I = 1.0\ M$, the forward rate constant for the reaction is 8.96 $(\pm 0.38)\ M^{-1}\ s^{-1}$. The rate of the U(III)–Ru(en)$_3^{3+}$ reaction is about 105 times faster, in accord with the Marcus cross-relations. The slower rate of the Np(III) reaction is reflected in less favorable ΔH^* and ΔS^* values. The rate constant for the self-exchange of Ru(en)$_3^{3+/2+}$ is calculated to be five times greater than that for Ru(NH$_3$)$_6^{3+/2+}$ in contrast to an earlier estimate.[9]

B. Np(IV) oxidation

An additional uncertainty in any mechanistic interpretation of redox reactions of the tetravalent actinide is caused by the tendency of these highly charged ions to undergo hydrolysis and/or complexation reaction. Such problems are reflected in the empirical forms of the rate laws in the following studies.

In a study of the oxidation of Np(IV) by persulfate[42] using the method of initial rates, it was found necessary to invoke three parallel paths, two of which reflected the complex formation between Np^{4+} and SO_4^- and $S_2O_8^{2-}$, respectively. The empirical form of the rate law for the oxidation of Np^{4+} by H_2O_2[43] in perchloric acid is

$$-d\text{Np(IV)}/dt = k[\text{Np(IV)}][H_2O_2]^{0.7}[H^+]^{0.1}$$

At 10°C and $I = 1.0\ M$, the activation enthalpy and entropy are 38 kJ mol^{-1} and $\Delta S^* = -104$ J mol^{-1} K^{-1}, respectively. The empirical form of the rate law determined for the oxidation of Np^{4+} by VO_2^+ is[44]

$$-d\text{Np(IV)}/dt = k[\text{Np(IV)}][V(V)]^{1.3}/[H^+]^{0.8}$$

where $k = 436\ (\pm 22)\ M^{0.5}\ \text{s}^{-1}$ at 25°C, $I = 2.0\ M$, HClO$_4$–NaClO$_4$. Values are reported for the activation enthalpy ($+106$ kJ mol^{-1}) and the activation entropy ($+108$ J mol^{-1} K^{-1}).

The approach to equilibrium

$$\text{NpO}_2^+ + \text{Fe}^{2+} + 4\text{H}^+ \underset{k_2}{\overset{k_1}{\rightleftharpoons}} \text{Np}^{4+} + \text{Fe}^{3+} + 2\text{H}_2\text{O}$$

was initially investigated in perchlorate media[4] and has been reinvestigated in nitric acid media by spectrophotometric techniques.[45] Using the method of initial rates the empirical form of the rate law is

$$d[\text{Np(V)}]/dt = k_2[\text{Np(IV)}][\text{Fe(III)}][H^+]^{-3}$$
$$- k_1[\text{Np(V)}][\text{Fe(II)}][H^+]^{1.38}$$

Values of the rate parameters were reported as $k_1 = 0.262\ M^{-2.38}\ \text{s}^{-1}$ and $k_2 = 0.028\ M^2\ \text{s}^{-1}$ at 25°C and $I = 1.5\ M$, respectively. Activation energies reported for the forward and reverse reactions are 42 kJ mol^{-1} and 146 kJ mol^{-1}, respectively. An unusual activated complex containing a NO$_3^-$ is postulated.

C. Np(V), Np(VI), Np(VII) reactions

The reduction of dioxoneptunium(V) by the hydrated electron has been found to proceed at a rate which is not phenomenologically undistinguishable from a diffusion-controlled process.[36] This result prompted speculation that the electron transfer reaction proceeded via a nonadiabatic path. The initial product of such a path would be a Np(IV) species in an excited electronic state with an inappropriate coordination configuration. Support for this postulate is obtained from a pulse radiolysis study using the transient conductivity detection[46] system. The rate constant remeasured for $e_{aq}^- + \text{Np(V)}$ was

$2.45\ (\pm 0.25) \times 10^{10}\ M^{-1}\ s^{-1}$ and a value of $k = 1.5 \times 10^{6}\ s^{-1}$ was calculated for the rearrangement of $NpO_2 \cdot 6H_2O$, the postulated initial product, to the fully hydrolyzed $Np(OH)_4 \cdot 4H_2O$.

Two pulse radiolysis studies of the oxidation of Np(V) to Np(VI) have been reported. The reaction[37] of Np(V) with CO_3^- will be discussed later. For the $(OH + NpO_2^+)$ reaction, a value of $k = 4.3\ (\pm 0.1) \times 10^{7}\ M^{-1}\ s^{-1}$ at 25°C and $I = 0.01\ M$ has been reported.[47]

The Marcus cross-relations[8] have been used in kinetic studies of a series of Np(VI) reductions. For a series of seven N-alkylphenothiazines[48] (which are oxidized to a cation radical) a linear correlation was observed between ΔG^* and ΔG° over a change in potential of the reductants of 0.27 V. The second-order rate parameters ranged from $3.6\ (\pm 0.1) \times 10^{6}\ M^{-1}\ s^{-1}$ to $1.91\ (\pm 0.06) \times 10^{5}\ M^{-1}\ s^{-1}$ at 25°C and $I = 1.0\ M$, $HClO_4$. The calculated rate parameters are 65–160 times higher than the experimentally determined values. There are also a marked discrepancy between experimental and calculated activation enthalpies and entropies with $\Delta H_E^* - \Delta H_C^* = 26\text{–}30\ kJ\ M^{-1}$ and $\Delta S_E^* - \Delta S_C^* = 50\text{–}63\ J\ M^{-1}\ K^{-1}$.

In a subsequent study[49] four substituted, 1,10-phenanthroline and 2,2′-bipyridine complexes of Fe(II) were used as reductants. In these systems a linear relation was observed between ΔG° (E° of reductants ranging from 1.12 to 0.93 V) and ΔG^* [rate parameters ranged from $4.34\ (\pm 0.03) \times 10^{3}\ M^{-1}\ s^{-1}$ to $68.8\ (\pm 2.1) \times 10^{3}\ M^{-1}\ s^{-1}$] at 25°C, $I = 1.0\ M$, $HClO_4$. There was sufficient auxiliary data to calculate $\Delta H^* = 27\ kJ\ mol^{-1}$ and $\Delta S^* = -63\ J\ mol^{-1}\ K^{-1}$ for the case of $Fe(phen)_3^{2+}$, which compares well with the respective experimental values of $28\ kJ\ mol^{-1}$ and $-83\ mol^{-1}\ K^{-1}$.

The reduction of Np(VI) by thioglycolic acid and coordinated thiolato ligands has been the subject of two mechanistic studies. The reduction of Np(VI) by free thioglycolic acid[50] has been investigated under conditions such that the reaction involves a 1-equivalent oxidation at the sulfur atom. The rate law is $-d(\ln[Np(VI)])/dt = (a + b/[H^+])[tga]$ where $a = 3.5\ (\pm 0.7)\ M^{-1}\ s^{-1}$ and $b = 34.2\ (\pm 0.3)\ s^{-1}$. With methyl thioglycolate the parameters are $58\ (\pm 2)\ M^{-1}\ s^{-1}$ and $2.7\ (\pm 0.1)\ s^{-1}$, respectively. Qualitative observations were also made on the rate of Np(VI) reduction by a series of Cr and Co thiolato complexes.

A study of the oxidation of (2-mercaptoethylamine-N,S)bis(ethylene-diamine)cobalt(III), (I), by Np(VI) or Co(III)[51] in perchloric acid media reported that a coordinated disulfide was produced. For Np(VI) as the oxidant the rate law is $-d[I]/dt = k[I][Np(VI)]$ where $k = 2842\ (\pm 15)\ M^{-1}\ s^{-1}$ at 25°C and $I = 1.00\ M$. It is of interest to note that the thermodynamically more powerful oxidant, Co(III), reacts more slowly than does Np(VI) (for a comparable acid-independent path) by a factor of ~ 3 which reflects the slower self-exchange rate of the Co(II)–Co(III) reaction.

The reduction of Np(VI) by a series of dicarboxylic acids was studied[52] at the tracer level using a solvent extraction technique. The results were verified for selected systems using an actinyl(VI) specific ion electrode. At 23°C, $I = 0.1\ M$, the rate parameters for reduction of Np(VI) in the NpO_2^{2+} complex were in the order of oxalate < malonate ≥ alkylmalonate ≃ succinate ≃ phthalate ≥ fumarate > glutarate where the values of the rate parameters range from 2.5 (± 0.3) × $10^{-4}\ s^{-1}$ to < $1.7 \times 10^{-6}\ s^{-1}$. When ascorbic acid is used as the reductant, the[53] reaction proceeds via an outer-sphere activated path. The value of the rate parameter at 25°C, $I = 1.0\ M$, is 1.23 (± 0.02) × $10^5\ M^{-1}\ s^{-1}$, reflecting the increased reducing power of ascorbic acid.

The Np(VII) ion is a powerful oxidant[54] capable of oxidizing Ag(I) to Ag(II), Co(II) to Co(III), and V(IV) to V(V).[55] The empirical form of the rate law for Ag(I) oxidation is $-d\text{Np(VII)}/dt = k_1[\text{Np(VII)}][\text{Ag(I)}][\text{H}^+]$, where $k_1 = 3.54$ (± 0.02) × $10^3\ M^{-2}\ s^{-1}$ at 25°C, $I = 1.0\ M$. Values for $\Delta H^* = 3.4\ \text{kJ mol}^{-1}$ and $\Delta S^* = -172\ (\pm 13)\ \text{J mol}^{-1}\ \text{K}^{-1}$ are reported. The rate law for the Co(II) oxidation is $-d\text{Np(VII)}/dt = k_2 K_3\ [\text{Np(VII)}][\text{Co(II)}][\text{H}^+]/(1 + K_3[\text{H}^+])$ where at 25°C, $I = 1.0\ M$, $k_2 = 5.4$ (± 0.2) × $10^4\ M^{-1}\ s^{-1}$ $K_3 = 1.83\ (\pm 0.09)\ M^{-1}$.

The rate law for the oxidation of V(IV) by Np(VII) is $-d\text{Np(VII)}/dt = k_1[\text{Np(VII)}][\text{VO}^{2+}]$, where k_1 exhibits only a weak acid dependence. At 25°C, $I = 1.0\ M$, the rate constant is 1.442 (± 0.001) × $10^3\ M^{-1}\ s^{-1}$. The reactivity patterns of these reactions are consistent with a mechanism[54] wherein the Np(VII) forms a protonated binuclear intermediate.

The empirical form of the rate law with Se(IV) as the reductant is[56]

$$-d[\text{Np(VII)}]/dt = (k_1[\text{Se(IV)}] + k_2[\text{Se(IV)}]^2)[\text{Np(VII)}]$$

The rate law is accounted for by a mechanism involving the known dimerization of Se(IV) in acid medium and parallel paths for oxidation of the monomer and dimer. At 25°C, $I = 1.0\ M$, the rate constants for oxidation of the monomer and dimer are 1.07 (± 0.02) × 10^3 and 4.37 (± 0.17) × $10^3\ M^{-1}\ s^{-1}$, respectively. The rate law and rate parameters are consistent with the established reactivity pattern exhibited by Np(VII) when oxidizing aquo metal ions.

For the reaction Np(VII) + Am(V) → Np(VI) + Am(VI)[57] the empirical form of the rate law is

$$-d[\text{Np(VII)}]/dt = \frac{\beta C[\text{Np(VII)}][\text{Am(V)}][\text{H}^+]}{1 + C[\text{H}^+]}$$

where at 25°C, $I = 1.0\ M$, $\beta = 4.22$ (± 0.10) × $10^5\ M^{-1}\ s^{-1}$, $C = 1.30 \pm 0.12\ M^{-1}$. Values are presented for activation parameters. The mechanism presented has the formation of a binuclear Np(VII) · Am(V) immediately followed by a rapid protonation equilibrium and rate-determining decomposition of this intermediate.

IV. PLUTONIUM OXIDATION–REDUCTION KINETICS

A study of the kinetics of the oxidation of Pu(III) by VO_2^+ has been reported in $HClO_4$ and HNO_3 solutions.[58] The rate law of the forward reaction is reported as

$$-d[\text{Pu(III)}]/dt = \frac{[\text{Pu(III)}][\text{V(V)}]}{1 + K_1[\text{H}^+] + K_2[\text{H}^+]^2}(k_1 + k_2[\text{H}^+]^2)$$

(K_1 and K_2 are equilibrium constants for the protonation of VO_2^+ ions). At 25°C, $I = 1.6\ M$, $HClO_4$, $k_1 = 3.38\ M^{-1}\ s^{-1}$ and $k_2 = 3.73\ M^{-3}\ s^{-1}$. In HNO_3 solutions at 25°C, $I = 2.0\ M$, $k_1 = 58.3\ M^{-1}\ s^{-1}$ and $k_2 = 267\ M^{-3}\ s^{-1}$. The marked catalysis by NO_3^- is not readily explicable.

In an accompanying publication, the same investigators[59] reported the kinetic results on the back reaction, where the empirical form of the rate law presented is

$$-d\,\text{Pu(IV)}/dt = [\text{Pu(IV)}][\text{V(IV)}](k_1 + k_2/[\text{H}^+]^2)$$

In perchloric acid–sodium perchlorate media at $I = 1.6\ M$, values reported for $k_1 = 0.497\ (\pm 0.027)M^{-1}\ s^{-1}$ and $k_2 = 0.48\ (\pm 0.03)M^{-1}\ s^{-1}$. Values of the rate parameters determined experimentally are consistent with those calculated from the forward reaction and the equilibrium constant.

The oxidation of Pu^{4+} by Ce(IV) in nitric acid solutions has been reported[60] to obey the empirical rate law

$$-d[\text{Pu(IV)}]/dt = k\,\frac{[\text{Pu(IV)}][\text{Ce(IV)}]}{a_{\pm}(\text{HNO}_3)}$$

where at 25°C, 1 M HNO_3, $k = 3.4\ M^{-1}\ s^{-1}$. The reaction was studied over a wide range of $HNO_3 + NaNO_3$ concentrations (1.0–5.7 M) and over a limited temperature range, from which the activation enthalpy ($\Delta H^* = 91.2$ (± 0.8) kJ mol^{-1}, $I = 1.0\ M$, was calculated. Mechanistic conclusions drawn from such studies are necessarily subject to reservations, but empirically it is to be noted that the presence of nitrate ions evidently slows the reaction rate as compared to perchlorate. This may be no more than a reflection of the change in values for the Pu(V)/(IV) and Ce(IV)/(III) couples due to the formation of nitrato complexes.

The reduction of Pu(IV) by H_2O_2 in nitric acid solutions can be described by an empirical rate law[61]

$$-d\text{Pu(IV)}/dt = k\frac{[\text{Pu(IV)}][H_2O_2]}{[H^+][NO_3^-]}$$

At 25°C, $I = 2$ M, $k = 0.198$ (± 0.017) M^{-1} s^{-1}, $(E_A = 95.5$ kJ mol$^{-1})$ and $\Delta S^* = 54$ kJ mol^{-1} K^{-1}. In contrast to results summarized earlier,[4] the authors present other evidence that the concentration of the Pu(IV)·H_2O_2 complex is minimized under the conditions of these experiments.

The reaction of Pu(VI) with HNO_2 in HNO_3–$NaNO_3$ solutions is reversible[61]:

$$2\text{PuO}_2^{2+} + HNO_2 + H_2O \underset{k_2}{\overset{k_1}{\rightleftharpoons}} 2\text{PuO}_2^+ + NO_3^- + 3H^+$$

The empirical forms of the rate laws for the forward and reverse reactions are

$$-d[\text{Pu(VI)}]/dt = k_1[\text{Pu(VI)}][HNO_2]/[H^+]$$

and

$$-d[\text{Pu(V)}]/dt = k_2[\text{Pu(V)}][HNO_2]^{0.5}[NO_3^-]^{0.4}[H^+]^{0.6}$$

where at 25°C, $I = 2.0$ M, $k_1 = 0.263$ (± 0.022) s^{-1}, $E_A = 111$ kJ mol^{-1} and $\Delta S^* = 109$ J mol^{-1} K^{-1}. These values can be qualitatively compared with those reported for Np(VI) and Am(VI) as oxidants,[63] with the caveat that the NO_3^- ions form only weak complexes. The ordering of the rate parameters follows the potentials for the An(VI)/(V) couples and the slower rate for the Pu system is reflected in greater values for ΔH^* and more positive values for ΔS^*.

The reduction of Pu(VI) by a series of phenothiazines has been reported[69] at 25°C, 1.0 M $HClO_4$. The rate laws are simple second order with rate parameters varying from 6×10^5 to 2×10^4 M^{-1} s^{-1}. A plot of ΔG^* vs $\Delta G°$ is linear over a range of 0.24 V for the reductants. A detailed calculation of rate parameters using the Marcus cross-relations again provided the correct ordering, but calculated values ranged from a factor of 18–40 higher than experimental. In comparison with the Np(VI) reactions with these reductants, the rates of Pu(VI) reduction are consistently slower.

The reduction of Np(VI) and Pu(VI) by I$^-$[65] is described by empirical rate laws that are first order in AnO_2^{2+} and I$^-$, and independent of acid. At 25°C, $I = 1.0$ M $(HClO_4 + LiClO_4)$ the values for the apparent second-order rate parameters are 530 (± 6) M^{-1} s^{-1} and 17.0 (± 0.1) M^{-1} s^{-1} for NpO_2^{+2} and PuO_2^{2+}, respectively. The apparent activation parameters for the reduction of NpO_2^{2+} are $\Delta H^* = 51.9$ (± 1.3) kJ mol^{-1} and $\Delta S^* = -25$ (± 4) J mol^{-1} K^{-1} and for the reduction of PuO_2^{2+} are $\Delta H^* = 57.7 \pm 0.4$ kJ mol^{-1} and $\Delta S^* = -33$ (± 2) J mol^{-1} K^{-1}. There is an adequate agreement between experimental values of the rate parameters and those calculated in terms of the Marcus cross-relations.

The results of a pulse radiolysis study[37] of the reactions between dioxoactinide(V) ions and the carbonate radical ions have been reported in 0.1 M Na_2CO_3 solutions. The rates for the reduction of the tris-(carbonato)actinide(VI) ions with e_{aq}^- have been determined as 1.18 ($\pm$0.6) $\times$ 10^{10}, 2.34 ($\pm$0.12) $\times$ 10^{10}, and 2.28 ($\pm$.11) $\times$ 10^{10} M^{-1} s^{-1}, respectively for U, Np, and Pu. The values determined for the rates of the reactions of CO_3^- with the dioxoactinide(V) ions are 4.88×10^8, 1.52×10^7, and 2.73×10^7 M^{-1} s^{-1}, respectively for U, Np, and Pu. The values calculated for the rate of the latter reactions using the Marcus cross-relations are in substantial agreement with the experimental data.

V. AMERICIUM OXIDATION–REDUCTION KINETICS

Americium(VI) is a potent oxidant having only marginal stability in aqueous solutions. The reduction of Am(VI) by nitrous acid has been reported to occur by two parallel paths involving reaction between AmO_2^{2+} and HNO_2 or NO_2^-.[63] A plausible mechanism has been proposed:

$$AmO_2^{2+} + HNO_2 \xrightarrow{k_1} AmO_2^+ + H^+ + NO_2^{\cdot} \qquad \text{slow}$$

$$AmO_2^{2+} + NO_2^- \xrightarrow{k_2} AmO_2^+ + NO_2^{\cdot} \qquad \text{slow}$$

$$H_2O + AmO_2^{2+} + NO_2^{\cdot} \longrightarrow AmO_2^+ + NO_3^- + 2H^+ \qquad \text{fast}$$

The correlation of log k with E^0 for the series Am(VI), Np(VI), Mn(III) (acid-independent path) is consistent with the Marcus formalism.

Cooper *et al.*[66] studied the kinetics of the reaction of Am(VI) with bromide. The stoichiometry of the reaction is

$$2Am(VI) + 2Br^- \rightarrow Br_2 + 2Am(V)$$

The reaction rate is independent of acidity over the range of 0.05 to 0.5 M. The rate law has terms proportional to $[Br^-]$ and $[Br^-]^2$. The suggested mechanism is

$$AmO_2^{2+} + Br^- \rightarrow AmO_2Br^+$$

$$AmO_2Br^+ \rightarrow AmO_2^+ + Br\cdot$$

$$AmO_2Br^+ + Br^- \rightarrow AmO_2^+ + Br_2^-$$

Oxidation of Br· and Br_2^- is presumed to follow rapidly. The inability of the Marcus relationship to reproduce the rate constants was taken as evidence that the reaction proceeded by an inner-sphere mechanism, although a contribution from outer-sphere electron exchange path could not be ruled out.

The stoichiometry of the reaction between V(IV) and Am(V) is[67]

$$Am(V) + 2V(IV) \rightarrow Am(III) + 2V(V)$$

The only reaction observed is the reduction of Am(V) to Am(IV), with the subsequent reduction of Am(IV) by V(IV) too fast to observe on the time scale of the stopped-flow device. The reaction is independent of acidity and is characterized by the simple rate law:

$$d[V(VI)]/dt = k[Am(V)][V(IV)]$$

The absence of acid dependence where one would normally be expected could be due to one of two phenomena:

1. Fortuitous cancellation of expected positive acid dependence for transformation of Am(V) to Am(III) and inverse acid dependence for the formation of VO^{2+} from VO_2^+.

2. An oxygen from AmO_2^+ replaces water as the second "yl" oxygen in vanadium is an atom transfer process.

The persulfate oxidation of Am(III) previously discussed by Newton[4] has been further explored in the intervening years, particularly by Russian researchers. This reaction has generally been reported to be rather slow at ambient temperatures, thus prompting studies of the catalytic effect of Ag^+ and reactions run at high temperatures. In general, the studies agree that the rate of thermal decomposition of $S_2O_8^{2-}$ to produce SO_4^- radical is the rate-determining step in the oxidation of Am(III). Because of the complex nature of this system, no single kinetic study is particularly illuminating as far as the redox kinetics of americium is concerned. No unifying rate law has been developed. References [68] and [69] are representative of a much larger body of data.

A similar situation exists regarding the many studies of the effects of radiolysis on the stability of americium in the higher oxidation states.[70–72] Since several different processes may occur simultaneously for net reduction of Am(IV), (V), or (VI) (disproportionation, reduction by H_2O_2), derivation of a mechanism is most often not particularly instructive.

A pulse radiolysis study of the dynamics of americium reaction addressed the rates of oxidation of Am(III) → Am(IV) by OH radical and of reduction of Am(III), Am(V), and Am(VI) by e_{aq}^-.[73] Americium(II) was produced by reaction with e_{aq}^- with a second-order rate constant $k = 1.55 \ (\pm 0.04) \times 10^8 \ M^{-1} \ s^{-1}$. Factors normally assumed to control such reactions (ionic radius, $E°$) are evidently not important since they are nearly identical for Am and Sm, but the rate of reduction of Sm(III) is 100 times faster. The difference is rationalized in terms of 5f orbital contribution to Am(III) hydration.

Likewise, the rates of reduction of Am(V) and Am(VI) by e_{aq}^- exhibits no discernible correlation with $E°$ when compared with the reduction of the other An(V) and An(VI) ions (all such reactions approach diffusion-controlled rates). Production of Am(IV) from the OH radical oxidation of Am(III) is characterized by a second-order rate constant of 4.1 (± 0.4) $\times 10^8 \, M^{-1} \, s^{-1}$. The rate constant for disproportionation of Am(IV) thus produced is $5(\pm 1) \times 10^6 \, M^{-1} \, s^{-1}$, indicating a trend of decreased dynamic stability of the tetravalent actinides toward disproportionation across the series.

VI. CURIUM, BERKELIUM, AND CALIFORNIUM OXIDATION–REDUCTION KINETICS

The short half-lives of the available nuclei of Cm, Bk, and Cf complicate studies of the aqueous chemistry of these elements. The most stable oxidation state of these elements in aqueous media is the trivalent state and as a result investigations of their solution chemistry have, for the most part, been concerned with equilibrium studies of complex ion formation.

Oxidation state IV of Bk, is readily available [$E°$(IV/III) = 1.54 V], and can be prepared electrolytically or with a variety of chemical oxidants.[74] Studies of the redox chemistry of Bk(IV) have been restricted to reduction by radiolytic decomposition products of water. That this is a readily available path is apparent when one notes that the most stable isotope available, ^{249}Bk, has a half-life of only 314 days.

The disappearance of Bk(IV) (monitored spectrophotometrically) obeys zero-order kinetics (dependent only on the total concentration of Bk) and is dependent on the composition of the medium (acidity and the anion). Studies by two different research groups are in remarkably good agreement.[75–77] In noncomplexing 2–3 M $HClO_4$ the rate of autoreduction of Bk(IV) is $8 \times 10^{-4} \, M \, h^{-1}$. The rate of reduction in 3 M HNO_3 is a factor of 2 faster $(1.6 \times 10^{-3} \, M \, h^{-1})$[77] while in 2–4.5 M H_2SO_4 it is at least 10 times slower $(1–7.4 \times 10^{-5} \, M \, h^{-1})$.[75,76] At lower concentrations of H_2SO_4 (0.05 M) the rate of autoreduction approaches that in $HClO_4$ ($3.8 \times 10^{-4} \, M \, h^{-1}$).[75] Detailed mechanistic schemes have been advanced but corroborative information from pulse radiolysis experiments is not available to provide supporting evidence.

Studies of the redox chemistry of Cm and Cf has been limited to pulse radiolysis investigations. The preliminary report of the reactions Cm(III) + $e_{aq}^- \rightarrow$ Cm(II) and Cm(III) + OH $\rightarrow$ Cm(IV) cited spectrophotometric evidence of these unusual oxidation states.[78] Half-lives estimated for the subsequent disappearance of Cm(II) and Cm(IV) (presumably by reaction with water) are 1.2 and 2.0×10^{-6} s, respectively. The rate constant for the reaction

$Cf(III) + e_{aq}^- \rightarrow Cf(II)$ has been estimated as $k \geq 3 \times 10^9$ M^{-1} s^{-1}.[79] The transient Cf(II) disappears via an apparent first-order process with half-life of 1×10^{-5} s. The lifetime for Am(II) in aqueous solution is reported as 7×10^{-6} s.[73] As the relative order for the half-lives of the divalent actinides in aqueous solution is Cf > Am > Cm, the stability of this state does not appear to be simply related to the electronic configurations of these elements.

VII. KINETICS OF COMPLEXATION REACTIONS

Because the actinides are hard acids, they form complexes mainly with the most electronegative elements. Bonding to fluoride- and oxygen-containing ligands are strongest while nitrogen and sulfur donors normally form strong bonds only when constrained by steric factors.

The bulk of the data available on rate of water exchange for the actinides indicates that this process is rapid and seldom affects the rate of complexation. The rate of exchange of water in the primary hydration sphere of U(IV) was reported[80] as being too fast to measure by the NMR technique (rapid relaxation resulted in no H NMR signal for these molecules). The rate of water exchange would therefore not normally be rate limiting. The rate of solvent water exchange on uranyl was reported[6] from NMR studies in d_6-acetone at $T = -70°C$. From activation parameters also reported in that study, the rate of water exchange of uranyl(VI) is calculated to be 1.1×10^6 s^{-1} at 25°C. Inasmuch as the actinides behave similarly in a given oxidation state, the rates of water exchange of the other hexavalent actinides would be expected to be of the same order of magnitude.

A study (T-jump) of the rates of formation and dissociation of uranyl(VI) complexes with sulfate, acetate, chloroacetate, and thiocyanate[81] showed that k_f/k_r ratios for the 1:1 complexes reproduce published values for the respective 1:1 stability constants (Table 3). The formation rate constants are in the order acetate > thiocyanate > sulfate > chloroacetate. With the exception of uranyl(VI) thiocyanate, these complexes are predominantly inner-sphere. The dissociation rates for the 1:1 inner-sphere complexes are ~ 4 s^{-1} while that for SCN^- is 45 s^{-1}. The difference between inner- and outer-sphere complex stability is reflected primarily in the faster rate of dissociation of the outer-sphere complex.

The T-jump relaxation method also was used to investigate[82] the rate of dimerization of uranyl monohydroxide. The reaction studied was the forward component of the equilibrium:

$$2UO_2^{2+} + 2H_2O \rightleftharpoons (UO_2)_2(OH)_2^{2+} + 2H^+$$

TABLE 3

Summary of kinetics results for actinide complexation[a]

Reactants[b]	Rate law	Parameters	Ref.	T (°C)	I (M)	Method[c]
U(VI), SCN$^-$	$R = k_1[\text{M}][\text{L}] - k_{-1}[\text{ML}]$	$k_1 = 290(\pm 30)$ $k_{-1} = 54(\pm 6)$	[81]	20	1.2	T-jump
U(VI), CH$_3$COO$^-$	1 for 1:1; 2 for 1:2; 3 for 1:3 complexes	$k_1 = 1050$ $k_{-1} = 4.4$ $k_2 = 800$ $k_{-2} = 8.4$ $k_3 = 200$ $k_{-3} = 2.1$	[81]	20	0.15	T-jump
U(VI), CH$_2$ClCOO$^-$	Same as above	$k_1 = 110$ $k_{-1} = 4.4$ $k_2 = 240$ $k_{-2} = 38$ $k_3 = 80$ $k_{-3} = 34$	[81]	20	0.15	T-jump
U(VI), SO$_4^{2-}$	Same as above	$k_1 = 180$ $k_{-1} = 3.6$ $k_2 = 300$ $k_{-2} = 43$ $k_3 = 160$ $k_{-3} = 22$	[80]	20	0.15	T-jump
U(VI) hydrolysis	$d[(\text{UO}_2)_2(\text{OH})_2]/dt = k[\text{UO}_2\text{OH}]^2$	$k = 116(\pm 6)$	[82]	25	0.5	T-jump
U(VI), H$_2$O$_2$	$d[\text{U(VI)L}]/dt = k[\text{U(VI)}][\text{H}_2\text{O}_2]$	$k = 1.31(\pm 0.16) \times 10^4$	[36]	25	$<10^{-3}$	stf, pr
U(VI), H$_2$O$_2$	$d[\text{U(VI)L}]/dt = k[\text{U(VI)}][\text{H}_2\text{O}_2]$ in Na$_2$CO$_3$	$k = 565(\pm 41)$ $\Delta G^* = +57.1(\pm 6.5)$ $\Delta H^* = +67.8(\pm 3.2)$ $\Delta S^* = +36(\pm 11)$	[83]	25	0.1	stf

206

System	Reaction / rate law	Parameters	Ref.	T	I	Method
$Np(VI), H_2O_2$	in $NaHCO_3$	$k = 254(\pm17)$ $\Delta G^* = +59.5(\pm4.1)$ $\Delta H^* = +53.5(\pm2.0)$ $\Delta S^* = -20(\pm7)$	[83]	25	0.1	stf
$Np(VI), H_2O_2$	$d[Np(VI)L]/dt = k[Np(VI)][H_2O_2]$ in Na_2CO_3	$k = 2.19(\pm0.01) \times 10^3$ $\Delta G^* = +54.4(\pm4.1)$ $\Delta H^* = +43.69(\pm2.0)$ $\Delta S^* = -36(\pm7)$	[83]	25	0.1	stf
$Pu(VI), H_2O_2$ $Am(III), dcta$	$d[Pu(VI)L]/dt = k[Pu(VI)][H_2O_2]$ $d[Am(dcta)]/dt = k[Am(Hdcta)^*]$ $k = k_f + k_f^h/[H]$	$k = 6.9(\pm0.8) \times 10^3$ $k_f = 76.4(\pm21.3)$ $E_{af}^* = +59.0(\pm5.9)$ $\Delta S_f^* = -19(\pm4)$ $k_f^h = 19.2(\pm1.9)$ $E_{af}^h = +65.7(\pm4.6)$ $\Delta S_f^h{}^* = -8.8(\pm3.3)$	[84] [85]	25 25	0.05 0.05	stf stf
$Am(dcta), Cu^{2+}$	$-d[Am(dcta)]/dt = k[Am(dcta)]$ $k = k_d + k_d^h[H]$	$k_d = 8(\pm2) \times 10^{-6}$ $k_d^h = 3.04(\pm0.03)$	[85]	25	0.05	spec, m-ex
$Eu(edta), Am^{3+}$	$R = (k_a[EuL][Am]/[Eu] - k_c[AmL])[H] +$ $k_b[EuL][Am] - k_d[Eu][AmL]$	$k_a = 187(\pm19)$ $k_b = 4.54(\pm0.65) \times 10^{-2}$ $k_c = 139(\pm13)$ $k_d = 3.19(\pm0.53) \times 10^{-2}$	[87]	25	0.1	i-ex, m-ex
$Eu(edta), Cm^{3+}$	Same as Am(III)	$k_a = 178(\pm5)$ $k_b = 4.623(\pm0.003) \times 10^{-2}$ $k_c = 119(\pm4)$ $k_d = 2.72(\pm0.11)$	[88]	25	0.1	i-ex, m-ex
$Eu(edta), Bk^{3+}$	Same as Am(III)	$k_a = 240(\pm24)$ $k_b = 8.81(\pm0.50) \times 10^{-2}$ $k_c = 56.6(\pm2.3)$ $k_d = 1.667(\pm0.18) \times 10^{-2}$	[88]	25	0.1	i-ex, m-ex

(continued)

TABLE 3 (continued)

Reactants[b]	Rate law	Parameters	Ref.	T (°C)	I (M)	Method[c]
Eu(edta), Cf^{3+}	Same as Am(III)	$k_a = 273(\pm 30)$ $k_b = 7.79(\pm 1.26) \times 10^{-2}$ $k_c = 25.4(\pm 0.7)$ $k_d = 0.774(\pm 0.02) \times 10^{-2}$	[88]	25	0.1	i-ex, m-ex
An(III)pdta	$-d[AnL]/dt = k[AnL]$ $k = k_1 + k_2[H]$ An = Am	$k_1 = 1.37(\pm 0.68) \times 10^{-2}$ $k_2 = 4.74(\pm 0.48) \times 10^{4}$	[89]	25	0.1	s-ex
	An = Cm	$k_1 = 3.5(\pm 1.3) \times 10^{-3}$ $k_2 = 3.22(\pm 0.42) \times 10^{4}$	[89]	25	0.1	s-ex
	An = Bk	$k_1 = 2.2(\pm 8.0) \times 10^{-4}$ $k_2 = 0.98(\pm 0.10) \times 10^{4}$	[89]	25	0.1	s-ex
	An = Cf	$k_1 = 9.4(\pm 5.1) \times 10^{-4}$ $k_2 = 0.43(\pm 0.01) \times 10^{4}$	[89]	25	0.1	s-ex
An(III)hedta, H^+	$-d[AnL]/dt = k[AnL][H]$ An = Am An = Cm An = Bk An = Cf	$k = 200(\pm 40)$ $k = 170(\pm 20)$ $k = 110(\pm 20)$ $k = 120(\pm 40)$	[89]	25	0.1	s-ex
An(IV)dcta, arsenazo III	$-d[AnL]/dt = k[H][AnL]$ An = Th	$k = 0.251(\pm 0.026)$ $\Delta G^* = +76.4(\pm 2.0)$ $\Delta H^* = +56.4(\pm 1.0)$ $\Delta S^* = -67.0(\pm 3.5)$	[90]	25	0.1	spec, l-ex
	An = U	$k = 0.266(\pm 0.050)$ $\Delta G^* = +76.2(\pm 5.6)$ $\Delta H^* = +50.6(\pm 2.8)$ $\Delta S^* = -85.7(\pm 9.4)$	[90]	25	0.1	spec, l-ex

		Ref	T (°C)	I	Method
An = Np	$k = 4.77(\pm 0.97) \times 10^{-2}$ $\Delta G^* = 80.3(\pm 5.3)$ $\Delta H^* = 47.8(\pm 2.6)$ $\Delta S^* = -109(\pm 9)$	[90]	25	0.1	spec, l-ex
An = Pu	$k = 1.02(\pm 0.15) \times 10^{-2}$ $\Delta G^* = +84.0(\pm 14.2)$ $\Delta H^* = +38.7(\pm 7.0)$ $\Delta S^* = -152(\pm 24)$	[90]	25	0.1	spec, l-ex
An(IV)dtpa, arsenazo III $-d[\mathrm{AnL}]/dt = k[\mathrm{AnL}][\mathrm{H}]^3$ An = Th	$k = 1640(\pm 400)$ $\Delta H^* = +15.5(\pm 12.1)$ $\Delta S^* = -131(\pm 17)$	[90]	25	0.1	spec, l-ex
An = U	$k = 2.3(\pm 0.6)$ $\Delta H^* = +33.9(\pm 5.0)$ $\Delta S^* = -124(\pm 15)$	[90]	25	≤0.77	spec, l-ex
An = Np	$k = 0.28(\pm 0.06)$ $\Delta H^* = +42.7(\pm 4.6)$ $\Delta S^* = -112(\pm 15)$	[90]	25	≤1.2	spec, l-ex
An = Pu	$k = 0.05(\pm 0.01)$ $\Delta H^* = +65.3(\pm 2.5)$ $\Delta S^* = -49.4(\pm 12.6)$	[90]	25	≤2.00	spec, l-ex
An^{4+}, arsenazo III $d[\mathrm{AnL}]/dt = k[\mathrm{An}][\mathrm{L}]$ An = Th	$k = 2.22(\pm 0.34) \times 10^6$ $\Delta H^* = +32.3(\pm 2.8)$ $\Delta S^* = -15.5(\pm 9.6)$	[91a]	25	2.0	stf
An = Np	$k = 3.30(\pm 1.6) \times 10^5$ $\Delta H^* = +32.6(\pm 2.7)$ $\Delta S^* = -29.0(\pm 14.1)$	[91a]	25	2.0	stf
An = U	$k = 3.76(\pm 0.46) \times 10^5$	[91a]	25	2.0	stf

(continued)

TABLE 3 (continued)

Reactants[b]	Rate law	Parameters	Ref.	T (°C)	I (M)	Method[c]
U(VI), arsenazo III	Rate $= k_f[\text{U(VI)}][\text{L}]$ $+ k_r[\text{U(VI)}]\text{L}$	$k_f = 2.14(\pm 5.2) \times 10^5$ $\Delta H^* = +30.1(\pm 5.2)$ $\Delta S^* = -42.2(\pm 17.8)$ $k_r = 8.86(\pm 0.42)$ $\Delta H^* = +48.2(\pm 5.1)$ $\Delta S^* = -65.3(\pm 4.2)$	[91b]	25	2.0	stf
U(VI), arsenazo III	$d[\text{U(VI)L}]/dt = k_f[\text{U(VI)}][\text{L}]$	$k_f = 2.1 \times 10^5$ $\Delta G^* = +42.7$ $\Delta H^* = +36.6$ $\Delta S^* = -20.0$	[92]	25	0.5	stf
U(VI), pyridylazoresorcinol,(L)	$d[\text{U(VI)L}]/dt = k[\text{U(VI)}][\text{L}]$	$k = 3.26 \times 10^4$ $\Delta H^* = +34.2(\pm 0.5)$ $\Delta S^* = -43.5(\pm 1.7)$	[93]	25	0.1	stf

[a] Units for rate constants determined by reaction order: for zero-order reactions (M s^{-1}), for first-order reactions (s^{-1}), for second-order reactions (M^{-1} s^{-1}), for third-order reactions (M^{-2} s^{-1}). Units for activation parameters: ΔH^* (kJ mol^{-1}) ΔS^* (J mol^{-1} K^{-1}).

[b] Short-hand notation used for chelating ligands: edta, ethylenediaminetetraacetic acid; dcta, trans-1,2-diaminocyclohexane tetraacetic acid; pdta, 1,3-propylenediaminetetraacetic acid; hedta, N-hydroxyethylethylenediaminetriacetic acid; dtpa, diethylenetriaminepentaacetic acid; arsenazo III, 3,6-bis(2-arsonophenylazo-4,5-dihydroxy-2,7-napthalenedisulfonic acid.

[c] Experimental methods and reaction type notation: T-jump, temperature jump; stf, stopped-flow spectrophotometry; pr, pulsed radiolysis; i-ex, ion exchange; m-ex, metal ion exchange; l-ex, ligand exchange; s-ex, solvent extraction; spec, conventional spectrophotometry.

Conditions were selected which favored the reaction written above over any of the many other potential hydrolysis reactions.

A series of reactions generically related to the hydrolysis of uranyl is the formation of actinyl(VI) peroxide complexes. The first report of the rate of complexation of U(VI) by hydrogen peroxide was conducted in pH 5.2 solution in which the rate determined for the formation of a $1:1$ U(VI)–H_2O_2 complex was the same when determined by either pulse radiolysis or stopped-flow techniques.[36] A similar U(VI)–peroxide complex was also observed in both dilute bicarbonate and carbonate solutions with the complex formed at a 20 times slower rate in carbonate and 40 times slower in bicarbonate than in perchlorate solutions.[83]

The rates of formation of transient peroxide complexes of Np(VI) in 0.05 M sodium carbonate[83] and Pu(VI) (in 0.05 M sodium bicarbonate[84]) have also been determined. Subsequent reduction of Np(VI) and Pu(VI) by H_2O_2 in carbonate media is complex. In general, the rate of complex formation increases with increasing atomic number. Speculation is that the complex is formed in one of three configurations, with peroxide replacing a coordinated carbonate ligand, coordinating equatorially without removal of carbonate, or interacting through a terminal oxo attached to the metal center.

Perhaps the most thoroughly studied complexation reactions of the actinides are those with chelating ligands. Both formation and dissociation reactions typically exhibit acid-dependent and -independent steps attributable to protonation equilibria. In contrast to simple ion combination reactions, the rate of complex formation in these systems can usually be related to some intramolecular process.

The rates of formation and dissociation of the Am(III) complex with dcta (*trans*-1,2-diaminocyclohexanetetraacetic acid) were determined using stopped-flow spectrophotometry to study the formation and conventional spectrophotometry for the decomposition reaction.[85] The experimental results are consistent with the interpretation that a precursor complex between Am(III) and the dcta ligand is formed. The rate-determining step in the reaction is postulated to be the formation of a bond between Am(III) and an imino nitrogen of dcta.

The dissociation of the Am(III)–dcta complex was studied by the metal ion exchange technique using Cu^{2+} as was reported in an analogous study of the lanthanide dcta chelates.[86] No dependence on the copper concentration was observed, implying that any reaction rates measured were pertinent to either acid-induced or spontaneous dissociation of the complex.

The rates of metal ion exchange for the trivalent actinides (Am, Cm, Bk, Cf) with Eu(edta) (edta = ethylenediaminetetraacetic acid) have been determined by an ion exchange technique.[87,88] The exchange reaction rate law indicates a dependence on acidity and in some cases dependence on free acetate concentration. The proposed mechanism includes steps analogous to those reported in the study of Am(dcta) (acid-catalyzed and non-acid-catalyzed dissociation reactions), and parallel steps involving a binuclear intermediate [An(edta)Eu] in which edta bridges the two metal ions. A mixed ligand edta–acetate complex is favored by the authors to explain the slight acetate dependence observed in the rate law.

The rate of dissociation of trivalent actinide (Am, Cm, Bk, Cf) complexes with edta analogs hedta(N-hydroxyethylethylenediaminetriacetic acid) and pdta (propylenediaminetetracetic acid) have been measured.[89] As is the case for the Am–Cf edta complexes, the rate of the acid-catalyzed dissociation decreases with increasing atomic number. A similar correlation was previously noted for a series of lanthanide–dcta complexes.[86] The observed correlations are consistent with a simple electrostatic model for the interactions of both lanthanides and actinides.

The results of a series of studies on the rates of dissociation of Th(IV), U(IV), Np(IV), and Pu(IV) dcta and dtpa (diethylenetriaminepentaacetic acid) complexes have been summarized.[90] The rates of dissociation of the dtpa complexes are greater than for the corresponding dcta complexes. There is no apparent correlation between rates of dissociation of these complexes and the radii of the tetravalent actinide ions as was noted in analogous studies with the trivalent transplutonium actinide ions.

The rate of formation of complexes between arsenazo(III) and Th(IV), Np(IV), and U(VI) has been interpreted[91] to involve the formation of a precursor complex (outer-sphere ion pair according to the Eigen mechanism). The rate-determining step is relaxation of the precursor complex to a more stable species. For the tetravalent actinides, the rate of formation of the complexes decreases with increasing atomic number with the rate decrease reflected in a less favorable activation entropy. The increased activation entropy with decreasing ionic radius is consistent with the argument of a precursor complex primarily influenced by electrostatic interaction. Two reports[91b,92] on the rate of formation of the U(VI)–arsenazo(III) complex differ in the range of pH investigated. The reported rate constants and activation parameters are in substantial agreement.

The rate of formation of a uranyl(VI) complex with a related ligand, 4-(2-pyridylazo)resorcinol (PAR),[93] was investigated by stopped-flow methods under second-order conditions and the data treated as a reversible second-order reaction. The reaction was found to be independent of ionic strength, which the authors interpret as an indication that the principal reacting species of the ligand is neutral (H_2L). The empirical rate law indicates a strong (not quite first-power) inverse dependence on acidity. Two very similar mechanisms were considered but neither was found to be acceptable, because they failed to reproduce the first ionization constant of the PAR ligand (the value derived from the kinetic expression is about an order of magnitude lower than the value determined by other, more conventional techniques). The authors suggest that the apparent complexity of the mechanism may be due to a preequilibrium between either the protonated or molecular form of the PAR ligand, or to the presence in the solution of a complexed species other than UO_2HR^+.

Acknowledgment

Work performed under the auspices of the Office of Basic Energy Sciences, Division of Chemical Sciences, U.S. Department of Energy, under contract number W-31-109-ENG-38, and the U.S. Geological Survey. The authors wish to thank Dr. Mary Woods for helpful discussions. We gratefully acknowledge the secretarial efforts expended by Janet Bergman and Laura Bowers.

References

[1] Choppin, G. R. *Radiochim. Acta* **1983**, *132*, 43.
[2] Keller, C. "The Chemistry of the Transuranium Elements"; Verlag Chemie: Weinheim, 1971.
[3] Fuger, J.; Oetting, F. L., "The Chemical Thermodynamics of Actinide Elements and Compounds, Part 2, The Actinide Aqueous Ions"; IAEA: Vienna, 1976.
[4] Newton, T. W. ERDA Critical Review Series, "The Kinetics of the Oxidation-Reduction Reactions of Uranium, Neptunium, Plutonium, and Americium in Aqueous Solutions"; (TID-26506); NTIX: Springfield, VA, 1975.
[5] Myasoedov, B. F. "Actinides Perspective, Proceedings of Actinides Conference"; Edelstein, N., Ed; Pergamon: Oxford, 1982, p. 509.
[6] Tomiyasu, H.; Fukutomi, H. *Bull. Res. Lab. Nucl. React.* **1982**, *7*, 57.
[7] Lincoln, S. F. *Pure Appl. Chem.* **1979**, *51*, 2059.
[8] Marcus, R. A. *Annu. Rev. Phys. Chem.* **1964**, *15*, 155.
[9] Lavallee, C.; Lavallee, D. K.; Deutsch, E. A. *Inorg. Chem.* **1978**, *17*, 2217.
[10] Adegite, A.; Iyun, J. F.; Ojo, J. F. *J. Chem. Soc., Dalton Trans.* **1977**, 155.
[11] Loar, M. K.; Sens, M. A.; Loar, G. W.; Gould, E. S. *Inorg. Chem.* **1978**, *17*, 330.
[12] Rajasekar, N.; Srinivasn, V. S.; Singh, A. N.; Gould, E. S. *Inorg. Chem.* **1982**, *21*, 3245.
[13] Srinivasn, V. S.; Singh, A. N.; Wieghardt, K.; Rajasekar, N.; Gould, E. S. *Inorg. Chem.* **1982**, *21*, 2531.
[14] Ekstrom, A.; McLaren, A. B.; Smythe, L. E. *Inorg. Chem.* **1975**, *14*, 2899.
[15] Ekstrom, A.; McLaren, A. B.; Smythe, L. E. *Inorg. Chem.* **1977**, *16*, 1032.
[16] Ekstrom, A.; McLaren, A. B.; Smythe, L. C. *Inorg. Chem.* **1975**, *14*, 1035.
[17] Ekstrom, A.; McLaren, A. B.; Smythe, L. E. *Inorg. Chem.* **1976**, *15*, 2853.
[18] Kojima, T.; Komatu, H.; Fukutomi, H. *Bull. Res. Lab. Nucl. React.* **1983**, *8*, 11.
[19] Adamcikova, L.; Treindl, L. *Chem. Zvesti* **1976**, *30*, 593.
[20] Adamcikova, L.; Treindl, L. *Coll. Czech. Chem. Commun.* **1978**, *43*, 1844 (*C. A.* 89: 136436s).
[21] Adamcikova, L.; Treindl, L. *Chem. Zvesti* **1980**, *34*, 145 (*C.A. 93*: 81217w).
[22] Adamcikova, L.; Lucivjansky, P.; Treindl, L. *React. Kinet. Catal. Lett.* **1980**, *15*, 451 (*C.A. 95*: 41830a).

[23] Adamcikova, L.; Treindl, L. *Coll. Czech. Chem. Commun.* **1980**, *45*, 26.
[24] Adamcikova, L.; Treindl, L. *Coll. Czech. Chem. Commun.* **1979**, *44*, 401.
[25] Adamcikova, L.; Treindl, L. *Coll. Czech. Chem. Commun.* **1974**, *39*, 1264.
[26] Gordon, G.; Taube, H. *J. Inorg. Nucl. Chem.* **1961**, *16*, 272.
[27] Silverman, R. A.; Gordon, G. *Inorg. Chem.* **1976**, *15*, 35.
[28] Adegite, A.; Ford-Smith, M. H. *J. Chem. Soc., Dalton Trans.* **1973**, 138.
[29] Adegites A.; Ford-Smith, M. H. *J. Chem. Soc., Dalton Trans.* **1973**, 134.
[30] Ekstrom, A. *Inorg. Chem.* **1977**, *16*, 845.
[31] Michaille, P.; Kikindai, T. *J. Inorg. Nucl. Chem.* **1977**, *39*, 859.
[32] Michaille, P.; Kikindai, T. *J. Inorg. Nucl. Chem.* **1977**, 493.
[33] Hassan, R. M.; Kojima, T.; Fukutomi, H. *Bull. Res. Lab. Nucl. React.* **1980**, *5*, 41.
[34] Hassan, R. M.; Kojima, T.; Fukutomi, H. *Bull. Res. Lab. Nucl. React.* **1981**, *6*, 35 (*CA. 95*: 50197b).
[35] Ekstrom, A. *Inorg. Chem.* **1974**, *13*, 2237.
[36] Sullivan, J. C.; Gordon, S.; Cohen, D.; Mulac, W.; Schmidt, K. H. *J. Phys. Chem.* **1976**, *80*, 1684.
[37] Mulac, W. A.; Gordon, S.; Schmidt, K. H.; Wester, D. W.; Sullivan, J. C. *Inorg. Chem.* **1984**, *23*, 1639.
[38] Nash, K.; Mulac, W.; Noon, M.; Fried, S.; Sullivan, J. C. *J. Inorg. Nucl. Chem.* **1981**, *43*, 897.
[39] Meisel, D.; Mulac, W.; Sullivan, J. C. *Inorg. Chem.* **1981**, *20*, 4247.
[40] Hinderberger, D. J., Jr.; Thompson, R. C. *Inorg. Chem.* **1975**, *14*, 784.
[41] Lavallee, C.; Lavallee, D. K. *Inorg. Chem.* **1977**, *16*, 2601.
[42] Chistyakov, V. M.; Ermakov, V. A.; Moknousov, A. D.; Rykov, A. G. *Sov. Radiochem.* **1975**, *17*, 401.
[43] Koltunov, V. S.; Kulikov, I. A.; Marchenka, V. I.; Milovanova, A. S. *Sov. Radiochem.* **1980**, *22*, 833.
[44] Koltunov, V. S.; Marchenka, V. I.; Shapovalov, M. P. *Sov. Radiochem.* **1977**, *19*, 56.
[45] Bond, W. D.; Jao, Y.; Peterson, J. R. *J. Inorg. Nucl. Chem.* **1977**, *39*, 1395.
[46] Schmidt, K. H.; Gordon, R. C.; Thompson, R. C.; Sullivan, J. C. *J. Inorg. Nucl. Chem.* **1980**, *42*, 611.
[47] Schmidt, K. H.; Gordon, S.; Thompson, M.; Sullivan, J. C.; Mulac, W. A. *Radiat. Phys. Chem.* **1983**, *21*, 321.
[48] Pelizzetti, E.; Woods, M.; Sullivan, J. C. *Inorg. Chem.* **1980**, *19*, 524.
[49] Pelizzetti, E.; Woods, M.; Sullivan, J. C. *Inorg. Chem.* **1981**, *20*, 3993.
[50] J. Weschler, C. J.; Sullivan, J. C.; Deutsch, E. *Inorg. Chem.* **1974**, *13*, 2360.
[51] Woods, M.; Karbwang, J.; Sullivan, J. C.; Deutsch, E. *Inorg. Chem.* **1976**, *15*, 1678.
[52] Rao, L. F.; Choppin, G. R. *Inorg. Chem.* **1984**, *23*, 2351.
[53] Woods, M.; Hoenich, C. A.; Sullivan, J. C. *J. Inorg. Nucl. Chem.* **1978**, *40*, 1907.
[54] Deutsch, E.; Sullivan, J. C.; Watkins, K. O. *Inorg. Chem.* **1975**, *14*, 550.
[55] Watkins, K. O.; Sullivan, J. C.; Deutsch, E. *Inorg. Chem.* **1974**, *13*, 1712.
[56] Cooper, J. N.; Woods, M.; Sullivan, J. C.; Deutsch, E. *Inorg. Chem.* **1976**, *15*, 2862.
[57] Copper, J. N.; Woods, M.; Sjoblom, R.; Sullivan, J. C. *J. Inorg. Nucl. Chem.* **1978**, *40*, 659.
[58] Koltunov, V. S.; Frolov, K. M.; Marchenko, V. I.; Tikhonov, M. F.; Shapovalov, M. P. *Sov. Radiochem.* **1981**, *23*, 103.
[59] Koltunov, V. S.; Frolov, K. M.; Marchenko, V. I.; Tikhonov, M. F.; Shopovalov, M. P. *Sov. Radiochem.* **1981**, *23*, 111.
[60] Nikitina, G. P.; Shumkov, V. G.; Egorova, V. P.; Ivanov, Yu. E. *Sov. Radiochem.* **1975**, *17*, 573.
[61] Koltunov, V. S.; Kulikov, I. A.; Keruranova, N. V.; Nikishova, L. K. *Sov. Radiochem.* **1981**, *23*, 462.
[62] Koltunov, V. S.; Ryabova, A. A. *Sov. Radiochem.* **1980**, *22*, 635.

[63] Woods, M.; Montag, T.; Sullivan, J. C. *J. Inorg. Nucl. Chem.* **1976,** *38*, 2059.

[64] Pelizzetti, E.; Woods, M.; Sullivan, J. C. *Inorg.Chim. Acta* **1983,** *76*, 163.

[65] Cooper, J.; Reents, W. D., Jr.; Woods, M.; Sjoblom, R.; Sullivan, J. C. *Inorg. Chem.* **1977,** *16*, 1030.

[66] Cooper, J. N.; Woods, M.; Sjoblom, R.; Sullivan, J. C. *Inorg. Chem.* **1977,** *16*, 2267.

[67] Woods, M.; Sullivan, J. C. *Inorg. Chem.* **1979,** *18*, 3317.

[68] Musikas, C.; Germain, M.; Bathellier, A. *Am. Chem. Soc. Symp. Ser.* **1980,** *117*, 157.

[69] Chistyakov, V. M.; Ermakov, V. A.; Mokruosov, A. D.; Rykov, A. G. *Sov. Radiochem.* **1974,** *16*, 793.

[70] a) Lebedev, I. A.; Frenkel, V. Ya.; Myasoedov, B. F. *Sov. Radiochem.* **1977,** *19*, 467.

b) Lebedev, I. A.; Milyukova, M. S.; Frenkel, V. Ya.; Litvina, M. N.; Myasoedov, B. F.; Mikhailov, V. M. *Sov. Radiochem.* **1976,** *18*, 561.

c) Myasoedov, B. F.; Milyukova, M. S.; Levedev, I. A.; Litvina, M. N.; Frenkel, V. Ya. *J. Inorg. Nucl. Chem.* **1975,** *37*, 1475.

[71] a) Frenkel, V. Ya.; Lebedev, I. A.; Myasoedov, B. F. *Sov. Radiochem.* **1980,** *22*, 53.

b) Kulikov, I. A.; Frenkel, V. Ya.; Vladimirova, M. V.; Lebedev, I. A.; Ryabova, A. A.; Myasoedov, B. F. *Sov. Radiochem.* **1979,** *21*, 719.

[72] Ermakov, V. A.; Frolov, A. A. *Sov. Radiochem.* **1979,** *21*, 58.

[73] Gordon, S.; Mulac, W. A.; Schmidt, K. H.; Sjoblom, R.; Sullivan, J. C. *Inorg. Chem.* **1978,** *17*, 294.

[74] Hobart, D. E.; Peterson, J. R. *Inorg. Radiochem.* **1984,** *28*.

[75] Gutmacher, R. G.; Bodie, D. D., Jr.; Longheed, R. W.; Hubt, E. K. *J. Inorg. Nucl. Chem.* **1973,** *35*, 979.

[76] Kulyaka, Yu. M.; Frenkel, V. Ya.; Lebedev, I. A.; Trofimov, T. I.; Myasoedov, B. F.; Mogilevskii, A. N. *Radiochim. Acta* **1981,** *28*, 119.

[77] Erin, E. A.; Vityatnev, V. M.; Kopytov; V. V.; Vasil'ev, V. Ta. *Radiokhimiya* **1982,** *24*, 179.

[78] Sullivan, J. C.; Gordon, S.; Mulac, W. A.; Schmidt, K. H.; Cohen, D.; Sjoblom, R. *Inorg. Nucl. Chem. Lett.* **1976,** *12*, 599.

[79] Sullivan, J. C.; Morss, L. R.; Schmidt, K. H.; Mulac, W. A.; Gordon, S. *Inorg. Chem.* **1983,** *22*, 2338.

[80] Kiener, C.; Folcher, G.; Rigny, P.; Virlet, J. *Can. J. Chem.* **1976,** *54*, 303.

[81] Hurwitz, P.; Kustin, J. *J. Phys. Chem.* **1967,** *71*, 324.

[82] Wittaker, M. P.; Eyring, E. M.; Dibble, E. *J. Phys. Chem.* **1965,** *69*, 2319.

[83] Thompson, M. E.; Nash, K. L.; Sullivan, J. C. *Isr. J. Chem.* **1985,** *25*, 155.

[84] Nash, K.; Noon, M. E.; Fried, S.; Sullivan, J. C. *Inorg. Nucl. Chem. Lett.* **1980,** *16*, 33.

[85] Sullivan, J. C.; Nash, K. L.; Choppin, G. R. *Inorg. Chem.* **1978,** *17*, 3374.

[86] Nyssen, G. A.; Margerum, D. W. *Inorg. Chem.* **1970,** *9*, 1814.

[87] Choppin, G. R.; Williams, K. W. *J. Inorg. Nucl. Chem.* **1973,** *35*, 4255.

[88] Williams, K. W.; Choppin, G. R. *J. Inorg. Nucl. Chem.* **1974,** *36*, 1849.

[89] Muscatello, A. C. Ph.D. dissertation, **1979,** Florida State University, Tallahassee.

[90] Nikitina, S. A.; Dem'yanova, T. A.; Stepanov, A. V.; Lipouskii, A. A.; Nemtsova, M. A. *Radioanal. Chem.* **1979,** *51*, 393.

[91] a) Pippin, C. G.; Sullivan, J. C.; Wester, D. W. *Radiochim. Acta* **1984,** *36*, 99.

b) Pippin, C. G.; Sullivan, J. C.; Wester, D. W. "Kinetics of the U(VI)-Arsenazo III Reaction"; to be published.

[92] Ishii, H.; Odashima, T.; Mogi, H. *Nippon Kagaku Kaishi* **1983,** 1442.

[93] Ekstrom, A.; Johnson, D. A. *J. Inorg. Nucl. Chem.* **1974,** *36*, 2549.

Solvation, Solvent Exchange, and Ligand Substitution Reactions of the Trivalent Lanthanide Ions

Stephen F. Lincoln

Department of Physical and Inorganic Chemistry
The University of Adelaide
Adelaide, South Australia, Australia

I. INTRODUCTION

Two decades or so ago it must have seemed that the 14 trivalent f-block ions $(Ce^{3+}, Pr^{3+}, Nd^{3+}, Pm^{3+}, Sm^{3+}, Eu^{3+}, Gd^{3+}, Tb^{3+}, Dy^{3+}, Ho^{3+}, Er^{3+}, Tm^{3+}, Yb^{3+}, Lu^{3+})$ plus the prototype La^{3+} ion, herein collectively referred to as trivalent lanthanides and denoted Ln^{3+}, offered a unique opportunity to study a smooth variation in ligand substitution rates on a wide range of metal ions untrammeled by the strong crystal field and directional bonding effects characterizing the first-row transition metal ions. After all, the relatively large size of Ln^{3+} and the $4f^n5s^25p^6$ electronic configuration ensured that the bonding and stereochemistry of Ln^{3+} complexes were largely determined by the variation of the predominantly ion–dipole bonding interaction between Ln^{3+} and the ligands and the interactions between coordinated ligands as the ionic radius $r(M^{3+})$ decreased monotonically from La^{3+} to Lu^{3+} with the lanthanide contraction. All that was needed was equipment capable of measuring the fast ligand substitution rates of Ln^{3+}. Not surprisingly, the first systematic kinetic study of ligand substitution on Ln^{3+} emanated from Göttingen in 1965 with Geier's publication of his study of the murexide monoanion substitution on Ln^{3+}, Y^{3+}, and Sc^{3+} in aqueous solution.[1] Instead of the smooth decrease in ligand substitution rate which might have been expected with the monotonic decrease in $r(M^{3+})$ from La^{3+} to Lu^{3+} Geier found that the substitution rate decreased precipitately from Sm^{3+} to Er^{3+} (as discussed in more detail in Section III,E). This observation stimulated considerable interest in Ln^{3+} ligand substitution processes in aqueous and nonaqueous solution; the accumulated studies of which, together with studies of the equilibrium, structural, bonding, and spectroscopic aspects of Ln^{3+} chemistry[2–6] and the employment of Ln^{3+} as shift reagents[7–10] and as spectroscopic probes in biological systems,[11–13] constitute a substantial part of present-day interest in Ln^{3+} chemistry.

This review[1] is predominantly concerned with solvation, solvent exchange, and ligand substitution processes on Ln^{3+} and also Y^{3+} and Sc^{3+}, which

---

[1] Abbreviations: $DCTA^{4-}$ = *trans*-1,2-diaminocyclohexane-N,N,N',N'-tetraacetate; DEA = N,N-diethylacetamide; DEF = N,N-diethylformamide; DMA = N,N-dimethylacetamide; DMF = N,N-dimethylformamide; DMMP = dimethyl methylphosphonate; DMSO = dimethyl sulfoxide; $DOTA^{4-}$ = 1,4,7,10-tetraazacyclododecane-N,N',N'',N'''-tetraacetate $DPTA^{5-}$ = diethylenetriamine-N,N,N',N',N''-pentaacetate; $EDTA^{4-}$ = 1,2-diaminoethane-N,N,N',N'-tetraacetate; $HEDTA^{3-}$ = 1,2-diaminoethane-N(2-hydroxyethyl)-N,N',N'-triacetate; $HIMDA^{2-}$ = hydroxyethyliminodiacetate; IDA^{2-} = iminodiacetate (the imidodiacetate group of $EDTA^{4-}$ is abbreviated as IDA); $MIDA^{2-}$ = methylaminodiacetate; NMA = N-methylacetamide; NMF = N-methylformamide; NTA^{3-} = nitrilotriacetate; OXS^{2-} = 8-hydroxyquinoline-5-sulfonate; $PDTA^{4-}$ = 1,3-diaminopropane-N,N,N',N'-tetraacetate; TMM^- = tetramethyl murexide; TMP = trimethyl phosphate; TMU = N,N,N',N'-

TABLE 1

Effective ionic radii of Ln^{3+}, Y^{3+}, and Sc^{3+} [a]

M^{3+}	$r(M^{3+})$ (pm)			M^{3+}	$r(M^{3+})$ (pm)		
	CN = 9	CN = 8	CN = 6		CN = 9	CN = 8	CN = 6
La^{3+}	121.6	116.0	103.2	Dy^{3+}	108.3	102.7	91.2
Ce^{3+}	119.6	143.0	101	Ho^{3+}	107.2	101.5	90.1
Pr^{3+}	117.9	112.6	99	Er^{3+}	106.2	100.4	89.0
Nd^{3+}	116.3	110.9	98.3	Tm^{3+}	105.2	99.4	88.0
Pm^{3+}	114.4	109.3	97	Yb^{3+}	104.2	98.5	86.8
Sm^{3+}	113.2	107.9	95.8	Lu^{3+}	103.2	97.7	86.1
Eu^{3+}	112.0	106.6	94.7	Y^{3+}	107.5	101.9	90.0
Gd^{3+}	110.7	105.3	93.8	Sc^{3+}	—	87.0	74.5
Tb^{3+}	109.5	104.0	92.3				

[a] From Shannon.[14] $r(M^{3+})$ values are based on $r(^{VI}O^{2-}) = 140$ pm for different coordination numbers (CN). These radii are smaller by a constant 14 pm than the crystal radii based on $r(^{VI}O^{2-}) = 126$ pm and $r(^{VI}F^{-}) = 119$ pm.

because of their trivalent state and lack of significant directional bonding effects are often grouped with Ln^{3+} as the pre- or pseudolanthanides. It is a personal view of this area of chemistry and does not claim to provide an exhaustive review of the literature; nevertheless, it is hoped that some of the many threads of current research are drawn together to present a reasonably comprehensive coverage of this area of Ln^{3+} chemistry to readers. It will be found that frequently the interpretation of variations in ligand substitution rates involves variations of $r(M^{3+})$. It is therefore appropriate to note the substantial change in $r(M^{3+})$ which occurs with change in coordination number as shown in Table 1 for the six-, eight-, and nine-coordinate states. These $r(M^{3+})$ are taken from Shannon's[14] compilation of effective ionic radii and are used exclusively in this review.

The precursor of metal complexes in aqueous solution is the hydrated metal ion, and in consequence a knowledge of the number and lability of water molecules in the first coordination sphere of this species is fundamental to an understanding of ligand substitution processes quite apart from its own intrinsic interest. Accordingly, and noting R. J. P. Williams' statement in a

Footnote 1 *(Continued)*
tetramethylurea; DPM = dipivalomethanate; FOD = 6,6,7,7,8,8,8-hepta-fluoro-2,2-dimethyl-3,5-octanedionate; LSR = lanthanide shift reagent; FT IR = Fourier transform infrared; NMR = nuclear magnetic resonance; RDF = radial distribution function. *N.B.* Ligand protonation is indicated by a proton in brackets in front of the ligand abbreviation, e.g., (H)EDTA^{3-}.

recent review of lanthanides in biochemistry that "the hydrates prove to be most intransigent complexes,"[13] discussion now turns to the hydrated trivalent lanthanide ions.

II. THE HYDRATED TRIVALENT LANTHANIDE ION

A. The hydrated trivalent lanthanide ion in the solid state

Early indications that the trivalent lanthanide ions could exist as nona-hydrates came in 1937 and 1939 with publications of the structures of $[Nd(OH_2)_9](EtOSO_3)_3$[15] and $[Nd(OH_2)_9](BrO_3)_3$[16] which were found to possess tricapped trigonal prismatic geometries with Nd–O distances of approximately 250 pm. Subsequently, a number of X-ray powder studies and molecular structure determinations indicated that all of the trivalent lanthanides[17–22] and yttrium(III)[17,21] can exist as the tricapped trigonal prismatic nonahydrates in the presence of trifluorosulfonate, ethylsulfate, and bromate anions. The metal to capping water and to prismatic water distances, M–O (capping) and M–O (prismatic), respectively, from these recent studies are collected in Table 2. It is seen that there is a tendency for both distances to decrease, as expected qualitatively on the basis of the lanthanide contraction; however, the decrease in the M–O (capping) distances is less than anticipated, as the van der Waals repulsions between all nine water molecules become particularly important for Ln^{3+} smaller than Gd^{3+}.[18] This is reflected in the increase in the ratio M–O (capping)/M–O (prismatic) as the radius of Ln^{3+} decreases (Table 1). In the extreme case, Lu–O (capping) exceeds Lu–O (prismatic) by 9%, and it has been commented[17] that the considerable increase observed in the thermal parameters characterizing the capping oxygens in these structures as Ln^{3+} decreases in size suggests that Ln^{3+} may tend toward six coordination with the progress of the lanthanide contraction.

It is pertinent to note, however, that the nature of the anion does have some effect on the cation stereochemistry. In the trifluorosulfonates and the ethylsulfates the angle of the six M–O (prismatic) bonds to the threefold axis is close to $45°$, whereas for the bromates this angle is close to $47°$. There are also differences in the orientation of the HOH planes of the coordinated waters in the presence of the different anions. These differences are small, however, by comparison to the change of coordination exhibited by La^{3+}, Tb^{3+}, and Er^{3+} in the presence of perchlorate, where the regular octahedral $[Ln(OH_2)_6]^{3+}$ is found[23] (Table 2). This major change in the coordination number of Ln^{3+} in the solid state must be substantially ascribed to packing forces but,

TABLE 2

First-coordination-sphere metal to oxygen distances determined for tricapped trigonal prismatic $[Ln(OH_2)_9]X_3$ and octahedral $[Ln(OH_2)_6]X_3$ in the solid state

Species	Ref.	M–O (capping) (pm)	M–O (prismatic) (pm)	M–O (capping)/ M–O (prismatic)	M–O (octahedral) (pm)	$r(M^{3+})$ (pm)
$[La(OH_2)_9](CF_3SO_3)_3$	[17]	261.9(5)	251.9(2)	1.04	—	121.6
$[Gd(OH_2)_9](CF_3SO_3)_3$	[17]	254.6(1)	240.2(2)	1.06	—	110.7
$[Lu(OH_2)_9](CF_3SO_3)_3$	[17]	250.3(8)	229.1(3)	1.09	—	103.2
$[Y(OH_2)_9](CF_3SO_3)_3$	[17]	252.5(6)	234.4(3)	1.08	—	107.5
$[Pr(OH_2)_9](EtOSO_3)_3$	[18]	259.2(3)	247.0(2)	1.05	—	117.9
$[Ho(OH_2)_9](EtOSO_3)_3$	[19]	247.4(15)	237.3(11)	1.04	—	107.2
$[Er(OH_2)_9](EtOSO_3)_3$	[20]	252	237	1.06	—	106.2
$[Yb(OH_2)_9](EtOSO_3)_3$	[18]	251.8(4)	232.1(3)	1.08	—	104.2
$[Y(OH_2)_9](EtOSO_3)_3$	[21]	251.8(2)	236.8(2)	1.06	—	107.5
$[Pr(OH_2)_9](BrO_3)_3$	[18]	252(1)	249(1)	1.01	—	117.9
$[Sm(OH_2)_9](BrO_3)_3$	[22]	255(1)	246(1)	1.04	—	113.2
$[Yb(OH_2)_9](BrO_3)_3$	[18]	243(1)	232(1)	1.05	—	104.2
$[La(OH_2)_6](ClO_4)_3$	[23]	—	—	—	248(3)	103.2
$[Tb(OH_2)_6](ClO_4)_3$	[23]	—	—	—	235(3)	92.3
$[Er(OH_2)_6](ClO_4)_3$	[23]	—	—	—	225(2)	89.0

nevertheless, it is unusual for hydrated metal ions to exhibit such a change of coordination number, and this is a significant demonstration of the facile nature of the first coordination sphere of the trivalent lanthanides.

Keppert[24,25], using a model in which the ligands are allowed to move on the surface of a sphere until interbond repulsion is minimized, has shown that the tricapped trigonal prism is the most stable geometry for a [M(unidentate ligand)$_9$] system and has predicted geometries very close to those observed for [Ln(OH$_2$)$_9$]$^{3+}$. This model also indicates that the capped square antiprism is only a slightly less stable geometry for [La(OH$_2$)$_9$]$^{3+}$ than is the tricapped trigonal prism and that in an unconstrained environment scrambling of all nine waters may readily occur, as any one of the three capping waters in the latter geometry may be selected to become the unique water of the capped square antiprism. On the basis of these theoretical calculations and the solid state structural data the following predictions may be made for the nonaaqualanthanide(III) ion in solution:

1. The structure of [Ln(OH$_2$)$_9$]$^{3+}$ will be tricapped trigonal prismatic where M–O (capping) > M–O (prismatic).

2. The interchange of capping and prismatic waters will be facile and proceed through a capped square antiprism intermediate. Should a water dissociate, then it is probable that this will also occur through this intermediate and thereby produce a square antiprism, which is likely to be the most stable geometry for [Ln(OH$_2$)$_8$]$^{3+}$.[24]

Eight coordination of Ln^{3+} has been observed with other monodentate oxygen donor ligands as exemplified by octakis(2,6-dimethyl-4-pyrone)-lanthanum(III) perchlorate,[26] octakis(pyridine-*N*-oxide)lanthanum(III) perchlorate,[27] as has seven coordination in heptakis(2,6-dimethyl-4-pyrone)-erbium(III) perchlorate,[28,29] and six coordination in hexakis(tetramethyl-urea)erbium(III) perchlorate[30] and hexakis(hexamethylphosphoramide)-neodymium(III) perchlorate.[31] The eight-coordinate [Ln(DMF)$_8$]$^{3+}$ cation is also well established in the solid state and solution and is discussed in more detail in Sections IV,A and B. The existence of these species and the aquo species clearly establishes the ability of the trivalent lanthanides to vary their coordination numbers considerably in the solid state in the presence of monodentate ligands. It will be found that the variation of the coordination number of Ln^{3+} is a pervasive theme throughout the succeeding sections.

There appears to be no report of a structure determination of a scandium(III) hydrate or indeed of a scandium(III) species with only monodentate oxygen donor ligands in the first coordination sphere. It is known, however, that scandium(III) does exhibit a variation of coordination number with multidentate oxygen donor ligands as exemplified by eight coordination in hydrogen tetrakis(tropolonate)scandium(III)[32,33] and six coordination in tris(tropolonate)scandium(III)[34] and tris-(acetylacetonate)scandium(III).[35]

B. The hydrated trivalent lanthanide ion in solution

An understanding of most of the properties of solutions of metal salts requires a knowledge of the composition of the first coordination sphere of the metal ion. This is particularly so in understanding the formation of metal complexes and related processes. Thus, considerable effort has been expended in determining the number of water molecules in the first coordination sphere of metal ions[36–38] and the stability constants governing the substitution of water by other ligands, as evidenced by the vast array of available data.[39] As a consequence, the composition of the first coordination spheres of all but the most labile metal ions down to the second transition metal series is reasonably well established in aqueous solution over a wide range of conditions. In the case of the trivalent lanthanides a large amount of equilibrium and thermodynamic data are available on the formation of polyaminocarboxylate complexes and related species,[2,40,41] largely as a consequence of their importance in lanthanide and actinide separation procedures.[42–44] However, in these cases the coordination number of Ln^{3+} is often uncertain, which is hardly surprising as there is still considerable debate about the value of n in the precursor of these complexes, $[Ln(OH_2)_n]^{3+}$. For this species there are protagonists for $n = 10$[45] and 9[46] for $Ln^{3+} = La^{3+}-Nd^{3+}$, for $n = 9$ for $Ln^{3+} = La^{3+}-Lu^{3+}$,[47–50] for $n = 8$ for $Ln^{3+} = Tb^{3+}-Lu^{3+}$,[51,52] for $n = 9$ for $Ln^{3+} = Tb^{3+}-Lu^{3+}$,[45] and for the coexistence of $n = 10$ and 9[45] and $n = 9$ and 8[52] for Ln^{3+} between Nd^{3+} and Tb^{3+}. Space does not permit specific mention of all of the material published concerning the value of n and only some of the salient publications are considered in the ensuing discussion. (Nevertheless, because of the vigorous debate in this area, the introductions to most of these publications contain substantial literature reviews which may be used by readers wishing for a more detailed coverage than is provided here.)

It is important at the outset to appreciate that the differentiation between $n = 8$ and 9, or 9 and 10 for $[Ln(OH_2)_n]^{3+}$ is inherently more difficult than distinguishing between $n = 6$ and 7, or 6 and 5 for the metal ions in the preceding periods, simply because the magnitude of the difference decreases with increase in n. (For the earlier metal ions $n = 6$ in most cases.[38]) The NMR methods[36,37] which worked quite well for the earlier metal hydrated metal ions have so far failed to provide definitive determinations of n for $[Ln(OH_2)_n]^{3+}$ although it will be seen in Section IV,A, B, and C that they are very effective for nonaqueous solvents. Thus for $[Ln(OH_2)_n]^{3+}$, the determination of n by low-angle X-ray[46,51,52] and neutron diffraction[53] methods has attracted considerable attention.

The measurement of the intensities of X-rays diffracted by a lanthanide salt solution and the bulk density of that solution leads to the determination of the radial distribution function (RDF) $4\pi r^2 \rho(r)$ where ρ is the diffracted intensity (e^2) as a function of distance r from Ln^{3+}. When specific solute–solute, solute–solvent, or solvent–solvent interactions occur they produce peaks in the RDF in contrast to the smooth parabola expected for a uniform liquid. In the normalized $RDF = G(r) = \rho(r)/\rho_0$, where ρ_0 is the diffraction intensity anticipated for a uniform liquid, the peaks arising from specific interaction in the lanthanide salt solution are clearly discerned, as is seen in Fig. 1. These $G(r)$

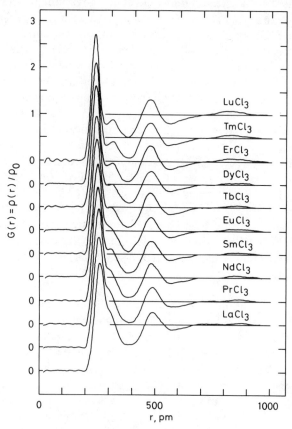

Fig. 1. The normalized radial distribution functions $G(r)$ for 10 concentrated aqueous $LnCl_3$ solutions. (Reproduced from Ref. [52] with permission.)

are taken from the extensive studies of Spedding's group[46,51,52] on aqueous solutions of $LnCl_3$ (in the concentration range 3.2–3.8 mol kg^{-1}) and are seen to exhibit a systematic variation as the lanthanide series is traversed. Three obvious peaks appear in the $G(r)$ of all 10 solutions and, taking $EuCl_3$ as an example, are identified as Eu^{3+}–OH_2 first-coordination-sphere interactions at 248 pm, unresolved H_2O–H_2O and H_2O–Cl^- interactions at 316 pm, and Eu^{3+}–Cl^- second-coordination-sphere interactions at 485 pm.

Assuming no significant entry of Cl^- into the first coordination sphere, a Gaussian analysis of each $G(r)$ produced the average n and Ln^{3+}–OH_2 distances for $[Ln(OH_2)_n]^{3+}$ given in Table 3. The decrease in Ln^{3+}–OH_2 from La^{3+} to Lu^{3+} is anticipated from the lanthanide contraction, and Spedding[52] interprets the variation of n in terms of a change in the coordination number of Ln^{3+} such that for $[Ln(OH_2)_n]^{3+}$ $n = 9$ for La^{3+}–Nd^{3+}, $n = 8$ for Tb^{3+}–Lu^{3+}, and an equilibrium between eight- and nine-coordinate species exists for Ln^{3+} between Nd^{3+} and Tb^{3+}. (A neutron diffraction study of a concentrated $NdCl_3$ solution determines n to be 8.5.[53]) From solid state data[17–22] and theoretical considerations[24,25] (Section II,A) $[Ln(OH_2)_9]^{3+}$ should exhibit two Ln^{3+}–OH_2 values which differ increasingly from La^{3+} to Lu^{3+}, such that for Lu^{3+} the difference between Ln^{3+}–OH_2 (capping) and Ln^{3+}–OH_2 (prismatic) approaches 20 pm. It would be interesting to know how a Gaussian resolution of $G(r)$ in Fig. 1 assuming two Ln^{3+}–OH_2 values similar to those observed in the solid state would affect n. Such a resolution has been reported[54] for the RDF characterizing tetragonally distorted $[Cu(OH_2)_6]^{2+}$. In this respect it is interesting to note that a preliminary report on a solution X-ray study of aqueous $La(ClO_4)_3$ and $Er(ClO_4)_3$ indicates that both Ln^{3+} exhibit two unequal Ln^{3+}–OH_2 distances (and also that in solution $n > 6$ observed in the solid state).[23] Solution X-ray studies by Wertz's group[55–58] on $LnCl_3$ (Ln = La, Nd, and Gd) solutions apparently similar to those studied by Spedding's group produce quite different RDFs which have been interpreted in terms of $[Ln(OH_2)_8]^{3+}$, $[Ln(OH_2)_7Cl]^{2+}$, and $[Ln(OH_2)_6Cl_2]^+$ being important species in solution; an earlier study reports $[Er(OH_2)_6Cl_2]^+$ as the predominant species in aqueous $ErCl_3$ solutions.[59] There are some variations in the concentrations of the $LnCl_3$ solutions used in these different studies, but the differences in the RDFs look rather large to have originated from this source. Spedding has discussed possible pitfalls in solution preparation[51] and has given substantial details of equipment alignment and calibration.[46]

The stability constants[60] governing the formation of Ln^{3+} chloro complexes in water range from 0.13 to 1.02 dm^3 mol^{-1}, depending on ionic strength, and it seems probable that some chloride enters the first coordination sphere of Ln^{3+} in concentrated Ln^{3+} solutions. (In the solid state the structures of $[Eu(OH_2)_6Cl_2]^+$ and $[Gd(OH_2)_6Cl_2]^+$ have been determined

TABLE 3

Mean $Ln^{3+}-OH_2$ and coordination numbers, n, obtained from X-ray studies of aqueous $LnCl_3$ solutions[a]

	La^{3+}	Pr^{3+}	Nd^{3+}	Sm^{3+}	Eu^{3+}	Tb^{3+}	Dy^{3+}	Er^{3+}	Tm^{3+}	Lu^{3+}
$Ln^{3+}-OH_2$ (pm)[b]	258.0	253.9	251.3	247.4	245.0	240.9	239.6	236.9	235.8	233.8
n^c	9.13	9.22	8.90	8.8	8.3	8.18	7.93	8.19	8.12	7.97
$[LnCl_3]$ (mol kg^{-1})	3.808	3.801	3.367	3.232	3.231	3.488	3.289	3.541	3.626	3.614

[a] From Refs. [46, 51, and 52].
[b] Uncertainty, ± 0.2 pm.
[c] Uncertainty, < 0.1.

by X-ray methods.[61,62]) Solution X-ray studies of Ln^{3+} in the presence of $CF_3SO_3^-$, $EtSO_4^-$, or BrO_3^-, which seem less likely to coordinate, would be interesting but it may be that the additional peaks arising from these anions in the RDF might further complicate the Gaussian resolution.

In the 3 mol kg^{-1} $LnCl_3$ solutions used in the X-ray studies there are ~ 6 water molecules available to separate $[Ln(OH_2)_n]^{3+}$ from Cl^-, which represents a situation very different from that applying in the dilute solutions in which the Ln^{3+} ligand substitution studies (discussed in Section III,A on) are usually carried out. While there are no reports claiming to detect a change in Ln^{3+} hydration as concentration is varied (except in glassy solution as discussed below), it is reported that the numbers of coordinated waters per Ca^{2+} detected in a neutron diffraction study are 10.0 ± 0.6 (246 ± 3), 7.2 ± 0.2 (239 ± 2), and 6.4 ± 0.3 (241 ± 3) in 1, 2.8, and 4.8 mol kg^{-1} $CaCl_2$ D_2O solutions.[63] (The figures in parentheses are the Ca^{2+}–OH_2 distances in pm.) As Ln^{3+} is used with increasing frequency to replace Ca^{2+} in investigations of biological molecules,[12,13] and as Ln^{3+} (although more strongly hydrated) does exhibit coordination numbers similar to Ca^{2+}, it may be that n could exhibit some variation between dilute and concentrated solution for $[Ln(OH_2)_n]^{3+}$ if the same applies for $[Ca(OH_2)_n]^{2+}$.

A study[64] of glassy aqueous solutions of $EuCl_3$ and $GdCl_3$ shows that substantial variations occur in the Raman spectra as $[LnCl_3]$ is increased. This is considered to indicate that $[Ln(OH_2)_9{}^{3+}]$ increases over $[Ln(OH_2)_8{}^{3+}]$ as $[LnCl_3]$ increases. As these are precisely the salts for which $[Ln(OH_2)_6Cl_2]^+$ has been shown to exist in the crystalline state,[61,62] it is surprising that these authors did not consider the possibility of chloride coordination in interpreting their spectra. Conversely, Karraker[65] has interpreted changes in the hypersensitive spectral bands of Nd^{3+} in the presence of HCl or LiCl varying from 0 to 12 mol dm^{-3} to indicate that $[Nd(OH_2)_9{}^{3+}]$ diminishes as $[Nd(OH_2)_8{}^{3+}]$ increases with increase in $[Cl^-]$. Similar spectral variations were not observed for Ho^{3+} and Er^{3+}, which was taken as an indication that the hydration of these ions is insensitive to $[Cl^-]$. A similar trend has been claimed in a ^{139}La NMR study where linewidth variations were considered to indicate that $[La(OH_2)_9]^{3+}$ converts to $[La(OH_2)_8]^{3+}$ as $[LaCl_3]$ increased.[66] Although there appears to be a consensus that $[Ln(OH_2)_n]^{3+}$ exists in concentrated solution with $n = 8$ or 9, the apparently contradictory interpretation of data arising from the same systems is disconcerting, as is the tendency of some authors to ignore the possibility of chloride coordination. It would be advantageous if future studies were carried out in the presence of anions with low coordinating abilities.

In the preceding discussion some emphasis has been placed on the coordinating ability of chloride. It is therefore appropriate to note that stability constants governing the formation of nitrate and thiocyanate Eu^{3+}

complexes in water have been reported in the respective ranges of 0.30–2.00 and 5.05–5.431 dm^3 mol^{-1},[60] and that the kinetics of $SO_4{}^{2-}$ substitution on Ln^{3+} has been extensively studied, as discussed in Section III,D. A fluorescence study[67] of aqueous $Eu(ClO_4)_3$ has established that $ClO_4{}^-$ does not coordinate to Eu^{3+}, and it seems probable that this is the case for other Ln^{3+}. While no studies of the coordinating ability of CF_3COO^-, $C_2H_5SO_4{}^-$ and $BrO_3{}^-$ in aqueous Ln^{3+} solutions appear to have been reported, their failure to coordinate to Ln^{3+} in the solid state suggests that they are unlikely to coordinate in aqueous solution. Thus the choice of one of these three anions or perchlorate appears to offer an opportunity to avoid unsought anion coordination in aqueous Ln^{3+} studies.

It is now appropriate to turn to less concentrated solutions. Spedding and co-workers[68] have shown that the apparent molar volume of Ln^{3+} ($LnCl_3$ solution) ϕ_v^0 (at 298.2 K) shows a smooth decrease from 14.51 to 10.18 cm^3 mol^{-1} for La^{3+} to Nd^{3+}, and an increase to 13.25 cm^3 mol^{-1} for Tb^{3+} followed by a smooth decrease to 9.26 cm^3 mol^{-1} for Yb^{3+}, and they suggest that this indicates a progressive change from $n = 9$ to $n = 8$ in $[Ln(OH_2)_n]^{3+}$ from La^{3+} to Lu^{3+}. Swaddle[69,70] has developed a model for computing absolute molar volumes of hydrated cations ($\bar{V}_{abs}^0$) through Eq. (1) in which r is the cationic radius of appropriate coordination number n, Δr is the diameter of a coordinated water molecule, and z is the cationic charge:

$$\bar{V}_{abs}^0 = 2.523 \times 10^{-6} (r + \Delta r)^3 - 18.07n - 417.5z^2/(r + \Delta r) \qquad (1)$$

Thus the effective volume of the hydrated cation is that of a sphere of radius $r + \Delta r$. Using this equation Swaddle has computed the variation of $\bar{V}_{abs}^0$ as r decreases from La^{3+} to Lu^{3+} for $[Ln(OH_2)_n]^{3+}$ for both $n=9$ and 8. The resulting two curves for $\bar{V}_{abs}^0$ lie close to the V_{abs}^0 data points for La^{3+}–Nd^{3+} and Tb^{3+}–Yb^{3+} calculated from the ϕ_v^0 data of Spedding and co-workers. It is concluded from this that a change in n probably occurs as the series La^{3+}–Lu^{3+} is traversed.[69,70]

Equation (1) also predicts $\bar{V}_{abs}^0$ values close to experimental values for the earlier hydrated metal ions for which together with Ln^{3+} $\Delta r = 238.7$ pm has been adopted. This value is the diameter of an electrostricted water molecule which is substantially less than the diameter of the bulk water molecule, which is estimated to be 276 pm. Intuitively this constancy of Δr seems a little unexpected for Ln^{3+} where $r(M^{3+})$ values for both nine and eight coordination decrease by 20 pm from La^{3+} (Table 1) and the crystal structure show that steric interactions in $[Ln(OH_2)_9]^{3+}$ cause an increasing inequality in Ln^{3+}–OH_2 from La^{3+} to Lu^{3+} as discussed in Section II,A. It would be very interesting to see if the Swaddle model could be developed to take these factors into account, particularly as earlier Geier and Karlen[49] contended that the variation of ϕ_v^0 with Ln^{3+} could be explained on the basis of the variation in the structure of $[Ln(OH_2)_9]^{3+}$ resulting from steric crowding. (The $[Ln(EDTA)(OH_2)_n]^-$ ϕ_v^0 variation[71] with Ln^{3+} is substantially greater than that for $[Ln(OH_2)_n]^{3+}$ and is discussed in Section III,F.)

Two spectroscopic studies appear to indicate that n is constant for $[Ln(OH_2)_n]^{3+}$ for all Ln^{3+}. In a study of the electron relaxation rates of $[Ln(OH_2)_n]^{3+}$ it was found that Pr^{3+} and Eu^{3+} exhibited temperature-independent electron relaxation rates consistent with n being an integer presumed to be 9.[72] The small uv-visible spectral temperature dependences exhibited by $Eu(ClO_4)_3$ solutions (by comparison to those exhibited by $[Eu(EDTA)(OH_2)_n]^-$ have also been interpreted as indicating $[Eu(OH_2)_9]^{3+}$ to be the only cationic species in solution.[48] Similarly, thermodynamic data for the formation of $[Ln(DPTA)(OH_2)_n]^{2-}$ are held consistent with n being identical in $[Ln(OH_2)_n]^{3+}$ for all Ln^{3+}.[73] A recently developed method for probing the hydration of Ln^{3+} is laser luminescence excitation spectroscopy.[12,45] Electronic transitions within a $4f^n$ configuration are Laporte forbidden and, in consequence, Ln^{3+} luminescence is weak by comparison to the fluorescence of organic molecules. This difficulty may be overcome by using a laser as the luminescence excitation source, which renders feasible the study of dilute solutions of Eu^{3+} and Tb^{3+}, which are the most strongly emitting members of the Ln^{3+} series. The observed luminescence decay constant, k_{lum}, is given by Eq. (2)

$$k_{lum} = k_{nat} + k_{nonrad} + k_{H_2O} \tag{2}$$

where k_{nat} is the rate constant for natural photon emission, k_{nonrad} is the rate constant for the radiationless deexcitation processes not involving water, and k_{H_2O} is the rate constant for energy transfer to the O–H vibrations of coordinated water. It appears that the magnitude of k_{H_2O} is directly proportional to the number of coordinated waters. In the solid state it is found that the change in k_{lum} found between Eu^{3+} and Tb^{3+} complexes containing no coordinated water and those containing n coordinated waters is linearly proportional to n. A similar correlation appears to exist in aqueous solution, and Horrocks and co-workers[45] have determined $n = 9.6 \pm 0.5$ and 9.0 ± 0.5 in $[Eu(OH_2)_n]^{3+}$ and $[Tb(OH_2)_n]^{3+}$, respectively. They conclude that in dilute solutions $n = 10$ for Ln^{3+}–Nd^{3+}, $n = 9$ for Tb^{3+}–Lu^{3+}, and that both stoichiometries exist for Ln^{3+} between Nd^{3+} and Lu^{3+}. In an earlier study also based on luminescence lifetime variations, however, Haas and Stein[74] concluded that $[Eu(OH_2)_9]^{3+}$ is the predominant species in solution.

C. Summary

It is clear from the preceding sections that despite many valiant efforts to determine a definitive value for n in $[Ln(OH_2)_n]^{3+}$ its value will probably continue to be a matter for debate for some time yet. The environment experienced by Ln^{3+} in the concentrated solution studies is such that second and subsequent hydration spheres do not exist as they do in dilute solution. With a species in which the lifetime of a single coordinated water is as short as 10^{-9} s[75] and which in the case of $[Ln(OH_2)_9]^{3+}$ probably interchanges prismatic and capping waters even more rapidly, it is probable that the second hydration sphere will be significant in determining the number of coordinated waters. Intuitively it seems unlikely that species exhibiting the stoichiometry $[Ln(OH_2)_9]^{3+}$ for La^{3+}–Lu^{3+} in the solid state will assume $n < 9$ in dilute aqueous solution. The possibility of $n = 10$ must be entertained for the lighter Ln^{3+}, as in addition to the recent suggestion of $[Eu(OH)_{10}]^{3+}$ in solution, La^{3+} is 10 coordinate in $[La((H)EDTA)(OH_2)_4]$ in the solid state.[76] On balance, however, the author is inclined to favor $[Ln(OH_2)_9]^{3+}$ for all Ln^{3+} in dilute solutions (an inclination which could change in the light of new data), and this stoichiometry will be assumed in subsequent discussion.

III. LIGAND SUBSTITUTION PROCESSES IN AQUEOUS SOLUTION

A. Reaction mechanisms

The relationship between solvent exchange and ligand substitution has been extensively reviewed[77–80] and accordingly this section is intended to provide only a brief mechanistic basis for subsequent sections. When solvent, S, in the first coordination sphere of a solvated metal ion $[MS_n]$ is substituted by another monodentate ligand, L, to form $[MS_{n-1}L]$ (or vice versa) as shown in Eq. (3), in which all charges are omitted, the transition or activated state may

$$[MS_n] + L \underset{k_b}{\overset{k_f}{\rightleftharpoons}} [MS_{n-1}L] + S \tag{3}$$

be produced through two activation modes. In the dissociative (**d**) activation mode the predominant process in the formation of the transition state is the breaking of a bond between M and S, whereas in the associative (**a**) activation mode the predominant process is the formation of a bond between M and L.

Two extreme mechanistic possibilities now arise. The first extreme possibility is the dissociative (**D**) mechanism in which the rate-determining step produces, through a **d** activation mode, a reactive intermediate of decreased coordination number, $[MS_{n-1}]$, which achieves thermal equilibrium with its environment before reaction either with L to produce $[MS_{n-1}L]$, or with another S such that solvent exchange results. If L competes very effectively with S such that $[MS_{n-1}]$ always reacts to form $[MS_{n-1}L]$, the rate of ligand substitution will be equal to the rate of solvent exchange in the absence of L. Thus, when the solvent exchange rate is expressed as in Eq. (4) where k_{ex}

$$\text{solvent exchange rate} = nk_{ex}[MS_n] \tag{4}$$

characterizes the exchange of a particular solvent molecule, then $k_f = nk_{ex}$. When the formation of $[MS_{n-1}L]$ is considered separately as expressed in Eq. (5) it may be shown, assuming the steady-state approximation for

$$[MS_n] + L \underset{k_{-1}}{\overset{k_1}{\rightleftharpoons}} [MS_{n-1}] + S + L \overset{k_2}{\longrightarrow} [MS_{n-1}L] + S \tag{5}$$

$[MS_{n-1}]$, that the observed rate constant for the formation of $[MS_{n-1}L]$ k_f varies with excess $[L]$ according to Eq. (6) in which $k_1 = nk_{ex}$. In the limit-

$$k_f = k_1 k_2 [L]/(k_{-1}[S] + k_2[L]) \tag{6}$$

ing condition $k_2[L] \gg k_{-1}[S]$, $k_f \approx k_1$ ($= nk_{ex}$). In strongly coordinating solvents both k_2 and $[L]$ will need to be substantial before such a limiting condition is achieved. At present there appears to be no well-established case of solvent substitution by another ligand occurring through a **D** mechanism. A substantial hindrance to the identification of such a mechanism is the tendency for positively charged $[MS_n]$ to form an ion pair with L which is often negatively charged and thereby presents the possibility of ligand substitution occurring through this ion pair (or encounter complex) in an interchange (**I**) mechanism as discussed below.

The second extreme possibility is the associative (**A**) mechanism in which the rate-determining step produces, through an **a** activation mode, an intermediate of increased coordination number $[MS_n L]$ which subsequently loses an S to produce $[MS_{n-1}L]$. In this case the rate of ligand substitution is linearly dependent on $[L]$ and k_f is a second-order rate constant. Solvent exchange may also proceed through an **A** mechanism but now there is no direct link between ligand substitution and solvent exchange, and the rate of the former may be faster or slower than that of the latter.

Between the two extremes is the interchange (**I**) mechanism, in which L and S interchange between the first and second coordination spheres of $[MS_n]$ in the encounter complex $[MS_n \cdots L]$ in which L resides in the second coordination sphere. The **I** mechanism is illustrated by Eq. (7) in which K_0

$$[MS_n] + L \overset{K_0}{\rightleftharpoons} [MS_n] \cdots L \underset{k_{-i}}{\overset{k_i}{\rightleftharpoons}} [MS_{n-1}L] \cdots S \tag{7}$$

characterizes the diffusion-controlled formation of the encounter complex, k_i the interchange of L and S between the second and first coordination spheres, and k_{-i} the reverse process. In this mechanism both M–L bond making and M–S bond breaking occur in the formation of the transition state leading to

$[MS_{n-1}L]$. When M–S bond breaking dominates the transition state energetics such that the activation mode is predominantly **d**, the mechanism is termed dissociative interchange (I_d). Solvent exchange also occurs through the I_d mechanism and a direct relationship exists between k_i in Eq. (7) and k_{ex} for solvent exchange. If in the encounter complex L is considered to occupy a triangular face delineated by three S of the polyhedron representing $[MS_n]$ and each S represents the apex of x such faces, then assuming L to compete equally with S for actual or incipient vacancies occurring in the first coordination sphere, $k_i = 3k_{ex}/x$ on a statistical basis. This estimate is a rather crude approximation but it is lent some credibility by the observation that for $[Ni(OH_2)_6]^{2+}$ and $[Co(OH_2)_6]^{2+}$ which are considered to undergo ligand substitution through an I_d mechanism, $k_i/k_{ex} \approx 1$ for a range of L. When M–L bond formation dominates the transition state energetics such that the activation mode is predominantly **a** the mechanism is termed associative interchange (I_a) and there exists no direct relationship between ligand substitution and solvent exchange such that $k_i < k_{ex} < k_i$ in principle. In theory a wide spectrum of **I** mechanistic variations exists between the **D** and **A** mechanistic extremes and the picture becomes a little wooly at the midpoint. It should also be noted that the I_d and I_a classifications have been used in significantly different ways by some authors[70,79] and that the framework erected here conforms to what appears to be the current usage of these classifications.

Regardless of the activation mode the observed rate constant k_{obs} for the approach to equilibrium of the interchange scheme shown in Eq. (7) is given by Eq. (8) where L is in excess concentration, [L]. Two limiting conditions apply.

$$k_{obs} = k_{-i} + k_i K_0[L]/(1 + K_0[L]) = k_b + k_f[L]/(1 + K_0[L]) \qquad (8)$$

The first occurs when $1 \gg K_0[L]$ and $k_{obs} \approx k_{-i} + k_i K_0[L]$. [This is a common situation with labile metal ions such as Ln^{3+} and k_f ($= k_i K_0$) has units of $dm^3 \ mol^{-1} \ s^{-1}$.] The second limiting condition occurs when $K_0[L] \gg 1$ and $k_{obs} \approx k_{-i} + k_i$. It is desirable to operate over a concentration range such that the complete variation of k_{obs} with [L] is seen, but often with labile species such as Ln^{3+} the magnitude of k_i is apparently so great that k_{obs} may only be determined in the first limiting condition. (The term encounter complex tends to the favored for $[MS_n] \cdots L$ by many authors, but it should be noted that the terms ion pair and outer-sphere complex are also applied to this species.)

It is evident from the above discussion that k_i should be invariant and variable with the nature of the ligand for I_d and I_a mechanisms, respectively. Such characteristics are observed for ligand substitution in $[Ni(OH_2)_6]^{2+}$, $[Co(OH_2)_6]^{2+}$, and $[Fe(OH_2)_6]^{2+}$ [81] for which I_d mechanisms operate, and for $[V(OH_2)_6]^{3+}$, $[Mo(OH_2)_6]^{3+}$,[82] and $[Cr(OH_2)_6]^{3+}$ [83] for which I_a mechanisms operate. However, for very reactive species which react every time a collision occurs, no discrimination is made between reactants and the selectivity for an incoming ligand L expected for an **a** activated ligand substitution vanishes. The $[Ln(OH_2)_9]^{3+}$ species do not undergo ligand substitution at such diffusion-controlled rates as will be seen below. Nevertheless, they are much more labile in water than the $[M(OH_2)_6]^{n+}$ species mentioned above and, accordingly, it is expected that their range of selectivity for entering ligands L, in the event that an **a** activated mechanism operated, would be compressed. Similarly the hard acid–hard base character of Ln^{3+}–L interactions should also serve to compress the selectivity range.

B. Water exchange on the trivalent lanthanide ions

Water exchange on the trivalent lanthanides has so far proved too fast to be quantitatively determined by the NMR line-broadening techniques[84] used to determine the water exchange rates of the labile six-coordinate di- and trivalent first-row transition metal ions, but it may be that with the advent of higher frequency NMR spectrometers this situation will change. Nevertheless, in the case of Gd^{3+} which has a symmetrical electronic ground state and long electronic relaxation times, T_{1s} and T_{2s}, Merbach's group found it possible to analyze the temperature variation of the bulk water ^{17}O longitudinal (T_1) and transverse (T_2) relaxation times and to obtain water exchange kinetic parameters from the scalar coupling interactions between the water ^{17}O and the unpaired electrons of Gd^{3+}.[75] The data obtained appear in Table 4 and are seen to be in good agreement with those obtained earlier by Marianelli.[85] It is important to notice that the derivation of these kinetic parameters assumes the solution species to be $[Gd(OH_2)_9]^{3+}$ such that the water exchange rate constant k_{ex} = water exchange rate/$9[Gd(OH_2)_9^{3+}]$. Thus, $[Gd(OH_2)_9]^{3+}$ is the only Ln^{3+} species for which k_{ex} has been directly determined in water, although lower limits of 0.3–2.6×10^7 s^{-1} have been estimated for k_{ex} for water exchange on Tb^{3+}, Dy^{3+}, Ho^{3+}, Er^{3+}, and Tm^{3+} from ^{17}O NMR line-broadening studies.[86] A more recent study has raised the lower limit for Dy^{3+} to 10^8 s^{-1}[87]; however it is important to note that these values are truly lower limits and not estimates of k_{ex}.

An examination of the data in Table 4 shows the k_{ex} value for $[Gd(OH_2)_9]^{3+}$ to be substantially greater than those for $[Ti(OH_2)_6]^{3+}$[88] and $[Mn(OH_2)_6]^{3+}$,[81] respectively the most labile tri- and divalent first-row transition ions for which k_{ex} has been determined for water. Similarly, $[Gd(OH_2)_9]^{3+}$ is seen to be much more labile than $[Al(OH_2)_6]^{3+}$ and its Ga^{3+} and In^{3+} analogs[89,90] and also $[Mg(OH_2)_6]^{2+}$.[91] The greater lability of $[Gd(OH_2)_9]^{3+}$ is probably a consequence of the larger ionic radius of

TABLE 4

Kinetic parameters for water exchange on selected metal ions

Metal ion	k_{ex} (298.2 K) (s^{-1})	$\Delta H^{\ddagger}$ $(kJ\,mol^{-1})$	$\Delta S^{\ddagger}$ $(J\,K^{-1}\,mol^{-1})$	$\Delta V_0^{\ddagger}$ $(cm^3\,mol^{-1})$	$r(M^{n+})^a$ (pm)	Ref.
$[Gd(OH_2)_9]^{3+}$	$(10.6 \pm 0.9) \times 10^8$	11.99 ± 1.40	-31.9 ± 4.26	—	110.7	[75]
$[Gd(OH_2)_9]^{3+}$	9×10^8	13	-29	—	110.7	[85]
$[Gd(PDTA)(OH_2)_2]^-$	$(3.3 \pm 0.4) \times 10^8$	9.59 ± 1.44	-49.80 ± 4.19	—	105.3	[74]
$[Ti(OH_2)_6]^{3+}$	$(1.81 \pm 0.03) \times 10^5$	43.4 ± 0.7	1.2 ± 2.2	-12.1 ± 0.4	67.0	[88]
$[Mn(OH_2)_6]^{2+}$	$(2.1 \pm 0.1) \times 10^7$	32.9 ± 1.3	5.7 ± 5.0	-5.4 ± 0.1	67.0	[81]
$[Al(OH_2)_6]^{3+}$	1.29 ± 0.04	84.7 ± 0.3	41.6 ± 0.9	5.7 ± 0.2	53.5	[89a]
$[Ga(OH_2)_6]^{3+}$	402.7 ± 38.7	67.1 ± 2.5	30.1 ± 7.5	5.0 ± 0.5	62.0	[89b]
$[In(OH_2)_6]^{3+}$	4.0×10^4	4.6	-23	—	80.0	[90]
$[Mg(OH_2)_6]^{2+}$	5.3×10^5	10.2	2	—	72.0	[91]

^{a} Ionic radius for the appropriate coordination number from Ref. [14].

Gd^{3+} and the increased steric crowding in the first coordination sphere (Section II,A). Nevertheless, the lability of water in $[Ga(OH_2)_9]^{3+}$ is substantially less than that expected for a diffusional process for which $k_{ex} \approx 5 \times 10^{11}$ s^{-1} [75] or for $k_{ex} \approx 10^{10}$ s^{-1} obtained for water exchange from the second coordination sphere of $[Cr(OH_2)_6]^{3+}$.[92]

The mechanism of water exchange on $[Gd(OH_2)_9]^{3+}$ cannot be deduced from the available data, but if the intramolecular interconversion of capping and prismatic waters occurs much more rapidly than intermolecular exchange as seems probable (see Section II,A and C), then it may be that the intermolecular exchange of all nine waters occurs predominantly through the three tricapped sites which have the greater Gd–OH_2 bond distance, or through the capped square antiprisms proposed for interchange of capping and axial sites, or both. (This process is similar to that proposed for water and methanol exchange on six-coordinated copper(II).[93,94] Here the rapid alternation of the tetragonal distortion over the x, y, and z axes occurs at a rate in excess of that of solvent exchange and is considered to be the source of the enhanced lability of $[Cu(OH_2)_6]^{2+}$. In the case of $[Cu(MeOH)_6]^{2+}$ $\Delta V^{\ddagger} = +8.3$ cm^3 mol^{-1} suggests a **d** activation mode for methanol exchange.[95]) This suggests the possibility of the operation of a **d** activation mode for water exchange on $[Gd(OH_2)_9]^{3+}$. Activation volume determinations for solvent exchange on lighter metal ions have proved mechanistically diagnostic[96] and a determination of $\Delta V^{\ddagger}$ for $[Gd(OH_2)_9]^{3+}$ could prove informative.

The only other directly determined water exchange rate for Ln^{3+} is that for assumed eight-coordinate $[Gd(PDTA)(OH_2)_2]^-$ in which k_{ex} (Table 4) is decreased by a factor of 3 by comparison to that for $[Gd(OH_2)_9]^{3+}$ [75] ($PDTA^{4-}$ is 1,3-diaminopropane-N,N,N',N'-tetraacetate). By contrast, the closely related $EDTA^{4-}$ (1,2-diaminoethane-N,N,N',N'-tetraacetate), which acts as a pentadentate ligand in $[Ni(EDTA)(OH_2)]^{2-}$ and a hexadentate ligand in $[Mn(EDTA)(OH_2)]^{2-}$ (Mn^{2+} is seven coordinated), labilizes the coordinated water in these species.[97,98] A combination of the smaller size of Ni^{2+} and Mn^{2+} and their strong directional bonding characteristics probably produces substantial steric strain in $[M(EDTA)(OH_2)]^{2-}$ which results in the observed labilization. The larger size of Gd^{3+} and the absence of significant directional bonding effects produce a relatively unstrained $[Gd(PDTA)(OH_2)_2]^-$ species, and it has been suggested that the decreased lability of coordinated water is a consequence of hydrogen bonding between the oxygen of $PDTA^{4-}$ and the two coordinated waters.[75]

In the absence of directly determined k_{ex} for Ln^{3+} other than Gd^{3+} a case will be made in Section III,D for using the first-order rate constant k_{34} for SO_4^{2-} substitution on Ln^{3+} determined by ultrasonic absorption techniques as an approximation to k_{ex}. The protolysis of coordinated water which complicates the interpretation of mechanistic studies of the lighter trivalent ions only occurs at moderate pH for Ln^{3+} ($pK_a = 9.33$ and 8.17 for $[La(OH_2)_9]^{3+}$ and $[Lu(OH_2)_9]^{3+}$, respectively[138]) and usually does not present significant problems in Ln^{3+} mechanistic studies.

C. Sulfate substitution on scandium(III)

Ligand substitution on Ln^{3+}, Y^{3+}, and Sc^{3+} in water is very rapid and the experimental data are mostly available in two forms. The first form is usually obtained with the temperature- or pressure-jump or stopped-flow techniques[99,100] and produces a linear variation of the observed rate constant for the approach to equilibrium, k_{obs}, with the excess concentration of a substituting ligand $[X]$, or the excess concentration of the metal ion. Such data are usually interpreted in terms of the operation of an **I** mechanism in the limiting condition: $1 \gg K_0[X]$ such that $k_{obs} = k_{-i} + k_i K_0[X] = k_b + k_f[X]$ as discussed in Section III,A. Substitution of an appropriate value of K_0 then produces k_i (which is not necessarily characteristic of the first bond formation when X is multidentate, as will be seen in Section III,E). The second form of experimental data is obtained from the very much faster ultrasonic absorption technique[99–101] and produces not only a rate constant equivalent to k_i directly but also rate constants for the formation of the encounter complex discussed in Section III,A. It is convenient to discuss ultrasonic studies of sulfate substitution on Ln^{3+} first because this ligand appears to substitute as a monodentate species, and also because a recent and particularly thorough study of the Sc^{3+} system[102] provides a convenient vehicle for a brief discussion of the ultrasonic method and some of the divergence of opinion which has surrounded the data obtained by this method.

Ultrasonic absorption involves oscillating perturbations of a chemical equilibrium such that when the reciprocal of the relaxation time of the reaction, $1/\tau$, is comparable to the ultrasonic frequency concentration changes lag behind the perturbation changes. The resultant ultrasonic absorption forms the basis for the determination of the relaxation times of the chemical equilibria. In order to explain the frequency variation of the ultrasonic absorption exhibited by aqueous solutions of $MgSO_4$ and other electrolytes, Eigen and Tamm[103] proposed a three-step mechanism which is shown in Eq. (9) for aqueous $Sc_2(SO_4)_3$.

$$Sc^{3+} + SO_4^{2-} \underset{k_{21}}{\overset{k_{12}}{\rightleftharpoons}} Sc^{3+} \cdot OH_2 \cdot OH_2 \cdot SO_4^{2-} \underset{k_{32}}{\overset{k_{23}}{\rightleftharpoons}} Sc^{3+} \cdot OH_2 \cdot SO_4^{2-} \tag{9}$$

$$\underset{k_{43}}{\overset{k_{34}}{\rightleftharpoons}} ScSO_4^{+}$$

The choice of this system is particularly appropriate as it provides some of the few data available for ligand substitution on Sc^{3+} and it has been studied over the wide frequency range 0.23–36 MHz.[102] In Eq. (9) only water molecules directly interposed between Sc^{3+} and SO_4^{2-} are shown, and in any event n for $[Sc(OH_2)_n]^{3+}$ is not known although it is probable that $8 \geq n \geq 6$. In the first step Sc^{3+} and SO_4^{2-} in their normal hydrated states form, at diffusion-controlled rates, an ion pair $(k_{12}/k_{21} = K_1)$ in which they are separated by two water molecules. In the second step SO_4^{2-} enters the second coordination sphere of Sc^{3+} to form a second ion pair $(k_{23}/k_{32} = K_2)$ analogous to the encounter complex of the **I** mechanism in Section III,A. The third step is the coordination of SO_4^{2-} to Sc^{3+} $(k_{34}/k_{43} = K_3)$, presumably as a monodentate ligand.

Provided that each of these steps produces a volume change and occurs with an appropriate relaxation time three maxima should be observed in the ultrasonic absorption over and above the normal absorption of water as the ultrasonic frequency f is varied. This excess absorption, $\alpha\lambda_{chem}$, is given by Eq. (10) where α is the absorbance coefficient, λ is the wave length, τ_i is a

$$\alpha\lambda_{chem} = \alpha\lambda - Bf = \sum_{i=1}^{n} \frac{A_i 2\pi f \tau_i}{1 + 4\pi^2 f^2 \tau_i^2} \tag{10}$$

chemical relaxation time, n is the number of chemical relaxations, A and B are constants, and Bf constitutes the background absorbance. Three relaxations may be seen in $\alpha\lambda_{chem}$ for a 0.033 dm^3 mol^{-1} aqueous $Sc_2(SO_4)_3$ solution as shown in Fig. 2, which also shows a resolution into three Gaussian curves characterized by $\tau_I = 5.5 \times 10^{-10}$ s, $\tau_{II} = 3.2 \times 10^{-8}$ s, and $\tau_{III} = 2.1 \times 10^{-7}$ s. The high-frequency absorbance and τ_I arise from the first step in Eq. (9), and the two lower frequency absorbances and τ_{II} and τ_{III} are coupled and contain components from the second and third steps. The derived rate

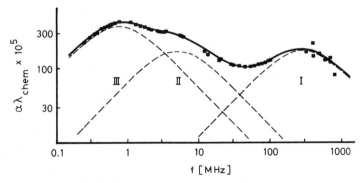

Fig. 2. The variation of ultrasound absorption $\alpha\lambda_{chem}$ with frequency for a 0.033 mol dm^{-3} aqueous solution of $Sc_2(SO_4)_3$. The solid curve represents the best fit of the data to three relaxations. The broken curves represent the three relaxation contributions. (Reproduced from Ref. [102] with permission.)

constants appear in Table 5. (The lower frequency relaxation is absent from a similar study of aqueous $Al_2(SO_4)_3$, consistent with the third step being too slow for ultrasonic studies. However, the two higher frequency relaxations are present, consistent with the first and second steps occurring.[102]) Strictly for an accurate derivation of the six rate constants of Eq. (9) from the ultrasonic absorption data, the inequality $\tau_{III} \gg \tau_{II} \gg \tau_I$ should hold. It is seen that τ_{III} and τ_{II} differ by a factor of ~ 6 only, but this is estimated to introduce an error $< 6\%$ in the derived rate constants.[102]

A significant problem in deriving $\alpha\lambda_{chem}$ in Eq. (10) is that the background absorption of water, Bf, varies with the electrolyte concentration. It has been argued[104] that incorrect values of Bf in earlier studies produced the absorbance characterized by τ_{II} and that a two-step mechanism, analogous to the **I** mechanism in Section III,A, operates such that the high-frequency relaxation encompasses the species analogous to $Sc^{3+} \cdot OH_2 \cdot OH_2 \cdot SO_4{}^{2-}$ and $Sc^{3+} \cdot OH_2 \cdot SO_4{}^{2-}$. In the $Sc_2(SO_4)_3$ study the three relaxations are seen for five solutions in the concentration range 0.0033–0.1 mol dm^{-3}, whereas in earlier studies carried out over much narrower frequency ranges three relaxations were more often inferred than seen. Thus, one is faced on the one hand with the possibility that the three-step mechanism usually operates but three relaxations are not observed because the frequency range studied was too narrow (or possibly ΔV for one step was too small), and on the other with the possibility that one of the three ultrasound absorbances is an experimental artifact. As a consequence, the treatment of ultrasonic data tends to be subdivided between that treated according to the three-step mechanism and that treated according to its truncation, the two-step mechanism, which can

TABLE 5

Kinetic and thermodynamic parameters for $SO_4{}^{2-}$
substitution on $[Sc(OH_2)_n]^{3+}$ at 298.2 K[a]

k_{12} (dm^3 mol^{-1} s^{-1})	$(2 \pm 0.5) \times 10^{11}$
k_{21} (s^{-1})	$(9 \pm 3) \times 10^8$
k_{23} (s^{-1})	$(1.4 \pm 0.5) \times 10^7$
k_{32} (s^{-1})	$(2.0 \pm 0.7) \times 10^7$
k_{34} (s^{-1})	$(9.9 \pm 3) \times 10^6$
k_{43} (s^{-1})	$(3.3 \pm 1) \times 10^6$
K_1 (dm^3 mol^{-1})	(220 ± 100)
K_2	(0.7 ± 0.5)
K_3	(3.0 ± 0.7)
ΔV_1 (cm^3 mol^{-1})	(13 ± 3)
ΔV_2 (cm^3 mol^{-1})	(7 ± 2)
ΔV_3 (cm^3 mol^{-1})	(7 ± 2)

[a] From Bonsen *et al.*[102]

produce a factor of ~ 2 difference in k_{34}.[78,101,104] Fortunately, this difference is not critical to the mechanistic arguments in this section or those which follow.

It is now appropriate to consider the kinetic data for Sc^{3+} in Table 5 in conjunction with the reaction scheme shown in Eq. (9). The diffusion-controlled formation (k_{12}) of $Sc^{3+} \cdot OH_2 \cdot OH_2 \cdot SO_4{}^{2-}$ is not unexpected but the subsequent and much slower formation (k_{23}) of $Sc^{3+} \cdot OH_2 \cdot SO_4{}^{2-}$ implies an unexpected degree of structuring in the second coordination sphere of Sc^{3+}. By comparison to $k_{ex} \approx 10^{10}\,s^{-1}$ determined for water exchange from the second coordination sphere of $[Cr(OH_2)_6]^{3+}$,[92] both k_{23} and k_{32} seem very slow; and k_{23} is very similar in magnitude to k_{34}, which according to Eq. (9) characterizes the formation of $ScSO_4{}^+$. (It is tempting to speculate that the middle and lowest frequencies seen in Fig. 2 characterize either the formation of the first and second bonds of a chelating $SO_4{}^{2-}$, or the formation of $ScSO_4{}^+$ and $Sc(SO_4)_2{}^-$, respectively, but there appears to be no independent support for such hypotheses.) However, this does not appear to seriously affect the derivation of k_{34}, which is the rate constant of major mechanistic interest. It has been shown[78] for a number of bivalent metal ions that k_{34} for $SO_4{}^{2-}$ substitution is within a factor of 2 of k_{ex} determined directly for water exchange. If a similar assumption is made for Sc^{3+} (for which the six- and eight-coordinate ionic radii are 74.5 and 87.0 pm, respectively[14]), $k_{ex} \approx 10^7\,s^{-1}$, which from Table 4 is seen to be substantially greater than that characterizing the similar sized $[Mg(OH_2)_6]^{2+}$ and comparable to that of the smaller $[Mn(OH_2)_6]^{2+}$. While $[Sc(OH_2)_n]^{3+}$ is much more labile than the trivalent hexaaquo ions in Table 4 it is two orders of magnitude less labile than $[Gd(OH_2)_9]^{3+}$. Thus, the lability of $[Sc(OH_2)_n]^{3+}$ seems too high for $n = 6$ but insufficient for $n = 9$, leaving the possibility that $n = 7$ or 8. As both $[Ti(OH_2)_6]^{3+}$ and $[In(OH_2)_6]^{3+}$ exchange water through **a** activation modes it could be argued that because the six-coordinate radius of Sc^{3+} is similar to that of these two ions, $[Sc(OH_2)_{7\,or\,8}]^{3+}$ would be relatively sterically crowded and therefore tend toward a **d** activation mode for water exchange. Some support for this rationale may be drawn from the studies of Sc^{3+} in nonaqueous solution discussed in Section IV,A. In subsequent discussion k_{34} for $SO_4{}^{2-}$ substitution on Sc^{3+} will be considered approximately equal to k_{ex} for water exchange.

D. Sulfate substitution on the trivalent lanthanides

Sulfate appears to coordinate to Ln^{3+} as a monodentate ligand and in consequence should provide valuable information on the first bond formation in $[Ln(OH_2)_9]^{3+}$ ligand substitution processes. Several studies of $SO_4{}^{2-}$

substitution have been published containing k_{34} and k_{43} values derived using both the three-step mechanism of Eq. (9) and its truncation the two-step mechanism.[101,105–109] Purdie's group has examined the widest range of Ln^{3+}, and their k_{34} and k_{43} data[105] obtained by assuming the three-step mechanism are plotted in Fig. 3 and form the basis of discussion here. Rate constants determined from the same data using the two-step scheme are in some cases smaller by 50% but in general show the same variation with Ln^{3+} as those derived using the three-step mechanism. Silber's[107] group has also reported k_{34} and k_{43} values for La^{3+}, Nd^{3+}, Gd^{3+}, and Dy^{3+} which while differing somewhat from the values obtained by Purdie are of the same order of magnitude and exhibit similar trends. In general terms these variations in magnitude of k_{34} and k_{43} between treatment and group do not seriously impair the ensuing mechanistic discussion.

It was noted in Section III,C that k_{34} for SO_4^{2-} substitution is often similar to k_{ex} for water exchange, and the same is also found to be the case for Gd^{3+}. Thus, k_{34} (at 298.2 K) is reported as $6.4 \times 10^{8\,[105]}$ and 8.2×10^8 s$^{-1[107]}$ and k_{ex} (298.2 K) is $10.6 \times 10^{8[75]}$ and 9×10^8 s^{-1}.[85] For SO_4^{2-} substitution $\Delta H^{\ddagger} = 9.2 \pm 3.8$ kJ mol^{-1} and $\Delta S^{\ddagger} = -46 \pm 13$ J K^{-1} mol^{-1},[105] which

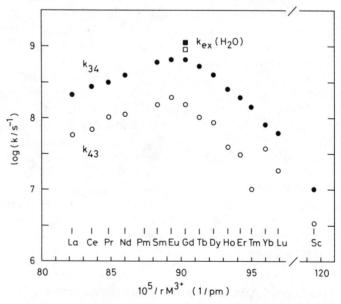

Fig. 3. Sulfate substitution on Ln^{3+} and Sc^{3+}. Log k_{34} (●) and log k_{43} (○) are plotted against the nine-coordinate $1/r(M^{3+})$. These data refer to 298.2 K and are taken from Refs. [101, 105, and 106]. Log k_{ex} for water exchange on $[Gd(OH_2)_9]^{3+}$ at 298.2 K from Refs. [85 and 75] are shown as (□) and (■), respectively.

though subject to substantial experimental error are values similar to those given for water exchange on $[Gd(OH_2)_9]^{3+}$ in Table 4. These data are consistent with $SO_4{}^{2-}$ substitution occurring on $[Gd(OH_2)_9]^{3+}$ through an I_d mechanism. However, it should not be forgotten that k_{34} for $SO_4{}^{2-}$ substitution[78] is close to k_{ex} for water exchange on $[Mn(OH_2)_6]^{3+}$, which a negative $\Delta V^{\ddagger}$ indicates exchanges through an I_a mechanism.[81] This assignment of a **d** activation mode to $[Gd(OH_2)_9]^{3+}$ is rather tentative and it is probably wiser to leave a final decision on the activation mode of water exchange and monodentate substitution on Ln^{3+} until higher frequency NMR spectrometers have provided an extended range of k_{ex} and $\Delta V^{\ddagger}$ data. (Those readers hungering for an unequivocal mechanistic statement should turn to Section IV,B in which such statements appear for nonaqueous solvent exchange on Ln^{3+}.) Nevertheless, it seems reasonable to assume that k_{34} for $SO_4{}^{2-}$ substitution is a close reflection of k_{ex} for $[Ln(OH_2)_9]^{3+}$, and in subsequent sections k_{34} for $SO_4{}^{2-}$ will be considered $\approx k_{ex}$ except in the case of $[Gd(OH_2)_9]^{3+}$ where k_{ex} is known from direct measurements.

If variation of the Ln^{3+} ionic radius was the sole determinant of lability the k_{34} values in Fig. 3 should decrease monotonically as the surface charge density increases from La^{3+} to Lu^{3+}. That this does not occur is a clear indication of a structural factor superimposing on the effect of variation in ionic radius. It was seen in Section II,A that while $Ln-OH_2$ distances decreased generally for $[Ln(OH_2)_9]^{3+}$ for La^{3+} to Lu^{3+}, the $Ln^{3+}-OH_2$ (capping)/$Ln^{3+}-OH_2$ (prismatic) ratio tended to increase as steric crowding increased. The labilizing effect of steric crowding (for a **d** activation mode) should grow as the ionic radius decreases and thus the variation of k_{34} (and k_{ex}) seen in Fig. 3 may be explained in terms of the labilizing effect of steric crowding superimposing on a general decrease in lability as ionic radius decreases. This rationalization is applicable only to the operation of a **d** activation mode for $[Ln(OH_2)_9]^{3+}$. A similar explanation is applicable to the variation of k_{43}. The k_{34} and k_{43} values for Sc^{3+} are also shown in Fig. 3 and are seen to be much less than those for Ln^{3+}, commensurate with the much smaller radius of this ion. (The variation of k_{34} could also be explained in terms of a change in coordination number from 9 to 8 occurring between Nd^{3+} and Tb^{3+} if the lability of $[Ln(OH_2)_9]^{3+}$ generally increased with decrease in ionic radius and vice versa for $[Ln(OH_2)_8]^{3+}$.)

Substitution of $NO_3{}^-$ on La^{3+}, Ce^{3+}, Pr^{3+}, Nd^{3+}, Gd^{3+}, Dy^{3+}, Er^{3+}, Yb^{3+}, and Y^{3+} has also been studied using ultrasonic absorption techniques.[110–113] Data treatment varies substantially such that a direct comparison of rate constants is difficult; however, for Nd^{3+} and Er^{3+} where k_{34} (at 298.2 K) are reported, the values of 1.8×10^6 and 6×10^7 s^{-1}, respectively,[112] are lower than the values for $SO_4{}^{2-}$ plotted in Fig. 3.

E. Multidentate ligand substitution on the trivalent lanthanides

The complexation of metal ions by multidentate ligands introduces the possibility that the rate-determining step in the formation reaction may not be the formation of the first metal to ligand bond, but may instead be the formation of a second or subsequent bond. Therefore, in the simplest case of a bidentate ligand, L–L, complexing $[Ln(OH_2)_9]^{3+}$ as shown in Eq. (11), the mechanistic interpretation of the observed k_f and k_b must be made with caution.

$$[Ln(OH_2)_9]^{3+} + L-L \underset{k_b}{\overset{k_f}{\rightleftharpoons}} \left[(H_2O)_7Ln\overset{L}{\underset{L}{\diagup\diagdown}}\right]^{3+} \qquad (11)$$

This is illustrated by the scheme shown in Eq. (12), in which reaction (11) is

$$[Ln(OH_2)_9]^{3+} + L-L \underset{fast}{\overset{K_0}{\rightleftharpoons}} [Ln(OH_2)_9]^{3+}\cdots L-L \underset{k_{-i}}{\overset{k_i}{\rightleftharpoons}}$$

$$[(H_2O)_8Ln-L-L]^{3+} \underset{k_{-r}}{\overset{k_r}{\rightleftharpoons}} \left[(H_2O)_7Ln\overset{L}{\underset{L}{\diagup\diagdown}}\right]^{3+} \qquad (12)$$

considered to proceed through an **I** mechanism in which k_i characterizes the first bond formation, and k_r the second in the ring closure or chelation step. By assuming $[(H_2O)_8Ln-L-L]^{3+}$ is a steady-state intermediate (and therefore does not exist in significant concentrations) and considering the case where $[Ln(OH_2)_9^{3+}\cdots L-L]$ is also small relative to the total Ln^{3+} concentration such that k_f is an observed second-order rate constant (as discussed for a similar case in Section III,A), it may be shown that

$$k_f = \frac{K_0 k_i k_r}{k_{-i} + k_r} \quad \text{(a)} \qquad k_b = \frac{k_{-i} k_{-r}}{k_{-i} + k_r} \quad \text{(b)} \qquad (13)$$

Thus, when $k_r \gg k_{-i}$, $k_f \approx K_0 k_i$ and the measured k_f refers to the first bond formation. However, when $k_r \ll k_{-i}$, $k_f \approx K_0 k_i k_r / k_{-i}$ and may be significantly less than expected for first bond formation. Similar considerations apply to k_d.

The substitution of $[LnOH_2)_9]^{3+}$ by bidentate ligands has generally been studied by temperature- or pressure-jump or stopped-flow techniques and the kinetic data have usually been in the form of a second-order k_f and a first-order k_b pertaining to Eq. (11). This is consistent with the operation of an **I** mechanism as shown in Eq. (12), and the data from several studies will be analyzed accordingly.

The study by Geier[1] of the substitution of the murexide monoanion (Fig. 4) on Ln^{3+}, Y^{3+}, and Sc^{3+} using a temperature-jump spectrophoto-

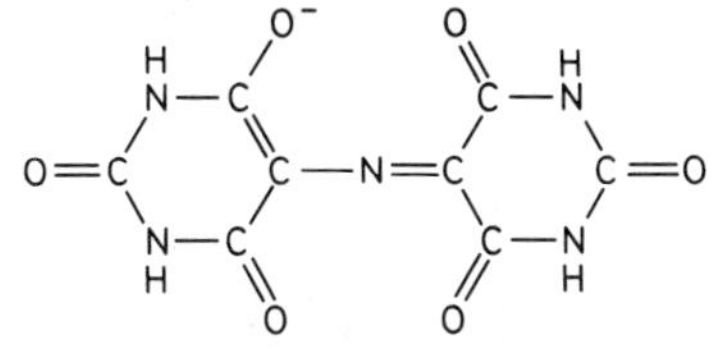

Fig. 4. Murexide.

metric method is probably the most discussed of all Ln^{3+} kinetic studies. The k_f and k_b data from this study are plotted in Fig. 5. Initially these data were interpreted in terms of first bond formation being the rate-determining step. The decrease in k_f from Sm^{3+} to Ho^{3+} was considered to be coincident with a change in stoichiometry from $[Ln(OH_2)_9]^{3+}$ to $[Ln(OH_2)_8]^{3+}$ with an equilibrium mixture of nine- and eight-coordinate species existing for the intervening Ln^{3+}. Thus, the effect of the coordination number change was thought to be superimposed on the variation of the lability of the nine- and eight-coordinate species with ionic radius. Subsequently, Geier[47] presented a revised interpretation based on $[Ln(OH_2)_9]^{3+}$ being the sole Ln^{3+} species, such that the variation in k_f arose from the variation of lability of co-ordinated water in tricapped trigonal prismatic $[Ln(OH_2)_9]^{3+}$ as steric crowding increased with decreasing ionic radius.

A pressure-jump conductometric study of substitution by the oxalate dianion[114] showed a trend in k_f similar to that for murexide, and a temperature-jump spectrophotometric study of substitution by the *o*-aminobenzoate monoanion[115–116] showed a less systematic variation of k_f, as shown in Fig. 5. If first bond formation is the rate-determining step in the substitution of murexide, *o*-aminobenzoate and oxalate on $[Gd(OH_2)_9]^{3+}$ in an I_d mechanism $k_f/k_{ex} \approx K_0 = 0.062$, 0.070, and 0.043 dm^3 mol^{-1} for murexide, *o*-aminobenzoate, and oxalate, respectively. [For $[Gd(OH_2)_9]^{3+}$ k_{ex} values were calculated from the data in Table 4 to be compatible with murexide k_f (285.2 K) and *o*-aminobenzoate k_f (285.7 K)]. At zero ionic strength and an interaction distance of 500 pm, the Fuoss[117] equation predicts $K_0 = 23$ dm^3 mol^{-1} for a $3+,1-$ ionic interaction and 1.7×10^3 dm^3 mol^{-1} for a $3+,2-$ ionic interaction. Perusal of Beck's compilation[118] of experimental K_0 values suggests that under the conditions of the studies considered here K_0 values of ~ 15 and ~ 200 dm^3 mol^{-1}, respectively, are appropriate for $3+,1-$ and $3+,2-$ interactions. However, murexide seems rather large for some of the assumptions inherent in the Fuoss calculations and it is substantially larger and more complex than the species whose K_0 values appear in Beck's compilation. The charge distribution on the murexide monoanion will be substantially delocalized and a K_0 value of ~ 1 dm^3 mol^{-1} may be more appropriate, and a similar value may be

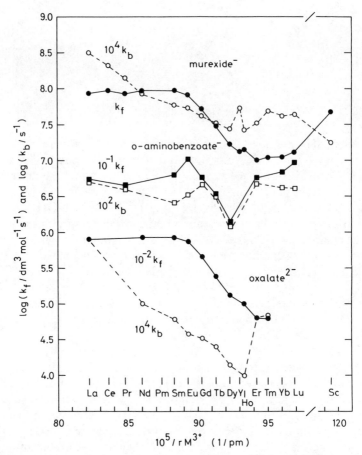

Fig. 5. Variation of multidentate ligand substitution rates with nine-coordinate $1/r(M^{3+})$ for Ln^{3+} and Y^{3+} and eight-coordinate $1/r(M^{3+})$ for Sc^{3+}. Log k_f (●) and log $10^4 k_b$ (○) at 285.2 K for murexide⁻ substitution[1] are shown in the upper data set. Log($10^{-1} k_f$) (■), and log($10^2 k_b$) (□) at 285.7 K for o-aminobenzoate⁻ substitution[116] are shown in the middle data set. Log($10^{-2} k_f$) (●) and log($10^4 k_b$) (○) at 298.2 K for oxalate²⁻ substitution[114] are shown in the lower data set.

appropriate for o-aminobenzoate. The oxalate dianion is relatively compact and a K_0 value ≥ 100 dm³ mol⁻¹ seems reasonable. On this basis the k_f/k_{ex} values indicate that k_f for murexide and o-aminobenzoate is ~20 times too small and for oxalate ~2000 times too small for first bond formation to be rate determining in an I_d mechanism. Therefore, the rate-determining step in the formation of the murexide, o-aminobenzoate, and oxalate Gd^{3+} complexes is ligand ring closure. A comparison of the k_f data in Fig. 5 with the $k_{43} \sim k_{ex}$ data for SO_4^{2-} substitution in Fig. 3 indicates that this is probably the case for all Ln^{3+}. Thus the variation of k_f (and k_b) in Fig. 5 reflects steric interactions experienced by the partially (and completely) complexed ligand.

The substitution by murexide on the smaller Sc^{3+} probably indicates a marked change in mechanism. The ratio $k_f/k_{ex} = K_0$ is 4.8 dm^3 mol^{-1} for murexide, which is within the previously discussed range of anticipated K_0 values for a $3+,1-$ interaction. (As k_{ex} refers to 298.2 K and k_f to 286.2 K it is clear that $K_0 > 4.8$ dm^3 mol^{-1}. It seems from this that the K_0 value of ~ 1 dm^3 mol^{-1} assumed for the Ln^{3+}–murexide interaction may be too small.) These Sc^{+3} results are consistent with first bond formation being rate determining in an I_d mechanism. This change in mechanism may simply be a result of k_{ex} of $\sim 10^7$ s^{-1} for Sc^{3+} being less than the k_{ex} for Ln^{3+} such that Sc^{3+} exhibits a smaller k_i/k_r ratio [Eqs. (12) and (13)] and does not necessarily require ligand ring closure for Sc^{3+} to be faster than for Ln^{3+}.

Clearly, it is desirable to observe both the first bond formation and the ring closure in the complexation of Ln^{3+} by a bidentate ligand to consolidate the preceding mechanistic deductions. The ultrasonic absorption study of the complexation of Sm^{3+} by malonate reported by Farrow and Purdie[109] appeared to achieve this. This study was carried out in parallel with a study of the SO_4^{2-} complexation of Sm^{3+}, and in the malonate system an extra relaxation at lower frequency was attributed to the ring closure step, as shown in Eq. (14), in which $O–O^{2-}$ represents malonate and the first two

$$Sm^{3+} \cdot OH_2 \cdot O{-}O^{2-} \underset{k_{43}}{\overset{k_{34}}{\rightleftharpoons}} Sm{-}O{-}O^+ \underset{k_{54}}{\overset{k_{45}}{\rightleftharpoons}} Sm{\overset{\textstyle O^+}{\underset{\textstyle O}{\diagdown\;|}}} \qquad (14)$$

stages involving the formation of ion pairs are not shown. For malonate, k_{34}, k_{43}, k_{45}, and $k_{54} = 7.1 \times 10^8$, 7.1×10^7, 5.3×10^7, and 8.1×10^6 s^{-1}, respectively, and k_{34} and $k_{43} = 3.3 \times 10^8$ and 6.3×10^7 s^{-1}, respectively, for SO_4^{2-} in the same study at 298.2 K. (The magnitude of the rate constants can change to some extent with the method of data treatment[109] but not sufficiently to affect mechanistic deductions.) The salient point to emerge is that k_{34} characterizing first bond formation ($\sim 10\ k_{45}$ for ring closure for malonate) is similar to k_{34} for SO_4^{2-}. On the basis of the similarity of k_{43} and k_{45} (equivalent to k_{-i} and k_r, respectively) Eq. (13a) predicts that $k_f \approx K_0 k_i/2 \approx K_0 k_{ex}/2$ for the Sm^{3+} malonate system. Petrucci and co-workers[111] and Purdie and Farrow[101] using a somewhat different approach to that presented above also concluded that ring closure is the rate-determining step in the formation of the murexide, o-aminobenzoate, and oxalate Ln^{3+} complexes; and the latter authors estimate $k_{45} = k_r = 1.2 \times 10^6$, 1.3×10^6, and 4×10^5 s^{-1} at 298.2 K for these respective ligands in the Dy^{3+} complexes.

The acetate Ln^{3+} complex formation was also thought to involve chelation at one stage but it appears now that this is unlikely.[109,119,120]

The murexide studies were mainly carried out in 0.1 mol dm^{-3} KNO_3 and KCl solution and no significant variation in rate constants was observed.[1] It has been previously noted that both NO_3^- and Cl^- enter the first coordination sphere of $[Ln(OH_2)_9]^{3+}$ but this appears to occur to only a minor extent under the conditions of the murexide studies and does not significantly affect the murexide substitution kinetics. In the study of substitution of Ln^{3+} by o-aminobenzoate[115,116] it was found that both k_f and k_b in 0.1 mol dm^{-3} KNO_3 increased to some extent compared to the values obtained in 0.2 mol dm^{-3} $NaClO_4$. This variation, which could well reflect a variation in K_0 with ionic strength in combination with a general salt effect on the reaction rate, does not alter the foregoing mechanistic discussion. A deuterium isotope effect has been reported[116] for o-aminobenzoate substitution on some $[Ln(OH_2)_9]^{3+}$ in D_2O solution. Similar isotope effects have been reported[107] for substitution by SO_4^{2-} but there is some dispute over the presence of the latter isotope effect.[101,107] The change of solvent from H_2O to D_2O will result in hydrogen bonding changes from the first coordination sphere to bulk solvent. In consequence, the interpretations of rate variations resulting from such a solvent change in terms of mechanism seem unlikely to be definitive and are not explored here.

F. Polyaminocarboxylate ligand substitution on the trivalent lanthanides

The polyaminocarboxylates form Ln^{3+} complexes of high stability[2,40] [as exemplified by log K (dm^3 mol^{-1}) = 15.50 and 19.83 for the La^{3+} and Lu^{3+} complexes of $EDTA^{4-}$, respectively[41]] and have been much studied as a consequence of their importance in lanthanide and actinide separation procedures.[42–44] Most of the kinetic data for the formation of these complexes have been obtained through studies of metal ion exchange reactions[121–124] as shown in Eq. (15) in which *M may be an isotope

$$ML + {}^*M \rightleftharpoons {}^*ML + M \tag{15}$$

of M or a different metal, and in which no charges are shown. This section is predominantly concerned with formation and dissociation reactions of polyaminocarboxylate lanthanide complexes, and metal ion exchange processes are considered more specifically in Section III,G. As Ln^{3+}, Y^{3+}, and Sc^{3+} constitute 17 different ions it is perhaps not surprising that there is only one kinetic study,[121] that of complexes of trans-1,2-diaminocyclohexane-N,N,N',N'-tetraacetate ($DCTA^{4-}$), which covers this range (excluding Pm^{3+}).

The more familiar $EDTA^{4-}$ ligand or complexone is the next most studied in this respect.[122,123] There are several studies of other polyaminocarboxylate systems but their experimental conditions vary substantially and make direct comparisons difficult. This section is largely confined to a consideration of Ln^{3+} complexation of flexible $EDTA^{4-}$ and $DCTA^{4-}$ which as a consequence of the cyclohexane ring has a substantially decreased flexibility. Together, these two systems typify most mechanistic aspects of polyaminocarboxylate Ln^{3+} complexation processes.

The formation and dissociation rate constants[121–123] for the general equilibrium shown in Eq. (16) for some Ln^{3+} are listed in Table 6.

$$M^{3+} + (H)_iL \underset{k_{bH_i}}{\overset{k_{fH_i}}{\rightleftharpoons}} [ML]^- + iH \qquad (16)$$

(Coordinated water is not shown.) The $EDTA^{4-}$ system will be considered first. Taking Gd^{3+} as our model ion again it is found that when $i = 1$ and 2, respectively, $k_{fH_i}/k_{ex} = 0.28$ and $0.001 \text{ dm}^3 \text{ mol}^{-1}$ which is approximately 10^3 times smaller than the anticipated K_0 for $3+,3-$ and $3+,2-$ interactions. There can be little doubt that the large size of $(H)_iEDTA^{(4-i)-}$ and the distribution of the charge over the polyaminocarboxylate will affect the encounter complex interaction, but it seems unlikely that the intensity of this interaction will be decreased to the extent required for the observed k_{fH_i} to characterize a first bond rate-determining step of an I_d mechanism. Thus, a ring closure step is probably rate determining as also appears to be the case when $M^{3+} = Nd^{3+}$, Er^{3+}, and Y^{3+}.

The formation of $(H)_iEDTA^{(4-i)-}$ complexes is generally considered to proceed through the coordination of a carboxylate oxygen which is followed by the sequential coordination of the second carboxylate and the nitrogen of one imidodiacetate (IDA) group. The sequential coordination of the second IDA group follows. (This is sometimes referred to as the IDA mechanism.) It is probable that the nonprotonated IDA of $(H)EDTA^{3-}$ coordinates first but it is not obvious at what coordination stage the proton is lost. In $(H)_2EDTA^{2-}$ in the free state both protons are thought to be on the IDA nitrogens, and it has been suggested[124] that the ~ 300-fold difference between k_{fH} and k_{fH_2} may be a consequence of a rate-determining proton transfer from a $(H)_2EDTA^{2-}$ nitrogen. As K_0 for $(H)EDTA^{3-}$ and $(H)_2EDTA^{2-}$ may well differ by ~ 300, the postulation of such a process seems unnecessary in this case.

It is appropriate at this point to review briefly the structural possibilities for $[M((H)_iEDTA)(OH_2)_n]^{(4-i)-}$ in solution. Solid-state X-ray studies[125] show that La^{3+} is nine coordinate in $[La(EDTA)(OH_2)_3]^-$ with $EDTA^{4-}$ acting as a hexadentate ligand, and that the three waters occupy adjacent positions to form a triangular face. Solution thermodynamic[49] and spectroscopic[48] studies have been interpreted to mean that a similar nine-coordinate species is the sole species in solution for $Ln^{3+} = Ce^{3+}-Nd^{3+}$, and for $Dy^{3+}-Lu^{3+}$

TABLE 6

Rate constants for the equilibrium $M^{3+} + (H)_iL \overset{k_{fH_i}}{\underset{k_{bH_i}}{\rightleftharpoons}} [ML]^- + iH^+$ at 298.2 K

M^{3+}	L	k_{fH} (dm^3 mol^{-1} s^{-1})	k_{fH_2} (dm^3 mol^{-1} s^{-1})	k_b (s^{-1})	k_{bH} (dm^3 mol^{-1} s^{-1})	k_{bH_2} (dm^6 mol^{-2} s^{-1})
La^{3+}	EDTA^{4-}[a]	—	—	2×10^{-2}	3.7×10^3	1.9×10^6
	DCTA^{4-}[b]	0.8×10^7	—	—	1.29×10^2	—
Pr^{3+}	EDTA^{4-}	1.6×10^7	—	9×10^{-3}	1.6×10^3	1.0×10^6
	DCTA^{4-}	—	—	1×10^{-4}	35	—
Nd^{3+}	EDTA^{4-}	2.7×10^8	1.1×10^6	—	1.09×10^3	7.0×10^5
	DCTA^{4-}	2.1×10^7	—	9.2×10^{-5}	16.2	—
Eu^{3+}	EDTA^{4-}	3.4×10^7	—	—	4.3×10^2	2.8×10^6
	DCTA^{4-}	—	—	9.8×10^{-6}	2.2	—
Gd^{3+}	EDTA^{4-}	3.0×10^8	1.1×10^6	—	—	—
	DCTA^{4-}	2.2×10^7	—	1×10^{-5}	1.3	—
Er^{3+}	EDTA^{4-}	1.8×10^8	0.54×10^6	—	8.9	7.7×10^4
	DCTA^{4-}	—	—	6.1×10^{-7}	5.3×10^{-2}	—
Yb^{3+}	EDTA^{4-}	—	—	—	0.76	1.43×10^4
	DCTA^{4-}	3.9×10^7	—	5.4×10^{-7}	2.3×10^{-2}	—
Y^{3+}	EDTA^{4-}	0.9×10^8	0.18×10^6	—	—	—
	DCTA^{4-}	2.4×10^7	—	1.7×10^{-6}	0.36	—

[a] k_b, k_{bH}, and k_{bH_2} refer to 0.5 mol dm^{-3} KCl data from Ref. [123]. k_{fH} and k_{fH_2} refer to 1.0 mol dm^{-3} data from Ref. [122].
[b] k_{fH}, k_b, and k_{bH} refer to 0.1 mol dm^{-3} NaClO$_4$ data from Ref. [121]. The rate constants k_{fH} and k_{fH_2} for EDTA^{4-} were determined through the reaction [CeL]$^-$ + M^{3+} → [ML]$^-$ + Ce^{3+}. For the determination of k_b, k_{bH}, and k_{bH_2} for EDTA^{4-} the reaction [ML]$^-$ + Cu^{2+} → [CuL]$^{2-}$ + M^{3+} was monitored. All rate constants for DCTA^{4-} were determined through this reaction. The processes characterized by the rate constants in this table are not affected by the presence of the second metal ion. Metal ion exchange reactions are discussed in Section III,G.

eight-coordinate $[Ln(EDTA)(OH_2)_2]^-$ is the sole species. It was also determined that for Sm^{3+}, Eu^{3+}, Gd^{3+}, and Tb^{3+} the ratio [nine-coordinate]/[eight-coordinate] $= 0.2, 0.7, 1.4$, and 10, respectively, at $293.2\,K$ in $0.1\,mol\,dm^{-3}\,KNO_3$ solution. Other studies[45,50] are in agreement that a change in coordination occurs in the middle of the Ln^{3+} series. On the basis of the apparent molar volume ϕ_v^0 Brücher and co-workers[71] have concluded that although for $Ln^{3+} = Gd^{3+}-Lu^{3+}$ eight-coordinate $[Ln(EDTA)(OH_2)_2]^-$ persists, as the size of Ln^{3+} increases a carboxylate group dissociates to an increasing extent until nine-coordinate $[La(EDTA)(OH_2)_4]^-$ is reached. This progressive dissociation of a carboxylate group has also been suggested by Kostromina and co-workers.[126,127] (The ϕ_v^0 values for $K[Ln(EDTA)(OH_2)_n]$, which vary from $166.2-164.8$ to $175.1-170.5\,cm^3\,mol^{-1}$ when $Ln^{3+} = La^{3+}, Pr^{3+}, Gd^{3+}$, and Yb^{3+}, respectively, show larger changes than those for $LnCl_3$, which may indicate a more substantial structure change for $[Ln(EDTA)(OH_2)_n]^-$ as Ln^{3+} varies.)

NMR studies[128–131] show conformational change, nitrogen inversion, and carboxylate group dissociation and reassociation to occur with increasing rapidity from Lu^{3+} to La^{3+}. It was also concluded from one of these studies[130] that the coordination number of Ln^{3+} in $[Ln(EDTA)(OH_2)_n]^-$ and several other polyaminocarboxylate complexes was invariant at nine for $Ce^{3+}-Yb^{3+}$. With the exception of this last study a consensus emerges that as $r(M^{3+})$ increases from Lu^{3+} to Tb^{3+} eight-coordinate $[Ln(EDTA)(OH_2)_2]^-$ is the sole species in solution, but with further increase in $r(M^{3+})$ a change in coordination occurs as a consequence of either a carboxylate detachment or the coordination of further water molecules, or both.

In the acid-independent dissociation path characterized by k_b (Table 6), it is expected that the dissociation of a carboxylate group is followed by the sequential dissociation of a second carboxylate and a nitrogen such that the complete dissociation of an IDA group is followed by that of a second. It seems probable that the $Ln-N$ bond dissociation is the rate-determining step for the IDA group dissociation.[123] It is reasonable to assume that k_{bH} characterizes a similar sequence of events preceded by the rapid protonation of the free carboxylate group. It should be noted, however, that in the solid state the proton resides on the noncoordinating oxygen of a coordinated carboxylate group in 10-coordinate $[Ln((H)EDTA)(OH_2)_4]^{-}$[76] and so this possibility must be entertained in solution also. (The log K values for such protonated Ln^{3+} species vary between 2.5 and 2.8.[132]) The k_{bH_2} path characterizes a doubly protonated species but the site of this second protonation is unclear.

The best determined dissociation rate constant, k_{bH}, exhibits a 5×10^3-fold decrease from La^{3+} to Yb^{3+} (Table 6). This decrease is substantially greater than that expected simply from the increase in surface charge density over this range, but is coincident with the decrease in the mobility of $EDTA^{4-}$ on the Ln^{3+} surface and the tendency for a carboxylate group to dissociate as the series is traversed from La^{3+} to Lu^{3+}. The tendency for the coordination number of $[Ln(EDTA)(OH_2)_n]^-$ to decrease from La^{3+} to Lu^{3+} with a consequent decrease in steric crowding may also be a factor in the variation of k_{bH}.

The formation and dissociation rate constants [Eq. (16)] k_{fH}, k_b, and k_{bH} determined for $[M(DCTA)]^-$ (where $M^{3+} = Ln^{3+}$, Y^{3+}, and Sc^{3+} and coordinated water is not shown) are plotted in Fig. 6. For Ln^{3+} and Y^{3+} k_{fH} varies over one order of magnitude with a tendency toward larger values for the heavier Ln^{3+}, while log k_b and log k_{bH} vary linearly with $1/r(M^{3+})$ over four orders of magnitude. The ratio k_{fH}/k_{ex} for $Gd^{3+} = 0.022$ dm^3 mol^{-1} which compares with $K_0 = 100$ dm^3 mol^{-1} estimated for a $3+,3-$ interaction by Nyssen and Margerum.[121] Clearly k_{fH} is too small for first bond formation in an I_d mechanism to be rate determining, and the same situation prevails for the remaining Ln^{3+} and Y^{3+}. Once again the situation is different for Sc^{3+}. The k_{fH} value for Sc^{3+} is two to three orders of magnitude greater than that for Ln^{3+}, and $k_{fH}/k_{ex} = 600$ dm^3 mol^{-1}, which is of a magnitude expected for K_0 characterizing $3+,3-$ interactions and indicates that first bond formation is the rate-determining step for Sc^{3+}. This implies that the slowest ring closure step for Sc^{3+} is substantially faster than that for Ln^{3+}, which suggests that Sc^{3+} can enter the coordination cage of $(H)DCTA^{3-}$ with a minimum of steric hindrance whereas the larger Ln^{3+} cannot. (The coordination number of $[Ln(DCTA)]^-$ in solution is not known with certainty but a recent spectroscopic study[133] detects two $[Eu(DCTA)]^-$ species in solution, which indicates a variation in the number of coordinated waters or the coordination of $DCTA^{4-}$.)

A comparison of the data in Table 6 shows k_{fH} for $(H)EDTA^{3-}$ to be 4–14 times greater than k_{fH} for (H)DCTA, consistent with the latter ligand being less flexible in adopting the stereochemistry required for Ln^{3+} complexation. (The ionic strength of the $(H)EDTA^{3-}$ studies is 1.0 M while that of the $(H)DCTA^{3-}$ studies is 0.1 M, which should cause the K_0 value in the former system to be substantially less than that in the latter and the true difference in k_{fH} to be greater.) The k_{bH} values for $[Ln(EDTA)]^-$ are greater than those for $[Ln(DCTA)]^-$, which is also consistent with the lower flexibility of the latter system. It is, however, k_b which indicates that the dissociation and formation of $[Ln(DCTA)]^-$ are unlikely to proceed through a simple sequential unwrapping and wrapping of $DCTA^{4-}$. The magnitudes of k_b (Fig. 6) are such that the values of k_f calculated from the appropriate stability constants[2,134]

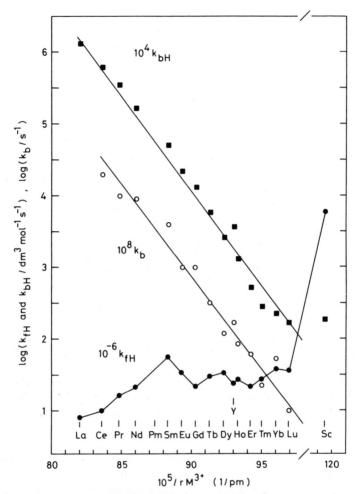

Fig. 6. Variation of $\log(10^{-6} k_{fH}; \text{dm}^3 \text{ mol}^{-1} \text{ s}^{-1})$ ($\bullet$), $\log(10^8 k_b; \text{s}^{-1})$ ($\bigcirc$), and $\log(10^4 k_{bH};$ $\text{dm}^3 \text{ mol}^{-1} \text{ s}^{-1})$ ($\blacksquare$) with nine-coordinate $1/r(M^{3+})$ for Ln^{3+} and Y^{3+}, and eight-coordinate $1/r(M^{3+})$ for Sc^{3+}, for the $[M(DCTA)]^-$ system. Data from Ref. [121].

substantially exceed the diffusion-controlled limit of $\sim 10^{10} \text{ dm}^3 \text{ mol}^{-1} \text{ s}^{-1}$, which indicates that k_f is unlikely to characterize the complete dissociation of $[Ln(DCTA)]^-$.

Nyssen and Magerum[121] have proposed that a rate-determining distortion of $[Ln(DCTA)]^-$ occurs to produce a reactive species which may then react rapidly with a proton or another metal ion to produce hydrated Ln^{3+}, as shown in Fig. 7. (This study was carried out in excess of Cu^{2+} scavenger and hence Cu^{2+} appears in Fig. 7.) In a fast preequilibrium M^{3+} and $(H)DCTA^{3-}$

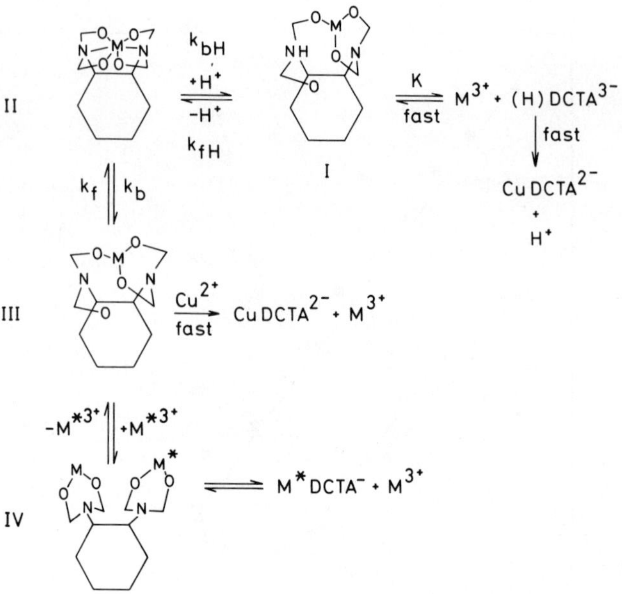

Fig. 7. Proposed reaction scheme[121] for [M(DCTA)]$^-$.

form I in which M^{3+} is coordinated by three oxygens but remains outside the DCTA^{4-} coordination cage. In the rate-determining step (k_{fH}) which may involve proton transfer from a nitrogen to a carboxylate oxygen, M^{3+} becomes fully coordinated to form II. For the paths characterized by k_b and k_{bH} the rate-determining deformation of II produces III and I, respectively. As both log k_b and log k_{bH} vary linearly with $1/r(M^{3+})$ it appears that the larger Ln^{3+} engender this deformation more effectively than the smaller Ln^{3+}. (To some extent this scheme resembles that proposed for the dissociation and metal exchange reactions of cryptates which are envisaged as proceeding through exclusive and inclusive forms of the cryptate.[135–137]) An alternative explanation for k_f has been proposed by Glentworth and co-workers[138–139] which envisages the protolysis of a coordinated water in [Ln(DCTA)(OH$_2$)$_n$]$^-$ followed by a rate-determining protonation of the DCTA^{4-} ligand, which subsequently rapidly dissociates from the complex as shown in Eq. (17):

$$[\text{Ln(DCTA)(OH}_2)]^- \underset{\text{fast}}{\overset{K_1}{\rightleftharpoons}} [\text{Ln(DCTA)(OH)}]^{2-} + \text{H}^+ \qquad (17)$$

$$[\text{Ln(DCTA)(OH)}]^{2-} + \text{H}^+ \xrightarrow{k_b'} [\text{Ln((H)DCTA)(OH)}]^-$$

$$[\text{Ln((H)DCTA)(OH)}]^- \xrightarrow[\text{fast}]{} \text{Ln}^{3+} + (\text{H})\text{DCTA}^{3-} + \text{OH}^-$$

Thus, $k'_b K_1 = k_b$, and it is estimated[121] that when $Ln^{3+} = Ce^{3+}$, $k'_b = 2.2 \times 10^5$ dm^3 mol^{-1} s^{-1}, which is much greater than the reported $k_{bH} = 67$[138,139] and 60 dm^3 mol^{-1} s^{-1}[121] values. [Such a labilization caused by coordinated OH^- is not unexpected as Ryhl[123,129] has reported k_b and k_{bOH} as 2×10^{-2} and $> 2 \times 10^5$ s^{-1}, respectively, for $[Ln(EDTA)]^-$ and $[Ln(EDTA)(OH)]^{2-}$ at 298.2 K. Southwood-Jones and Merbach[140] report the pK_a values for $[M(EDTA)(OH_2)_n]^-$ to be 10.6 (Sc^{3+}), 11.9 (Y^{3+}), > 11 (La^{3+}), 12.86 (Pr^{3+}), ~ 12.2 (Nd^{3+}), 12.48 (Eu^{3+}), 12.3 (Yb^{3+}), and 11.7 (Lu^{3+}).] Glentworth and Newton[139] rationalize the greater k'_b value largely on the basis of a more favorable electrostatic factor for the protonation of $[Ln(DCTA)(OH)]^{2-}$ by comparison to that for $[Ln(DCTA)]^-$. Though a choice between these two mechanistic proposals is not readily made from a comparison of the magnitudes of k_b, k'_b, and k_{bH}, the observation of direct exchange of Y^{3+} on $[Y(DCTA)]^-$ does require a binuclear intermediate bearing some similarity to form IV in Fig. 7.[78,139]

An extreme case of a polyaminocarboxylate decreased flexibility is that of 1,4,7,10-tetraazacyclododecane-N,N',N'',N'''-tetraacetate ($DOTA^{4-}$), which is also characterized by very slow formation reactions with Ln^{3+}. Nevertheless, the stability constant for $[Gd(DOTA)]^-$ is 10^{28}, some 10 orders of magnitude greater than that for $[Gd(EDTA)]^-$.[141]

G. Metal ion exchange reactions in polyaminocarboxylate complexes

The exchange of the same or different metal ions between a metal complex, ML, and the hydrated state, M, as shown in Eq. (18) (in which charges are

$$[ML] + {}^*M \xrightarrow{k_{Mex}} [{}^*ML] + M \qquad (18)$$

omitted) has been studied quite widely. In the case of Ln^{3+}, Y^{3+}, and Sc^{3+} such studies have largely been confined to polyaminocarboxylate complexes where the observed metal exchange rate constant k_M varies according to Eq. (19), in which $i = 1, 2$, and 3. The k_b and k_{bH_i} terms are the same as those

$$k_{Mex} = k_b + \sum_i k_{bH_i}[H^+]^i + k_M[M] + k_{MH}[H^+][M] \qquad (19)$$

discussed in Section III,F, and the new terms, k_M and k_{MH}, characterize the formation of a binuclear intermediate species. The relative importance of the terms in Eq. (19) varies with the nature of both M and L as may be seen from the rate constants obtained through ${}^{90}Y$, ${}^{144}Ce$,[139] ${}^{152-4}Eu$,[142] and ${}^{46}Sc$[143] isotope exchange studies listed in Table 7. Depending on the relative sizes of the terms in Eq. (19), some terms may not be observed and thus the

TABLE 7

Rate constants for metal ion exchange reactions in polyaminocarboxylate complexes

[ML]	k_b (s⁻¹)	k_{bH} (dm³ mol⁻¹ s⁻¹)	k_{bH_2} (dm⁶ mol⁻² s⁻¹)	k_M (dm³ mol⁻¹ s⁻¹)	k_{MH} (dm³ mol⁻¹ s⁻¹)	T (K)
[Ce(EDTA)]⁻ [a]	—	1.6 × 10³	—	—	—	293.2
[Y(EDTA)]⁻ [b]	—	30	5.2 × 10⁵	—	3.9 × 10³	298.2
[Ce(DPTA)]²⁻ [a]	—	—	5.0 × 10⁶	—	2.7 × 10⁻²	293.2
[Y(DPTA)]²⁻ [b]	—	—	1.8 × 10⁵	4 × 10⁻³	8.6 × 10²	298.2
[Ce(HEDTA)]ᵃ [a]	—	3.6 × 10³	—	—	—	293.2
[Y(HEDTA)]ᵇ [b]	—	72	—	—	—	298.2
[Sc(HEDTA)]ᶜ [c]	—	12.4	—	—	—	298.2
[Ce(DCTA)]⁻ [a]	7 × 10⁻⁵	65	—	9.3 × 10⁻²	—	293.2
[Eu(DCTA)]⁻ [d]	9 × 10⁻⁶	3.15	—	—	—	298.2
[Y(DCTA)]⁻ [c]	6.7 × 10⁻⁷	0.49	—	1.3 × 10⁻⁴	—	298.2

[a] Ref. [141]. Studies in 10⁻² mol dm⁻³ ammonium acetate buffer with ionic strength adjusted to 1.0 mol dm⁻³ with KNO₃.
[b] Ref. [139]. Same conditions as above.
[c] Ref. [143]. Studies in 2 × 10⁻² mol dm⁻³ sodium acetate buffer with ionic strength adjusted to 0.5 mol dm⁻³ with NaClO₄.
[d] Ref. [142]. Same conditions as above.

data sets in Table 7 are incomplete. Nevertheless, there is a general trend for k_b, k_{bH}, and k_{bH_2} to decrease as $r(M^{3+})$ decreases. However, in the case of k_M and k_{MH} the trends are variable, which is probably a reflection of the change in the nature of the binuclear intermediate as the polyaminocarboxylate changes.

It is generally considered that the most probable path for direct metal ion exchange in $[Ln(EDTA)]^-$ complexes is through a binuclear species in which Ln^{3+} and the exchanging M^{m+} are both coordinated by IDA groups of $EDTA^{4-}$. This exchange process is shown in Eq. (20) in which several steps are combined together. Thus, the first major stage in the exchange mechanism is for an IDA group of $EDTA^{4-}$ to become detached from $[Ln(EDTA)]^-$ to produce A (as in spontaneous dissociation of the complex; Section III,F).

$$\tag{20}$$

(A)   (B)   (C)

Subsequently, M^{m+} coordinates to the free IDA group to produce the binuclear intermediate (B) from which Ln^{3+} then dissociates to form C with the subsequent formation of $[M(EDTA)]^{(4-m)-}$. The two mononuclear (A and C) and the binuclear (B) intermediates are formed in very low concentration such that $k_{-1} \gg k_1$, $k_{-1} \gg k_2$, $k_4 \gg k_{-3}$ and $k_4 \gg k_{-4}$. If it is assumed that the relative stability constants of the mononuclear intermediates are similar to $K_{Ln(IDA)}$ and $K_{M(IDA)}$ and that water exchange on Ln^{3+} and M^{m+} proceeds through **d** activation modes, it may be shown that

$$\frac{k_{-2}}{k_3} \approx \frac{k_2 K_{Ln(IDA)}}{k_{-3} K_{M(IDA)}} \approx \frac{C_M k_{Mex} K_{Ln(IDA)}}{C_{Ln} k_{Lnex} K_{M(IDA)}} \tag{21}$$

where k_{Mex} and k_{Lnex} are the water exchange rate constants for M^{m+} and Ln^{3+}, respectively, and C_M and C_{Ln} are constants which include the encounter complex stability constant K_0 and also any effects of the rate-determining step being ring closure, as expressed through Eq. (13a). Returning to Eq. (20) it may be shown using the steady-state approximation that $[B] = k_2[A][M^{m+}]/(k_{-2} + k_3)$, and since A and $[Ln(EDTA)]^-$ are in equilibrium, the rate of direct metal ion exchange is given by Eq. (22):

$$\text{rate} = \frac{k_1 k_2 k_3}{k_{-1}(k_{-2} + k_3)}[Ln(EDTA)^-][M^{m+}] \tag{22}$$

Brücher and Laurenczy[144] have found a linear correlation between k_{ex} for $Ni^{2+}, Co^{2+}, Cu^{2+}$, and Nd^{3+} and k_M characterizing the direct displacement of Ce^{3+} from $[Ce(EDTA)]^-$ by these cations. This observation may be analyzed through Eq. (22). Since k_1/k_{-1} is constant it may be inferred that $k_3 \gg k_{-2}$ such that Eq. (22) approximates with appropriate substitutions to

$$\text{rate} = \frac{k_1}{k_{-1}} C_M k_{ex} [Ce(EDTA)^-][M^{2+}] \qquad (23)$$

Brücher and Laurenczy have inferred by substituting k_{Mex} and $K_{M(IDA)}$ values into Eq. (22) that $k_3 \gg k_{-2}$, and further have used the approximations $k_1/k_{-1} \approx K_{Ce(IDA)}/K_{Ce(EDTA)} \approx 10^{-9.9}$ and $k_2 \approx k_{ex}K_0$ where $K_0 \approx 1$ dm^3 mol^{-1}. Thus, the slope $\approx 10^{-9.1}$ obtained from a plot of k_M for $Ni^{2+}, Co^{2+}, Cu^{2+}$, and Nd^{3+} against k_{ex} is within an order of magnitude of $K_{Ce(IDA)}/K_{Ce(EDTA)}$ and is considered to represent the ratio of $[A]/[Ce(EDTA)^-]$ [see Eq. (20)]. A similar relationship between k_M and k_{ex} has been reported for the transition metal ions.[145] Similar mechanistic schemes may be proposed for other flexible polyaminocarboxylate complexes but in the case of the less flexible $DCTA^{4-}$ systems a more constrained mechanism has been proposed,[121] as discussed in Section III,F.

Acetate catalysis of metal ion exchange has been observed in the $[Sc(HEDTA)]$[143] and $[Ce(HEDTA)]$[138] systems. It is probable that a ternary complex is formed in which the $HEDTA^{3-}$ ligand is labilized. Smaller effects have been observed in the $[Eu(DCTA)]^-$,[142] $[Ce(DCTA)]^-$,[138] and $[Eu(EDTA)]^-$ [146] systems.

H. Ternary complex formation

The formation of ternary complexes is illustrated by Eq. (24) in which L is a multidentate ligand, Y is a y-dentate ligand, and charges are omitted. In the

$$[LnL(OH_2)_n] + Y \underset{k_b}{\overset{k_f}{\rightleftharpoons}} [LnLY(OH_2)_{n-y}] \qquad (24)$$

case of Ln^{3+}, L is inevitably a polyaminocarboxylate and Y contains oxygen donor atoms. Of the few reported kinetic studies of such ternary systems, those of Geier and co-workers have covered the widest range of Ln^{3+}. Some of their k_f data[147,148] for L = $EDTA^{4-}$ and $DPTA^{4-}$ and Y = tetramethyl-murexide monoanion (TMM^-) and 8-hydroxyquinoline-5-sulfonate dianion (OXS^{2-}) are shown in Fig. 8. Also shown in Fig. 8, largely for purposes of comparison, are k_f data for TMM^- substitution on $[Ln(OH_2)_9]^{3+}$. These latter k_f data are of similar magnitude to those for murexide (see Fig. 5). The ratio $k_f/k_{ex} \approx 0.07$ dm^3 mol^{-1} for $[Gd(OH_2)_9]^{3+}$, which is similar to that observed for murexide and which on the same basis as that discussed in

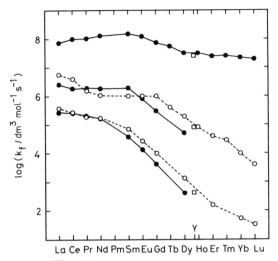

Fig. 8. The variation of k_f with M^{3+}. Data for substitution by TMM^- (tetramethylmurexide) (●) and OXS^{2-} (8-hydroxyquinoline-5-sulfonate) (○). The upper data set refers to TMM^- substitution on $[Ln(OH_2)_9]^{3+}$, the middle data set to TMM^- and OXS^{2-} substitution on $[Ln(EDTA)(OH_2)_n]^-$, and the lower data set to TMM^- and OXS^{2-} substitution on $[Ln(DPTA)(OH_2)_n]^-$. For the latter data set $k_f/10$ is plotted. (□) Data for TMM^- substitution on the analogous Y^{3+} species, positioned on the abscissa at a position appropriate to the relative magnitudes of Y^{3+} and Ln^{3+} eight- or nine-coordinate ionic radii. All data refer to 293.2 K and are from Ref. [147].

Section III,E is judged to be too small for k_f to characterize a first bond formation occurring through an I_d mechanism. Therefore the rate-determining step in the substitution of TMM^- on $[Gd(OH_2)_9]^{3+}$ and $[Ln(OH_2)_9]^{3+}$ generally is ring closure.

An inspection of Fig. 8 reveals that (1) k_f for $[Ln(OH_2)_9]^{3+}/TMM^-$ is greater than k_f for the formation of the ternary complexes; (2) k_f for the $[Ln(OH_2)_9]^{3+}/TMM^-$ varies over one order of magnitude; (3) k_f values for both ternary complexes vary over four orders of magnitude but show relatively small variations with L and Y; (4) k_f values for $[Y(OH_2)_9]^{3+}$ appear in positions predicted on the basis of $r(M^{3+})$ being a major determinant of k_f. The marked difference in the variation of k_f characterizing $[Ln(OH_2)_9]^{3+}$ and the ternary complexes is consistent with n changing in $[LnL(OH_2)_n]^-$ (as other evidence suggests; Section III,F) but remaining constant at 9 for $[Ln(OH_2)_9]^{3+}$. The variation of k_f for $[Ln(EDTA)(OH_2)_n]^-$ and $[Ln(DPTA)(OH_2)_n]^-$ may be perceived as a monotonic decrease with $1/r(M^{3+})$ with the effect of the coordination number change superimposed.

A semiquantitative analysis of mechanistic possibilities may be made starting from the water exchange data for $[Gd(PDTA)(OH_2)_2]^-$. At 293.2 K

$k_{ex} = 3.0 \times 10^8 \text{ s}^{-1}$ and the ratios k_f/k_{ex} for $[\text{Gd(PDTA)(OH}_2)_n]^-$ and TMM^- and OXS^{2-} are 1.4×10^{-4} and $3.3 \times 10^{-4} \text{ dm}^3 \text{ mol}^{-1}$, respectively, which if first bond formation occurred through an I_d mechanism may be equated to sK_0 where s is a statistical factor. Although both TMM^- and OXS^{2-} are negatively charged, the Fuoss equation predicts $K_0 \approx 0.2$ and $0.06 \text{ dm}^3 \text{ mol}^{-1}$ for interaction with $[\text{Gd(PDTA)(OH}_2)_2]^-$ at a distance of 600 pm under the conditions of the ternary complex formation studies ($0.1 \text{ mol dm}^{-3} \text{ KNO}_3$ solution). To an approximation, the geometry of $[\text{Gd(PDTA)(OH}_2)_2]^-$ will be either a square antiprism or a triangular dodecahedron,[24] and allowing TMM^- and OXS^{2-} an equal opportunity to occupy any face and an equal competitiveness with water for vacancies occurring in the first coordination sphere of Gd^{3+}, a statistical factor $s \approx 0.2$ is obtained. On this basis, the calculated $sK_0 \approx 4 \times 10^{-2}$ and 1.2×10^{-2}, respectively for TMM^- and OXS^{2-}, values which are significantly greater than the observed $k_f/k_{ex} = sK_0$. The large bulk and the delocalization of charge of TMM^- and OXS^{2-} may decrease sK_0 by a factor of 10 or so which still leaves the calculated sK_0 substantially greater than k_f/k_{ex}, thus suggesting that ring closure is determining the magnitude of k_f. On an intuitive basis this seems a reasonable deduction as ring closure also determines k_f for the substitution of TMM^- on $[\text{Gd(OH}_2)_9]^{3+}$. Nevertheless, semiquantitative estimates of this type are likely to involve substantial error. (It is appropriate to note at this point that should three H_2O be bound in $[\text{Gd(PDTA)(OH}_2)_3]^-$ the k_{ex} recalculated from the ^{17}O NMR data is $\sim 2/3 \, k_{ex}$ calculated for $[\text{Gd(PDTA)(OH}_2)_2]^-$ and so no significant changes in the preceding mechanistic deductions result.)

The deduction of the substitution mechanisms for the remaining $[\text{Ln(PDTA)(OH}_2)_n]^-$ is hampered by the absence of k_{ex} data. It is seen that for substitution by both TMM^- and OXS^{2-} k_f increases from Gd^{3+} to La^{3+} and decreases from Gd^{3+} to Ln^{3+}. If first bond formation proceeds through a **d**-activated pathway then the slower ring closure step which determines k_f will probably do likewise. On this basis the variation of k_f will reflect the variation of the lability of water in the Ln^{3+} complexes, although the magnitude of k_f will be decreased to some extent by the steric constraints imposed on ring closure. Using this rationale, it may be deduced that the lability of water decreases by three to four orders of magnitude from $[\text{La(PDTA)(OH}_2)_n]^-$ to $[\text{Lu(PDTA)(OH}_2)_n]^-$ as n decreases. (Such decreases in lability with decrease in coordination number have been observed in nonaqueous studies; Section IV,B and C.) A similar reasoning may be applied to the variation of k_f for $[\text{Ln(EDTA)(OH}_2)_n]^-$.

The variations of k_b with Ln^{3+} for the five systems are similar to those exhibited by k_f in Fig. 8, particularly in the case of the $[\text{Ln(DPTA)(OH}_2)_n]^-$ and $[\text{Ln(EDTA)(OH}_2)_n]^-/\text{OXS}^{2-}$ systems and the $[\text{Ln(OH}_2)_9]^{3+}/\text{TMM}^-$ system. In the case of the $[\text{Ln(DPTA)(OH}_2)_n]^-$ and

$[Ln(PDTA)(OH_2)_n]^-/TMM^-$ systems, the decrease in k_b with $1/r(M^{3+})$ is less marked than for k_f. Some k_b and equilibrium constants for the formation of the ternary complexes appear in Table 8.

Merbach and co-workers have demonstrated the utility of NMR in studying the exchange of multidentate ligands on Ln^{3+}. For the following exchange process,[149]

$$[Ln(EDTA)(L)] + {}^*L \underset{k_L}{\overset{}{\rightleftharpoons}} [Ln(EDTA)({}^*L)] + L \qquad (25)$$

it is found that where $L = IDA^{2-}$, k_L (293.2 K) is too large to measure for Pr^{3+} but values of 210 and 119 s^{-1} are obtained for Eu^{3+} and Yb^{3+}, respectively, consistent with the general decrease in lability of $[Ln(EDTA)]^-$ complexes as $1/r(M^{3+})$ increases. When $Ln^{3+} = Eu^{3+}$, k_L values (at 293.2 K) of 210, 250,

TABLE 8

Dissociation rate constants (k_b) and equilibrium constants (K) for the reactions:

$$[Ln(OH_2)_9]^{3+} + X^{x-} \underset{k_b}{\overset{k_f}{\rightleftharpoons}} [Ln(OH_2)_{9-x}X]^{(3-x)+}$$

$$[LnL(OH_2)_n]^- + X^{x-} \underset{k_b}{\overset{k_f}{\rightleftharpoons}} [LnLX]^{(1+x)-}$$

(where $X^{x-} = TMM^-$ or OXS^{2-} and $L = EDTA^{4-}$ or $DPTA^{4-}$)[a]

Complex	Leaving ligand	k_b (at 293.2 K) (s^{-1})	$\log(K; dm^3 \, mol^{-1})$ (at 293.5 K)
$[Ln(OH_2)_9]^{3+}$	TMM^-		
La		1.33×10^4	3.75
Dy		2.0×10^3	4.2
Lu		1.0×10^3	4.35
$[Ln(EDTA)(OH_2)_n]^-$	TMM^-		
La		8.7×10^3	2.45
Dy		4.7×10^3	1.0
	OXS^{2-}		
La		2.1×10^3	3.36
Dy		5.1	4.56
Lu		5.6×10^{-2}	4.82
$[Ln(DPTA)(OH_2)_n]^-$	TMM^-		
La		8.7×10^3	2.49
Dy		2.9×10^2	1.2
	OXS^{2-}		
La		9.1×10^2	3.60
Dy		0.74	4.23
Lu		1.8×10^{-2}	4.25

[a] All data obtained in 0.1 mol dm^{-3} KNO_3 solution; k_b values from Ref. [147] and K values from Refs. [47 and 147].

and 25 s^{-1} are observed for IDA^{2-}, MIDA^{2-} (methylaminodiacetate), and NTA^{3-} (nitrilotriacetate), respectively. In earlier studies[150,151] of NTA^{3-} exchange on $[Ln(NTA)_2]^{3-}$ the four exchange pathways shown in Eq. (26) were considered.

$$[Ln(NTA)(*NTA)]^{3-} \underset{k_{-0}}{\overset{k_0}{\rightleftharpoons}} [Ln(NTA)] + *NTA^{3-} \tag{26a}$$

$$[Ln(NTA)(*NTA)]^{3-} + NTA^{3-} \underset{k_{-1}}{\overset{k_1}{\rightleftharpoons}} [Ln(NTA)_2]^{3-} + *NTA^{3-} \tag{26b}$$

$$[Ln(NTA)(*NTA)]^{3-} + (H)NTA^{2-} \underset{k_{-2}}{\overset{k_2}{\rightleftharpoons}} [Ln(NTA)_2]^{3-} + (H)*NTA^{2-} \tag{26c}$$

$$[Ln(NTA)(*NTA)]^{3-} + H^+ \underset{k_{-3}}{\overset{k_3}{\rightleftharpoons}} [Ln(NTA)] + (*NTA)^{2-} \tag{26d}$$

According to Eq. (26) the observed exchange rate constant k_{obs} ($= 1/\tau_c$ = reciprocal of mean lifetime of $[Ln(NTA)_2]^{3-}$) should vary with pH according to Eq. (27).

$$1/\tau_c = k_0 + k_1[NTA^{3-}] + k_2[H(NTA)^{2-}] + k_3[H^+] \tag{27}$$

(Under the experimental conditions $\log K_H = pK_a = 9.55$ for (H)NTA^{2-}.) A study in the pH range 4–12 showed the contribution of the k_2 term to be negligible, consistent with (H)NTA^{2-} being relatively inactive in the exchange process.

The derived rate constants appear in Table 9 as do the independently determined equilibrium constants for $[Ln(NTA)_2]^{3-}$. The formation rate constants k_{-0} vary little from La^{3+} to Nd^{3+} and are comparable in magnitude to k_f observed for TMM$^-$ substitution on $[Ln(OH_2)_9]^{3+}$ (Fig. 8). If k_{ex} for $[Ln(NTA)(OH_2)_n]$ is $\sim 10^8$–10^9 s^{-1}, K_0 of ~ 0.4–0.04 would be required if first bond formation through an I_d mechanism is rate determining. This is not implausible as NTA^{3-} is a flexible species, but the difficulty of reliably calculating K_0 in the high [NTA^{3-}] concentrations employed in the study and the absence of directly determined k_{ex} values for $[Ln(NTA)(OH_2)_n]$ make a definitive mechanistic analysis difficult. The decrease in k_0 from La^{3+} to Sm^{3+} reflects the increase in $K_{[Ln(NTA)_2]^{3-}}$ as $1/r(M^{3+})$ increases. The markedly smaller value of k_{-3} by comparison to k_{-0} may be explained in terms of a fast first metal bond formation being followed by a rate-determining proton loss from the (H)NTA^{2-} nitrogen before coordination is completed. Alternatively, it has been estimated that the ratio of nitrogen protonated to carboxylate oxygen protonated (H)NTA^{2-} is $\sim 3 \times 10^4$, and thus if nitrogen bonds first, k_{-3} will therefore be small.[150,152] However, as Ln^{3+} exhibits a preference for oxygen donor atoms over nitrogen this latter proposal seems less acceptable for Ln^{3+}. The small values of k_1 probably reflect the instability of the $[Ln(NTA)_3]^{6-}$ intermediate for this exchange process.

TABLE 9

Rate constants for NTA^{3-} exchange on [M(NTA)$_2$]$^{3-}$ in the presence of excess [NTA^{3-}] and associated equilibrium constants determined at 293.2 K[a]

M^{3+}	k_0 (s^{-1})	$10^{-7} k_{-0}$[b] (dm^3 mol^{-1} s^{-1})	$10^{-4} k_3$ (dm^3 mol^{-1} s^{-1})	$10^{-3} k_{-3}$[c] (dm^3 mol^{-1} s^{-1})	k_1 (dm^3 mol^{-1} s^{-1})	$\log K_{[M(NTA)_2]^{3-}}$ (dm^3 mol^{-1})
La^{3+}	13 ± 1	5.7	500 ± 100	6.0	(6.7 ± 1.7) × 10^2	6.64
Ce^{3+}	1.7 ± 0.5	3.2	83 ± 4	4.4	65 ± 5	7.27
Pr^{3+}	0.8 ± 0.5	3.7	22 ± 2	2.8	37 ± 4	7.66
Nd^{3+}	0.5 ± 0.3	3.6	14 ± 1	2.9	17 ± 4	7.86
Sm^{3+}	<0.3	—	9 ± 0.7	3.1	6.5	8.09
Eu^{3+}	—	—	5.9 ± 0.4	2.3	5.6	8.15
Tb^{3+}	—	—	—	—	—	8.28
Dy^{3+}	—	—	—	—	—	8.36
Ho^{3+}	—	—	—	—	—	8.30
Er^{3+}	—	—	—	—	—	8.20
Tm^{3+}	—	—	2.8 ± 0.3	2.3	—	8.47
Yb^{3+}	—	—	2.1 ± 0.2	2.0	—	8.53
Lu^{3+}	—	—	1.9 ± 0.2	2.1	—	8.60
Y^{3+}	—	—	11 ± 2	3.2	—	8.02
Sc^{3+}	—	—	0.8 ± 0.1	—	—	—

[a] From Refs. [150 and 151].
[b] $k_{-0} = k_0 \cdot K_{[M(NTA)_2]^{3-}}$.
[c] $k_{-3} = k_3 \cdot K_{[M(NTA)_2]^{3-}} \cdot K_H$.

A similar study has detected several pathways for $HIMDA^{2-}$ (hydroxyethyliminodiacetate) exchange on $[Pr(HIMDA)_2]^-$.[153] The formation of the ternary complex $[Eu(DCTA)(IDA)]^{3+}$ as shown in Eq. (28) has been

$$[Eu(DCTA)]^- + IDA^{2-} \underset{k_b}{\overset{k_f}{\rightleftharpoons}} [Eu(DCTA)(IDA)]^{3-} \tag{28}$$

studied by a luminescence emission spectroscopy technique, and k_f and k_b values of 1.6×10^7 dm^3 mol^{-1} s^{-1} and 2.6×10^4 s^{-1}, respectively, have been obtained at 296.7 K.[133]

I. Summary

A knowledge of the mechanism of water exchange on $[Ln(OH_2)_9]^{3+}$ is essential to the interpretation of Ln^{3+} complexation rate data, but at this time the only accurately determined water exchange data are those available for $[Gd(OH_2)_9]^{3+}$. Even in this case a $\Delta V^{\ddagger}$ determination is not available to aid in mechanistic interpretation. Although it is hoped that high-field NMR spectrometers will soon provide accurate water exchange data for other $[Ln(OH_2)_9]^{3+}$, it is clear that in the meantime the k_{34} values obtained for SO_4^{2-} substitution from ultrasonic studies probably provide reasonable approximations to k_{ex}. Accordingly, it would be interesting to know if substitution by other monodentate species such as SCN^- or OCN^- (which so far seem not to have been investigated) yield k_{34} values similar to those for SO_4^{2-}. Given the high lability of $[Ln(OH_2)_9]^{3+}$ and its preference for hard-atom donor species it is likely that any selectivity exhibited for incoming ligands will be compressed into a small range.

The substitution of multidentate ligands onto Ln^{3+} appears to be ring closure controlled in most cases and yet such substitutions are substantially faster than those of early trivalent metal ions. This suggests that while ring closure rates reflect the general lability of Ln^{3+}, the major retardation of ring closure arises from steric effects within the entering ligand and also within the precursor complex. The same factors affect ternary complex formation, and ligand and metal ion exchange reactions.

The question of coordination number changes has inevitably arisen in this section. The overall observation is that monodentate (SO_4^{2-}) substitution rates for $[Ln(OH_2)_9]^{3+}$ vary within one order of magnitude as do the substitution rates for multidentate ligands where ring closure appears to be rate determining. In the case of ternary complex formation and related reactions the substitution rates for a given entering ligand vary over four or so orders of magnitude, a finding consistent with the general belief that coordination changes occur for species such as $[Ln(EDTA)(OH_2)_n]^-$ and $[Ln(DPTA)(OH_2)_n]^-$ as Ln^{3+} varies.

Where Y^{3+} and Sc^{3+} have been studied the former behaves much as expected from its $r(M^{3+})$, midway between that of Dy^{3+} and Ho^{3+}. Usually Sc^{3+} does not fit either the Ln^{3+} kinetic characteristics or those of the lighter M^{3+} ions, and in consequence appears to be in a class midway between Ln^{3+} and M^{3+}.

IV. NONAQUEOUS SOLUTION

A. Solvation

The study of solvation and solvent exchange processes of Ln^{3+}, Y^{3+}, and Sc^{3+} has proved rather more tractable in nonaqueous solution than is the case for water. The most extensive studies so far have been carried out on N,N-dimethylformamide (DMF) solutions, largely as a consequence of the large liquid temperature range (213–429 K) of this solvent. In the solid state $[Ln(DMF)_8](ClO_4)_3$ has been isolated for $Ln^{3+} = La^{3+}$–Lu^{3+},[154,155] and conductivity[156,157] and ^{35}Cl NMR[155] studies indicate that $ClO_4{}^-$ does not enter the first coordination sphere of Ln^{3+} to a detectable extent in DMF solution. Bünzli's spectroscopic studies[157] show that the predominant Eu^{3+} species in DMF solution has C_{2v} symmetry consistent with the stoichiometry $[Eu(DMF)_8]^{3+}$. The ^1H NMR and spectrophotometric studies of Merbach and co-workers[155,158–160] show that $[Ln(DMF)_8]^{3+}$ is the major species when $Ln^{3+} = Ce^{3+}$–Nd^{3+}, and that this becomes the only stable species when $Ln^{3+} = Tb^{3+}$–Lu^{3+}. Thus, in $[Yb(DMF)_8](ClO_4)_3/DMF/CD_2Cl_2$ solutions at 175 K the rate of DMF exchange was sufficiently slow for the ^1H NMR resonances of coordinated and free DMF to be well resolved, as seen in Fig. 9, and the mean coordination number to be determined from integration of these resonances. This value and that similarly determined for Tm^{3+} appear in Table 10 and are interpreted as indicating that $[Ln(DMF)_8]^{3+}$ is the greatly predominant species in solution.

However for the lighter Ln^{3+} ions the equilibrium shown in Eq. (29)

$$[Ln(DMF)_8]^{3+} + DMF \rightleftharpoons [Ln(DMF)_9]^{3+} \tag{29}$$

becomes important as exemplified by Nd^{3+}. The absorption band $^4I_{9/2} \rightarrow {}^2P_{1/2}$ in the electronic spectrum of Nd^{3+} is sensitive to environmental changes,[161] and it is found that the spectra of $[Nd(DMF)_8](ClO_4)_3/DMF$ solutions exhibit two temperature- and pressure-sensitive bands in the region 425–435 nm,[158] as shown in Fig. 10. The band at 429.3 nm is assigned to $[Nd(DMF)_8]^{3+}$ on the basis of the spectra of solutions of $[Nd(DMF)_8](ClO_4)_3$ in the noncoordinating solvents CH_3NO_2 and CH_2Cl_2.

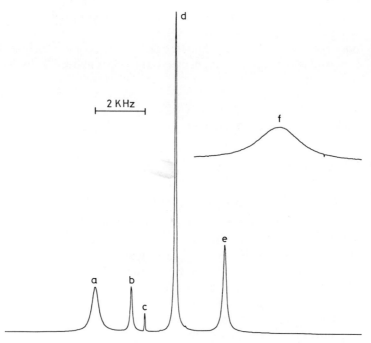

Fig. 9. The 1H NMR (360 MHz) spectrum of a $[Yb(DMF)_8](ClO_4)_3/DMF/CD_2Cl_2$ solution at 175 K. The field increases from left to right and the resonances are assigned as follows: (a) N–CH_3 *trans* to the formyl proton of coordinated DMF; (b) formyl proton of free DMF; (c) proton impurity in CD_2Cl_2; (d) both N–CH_3 resonances of free DMF; (e) N–CH_3 *cis* to formyl proton of coordinated DMF; (f) formyl proton of coordinated DMF. Resonance (f), which is displayed at a higher amplitude, appears 20.8 kHz upfield of (c). (From Ref. [158] with permission.)

The band at 428.2 nm is assigned to $[Ln(DMF)_9]^{3+}$, which as may be seen from Fig. 10 is favored at low temperature and high pressure. Variable temperature and pressure studies of this system yield the thermodynamic parameters shown in Table 10. No such spectral variations are observed for Nd^{3+} in aqueous solution. However, earlier visible spectroscopic studies[156] indicate the presence of both NdO_8 and NdO_9 chromophores in aqueous DMF solutions of $Nd(ClO_4)_3$. When the NMR and spectrophotomeric data are considered together with the differences observed in the temperature variations of the formyl proton chemical shift of bulk DMF in $[Ln(DMF)_8](ClO_4)_3/DMF$ solutions,[158,159] the conclusion is reached that $[Ln(DMF)_8]^{3+}$ is the only stable species when $Ln^{3+} = Tb^{3+}$–Lu^{3+} and that it is the major species present in the equilibrium shown in Eq. (29) when $Ln^{3+} = Ce^{3+}$–Nd^{3+}.

In contrast to ClO_4^-, conductivity and spectrophotometric studies[157,162] show NO_3^- to coordinate to Eu^{3+} in DMF solution to form the three species $[Eu(NO_3)(DMF)_{x-2}]^{2+}$, $[Eu(NO_3)_2(DMF)_{x-4}]^+$,   and

TABLE 10

Thermodynamic parameters for the reaction $[LnS_n]^{3+} + S \rightleftharpoons [LnS_{n+1}]^{3+}$ and average coordination numbers (CN)[a]

S	La³⁺ CN	Nd³⁺ CN	ΔH (kJ mol⁻¹)	ΔS (J K⁻¹ mol⁻¹)	ΔV (cm³ mol⁻¹)	Eu³⁺ CN	Tb³⁺ CN	Tm³⁺ CN	Yb³⁺ CN	La³⁺ CN
DMF	—	—	−14.9 ± 1.3	−69.1 ± 4.2	−9.8 ± 1.1[b]	—	—	7.7 ± 0.2	7.8 ± 0.2	—
TMP	6.8 ± 0.2	6.8 ± 0.3	−32.7 ± 3.3	−114 ± 11	−23.8 ± 1.5[c]	6.9 ± 0.3	5.9 ± 0.2	—	—	5.9 ± 0.2
TMU	—	—	—	—	—	—	—	—	—	6.0 ± 0.1

[a] All values from Ref. [158] except that for TMU, which is from Ref. [183].
[b] At 256 K and 231 K.
[c] At 298 K.

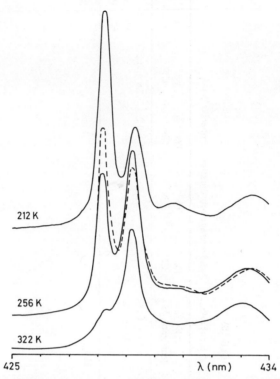

Fig. 10. The temperature and pressure variation of a $[Nd(DMF)_8](ClO_4)_3$/DMF solution. The spectra drawn in solid lines were observed at ambient pressure at the temperatures indicated. The peak at 429.3 nm arises from $[Nd(DMF)_8]^{3+}$ and that at 428.2 nm arises from $[Nd(DMF)_9]^{3+}$. The spectrum indicated by a broken line is that observed at 120 MPa. (From Ref. [158] with permission.)

$[Eu(NO_3)_3(DMF)_{x-6}]$, clearly indicating the greater coordinating ability of NO_3^- by comparison to that of ClO_4^-. (An earlier study[163] indicates that the interaction of NO_3^- with Tb^{3+} is stronger than with Eu^{3+} in DMF solution.) In the case of the less strongly coordinating solvent CH_3CN, however, FT IR and fluorometric studies[164] show that ClO_4^- coordinates to Eu^{3+} such that in 0.05 mol dm^{-3} solutions an equilibrium exists between the nine-coordinate species $[Eu(ClO_4)(CH_3CN)_8]^{2+}$ and $[Eu(CH_3CN)_9]^{3+}$. Nitrate is more competitive, and a sequential displacement of CH_3CN from the first coordination sphere is observed until the 10-coordinated $[Eu(NO_3)_5]^{2-}$ species is formed.[165] Similar studies[166] of dimethyl sulfoxide (DMSO) solutions indicate that $[Eu(DMSO)_9]^{3+}$ is formed in the presence of both ClO_4^- and NO_3^-. In CH_3CN solution DMSO readily displaces CH_3CN from the first coordination sphere, and NO_3^- is able to compete with DMSO for first-coordination-sphere sites. From an observation of such displacements the affinity sequence $NO_3^- > DMSO > H_2O \sim DMF > (CH_3)_2CO > CH_3CN$ has been proposed for Tb^{3+}.[163]

The trimethyl phosphate (TMP) molecule is more bulky than those of the solvents so far discussed and this causes the coordination number of Ln^{3+} to decrease. Thus, in $[Ln(TMP)_6](CF_3SO_3)_3/TMP/CD_2Cl_2$ solutions where $Ln^{3+} = La^{3+}, Nd^{3+}, Eu^{3+}, Tb^{3+}$, and Lu^{3+}, separate ^{31}P NMR resonances were observed for coordinated and free TMP such that the average coordination numbers shown in Table 10 were determined by integration.[158] It is found that the stoichiometries $[Ln(TMP)_7]^{3+}$ and $[Ln(TMP)_6]^{3+}$ are favored for the light and heavy Ln^{3+} ions, respectively. In neat TMP a similar temperature and pressure dependence of the Nd^{3+} spectrum to that observed in DMF (as discussed above) afforded the thermodynamic parameters for the equilibrium shown in Eq. (30) which appear in Table 10. The

$$[Nd(TMP)_6]^{3+} + TMP \rightleftharpoons [Nd(TMP)_7]^{3+} \tag{30}$$

larger ΔH characterizing the addition of TMP to $[Nd(TMP)_6]^{3+}$ (by comparison to that characterizing the addition of DMF to $[Nd(DMF)_8]^{3+}$) indicates the formation of a stronger Nd^{3+}–O bond in this system, which is a consequence of the lower coordination number facilitating the formation of a shorter Nd^{3+}–O bond than is possible in the more crowded $[Nd(DMF)_9]^{3+}$ species. The ΔV values in Table 10 are of considerable interest as they not only provide information about ground state volume changes but also provide valuable comparisons for activation volumes, $\Delta V^{\ddagger}$, as discussed in Section IV,B and C.

A number of studies have been reported for aqueous solvent mixtures. In aqueous acetone and acetonitrile the formation of 10-coordinate $[Eu(NO_3)_3(H_2O)_4]$ and nine-coordinate $[Tb(NO_3)_3(H_2O)_3]$ species has been reported from electronic absorption and emission spectroscopic studies.[163,165] (Similar stoichiometries have been observed in the solid state.[167–169]) Silber and co-workers[170] have used the variation of ultrasonic absorption arising from anion substitution on Ln^{3+} with variation of the composition of mixed aqueous solvent solutions in an attempt to probe solvation changes of Ln^{3+}. The difficulty with this approach is that it is probable that several Ln^{3+} species with different first-coordination-sphere compositions will exist in solution such that the absorptions characterizing each stage of complexation [see Eqs. (9) and (10)] for each Ln^{3+} species may superimpose. Nevertheless, interesting results have been obtained in aqueous DMF,[171] aqueous methanol,[172] and aqueous DMSO[173] for a range of lanthanide salts. It is noteworthy that in aqueous methanol, k_{34} (at 298.2 K) = $(9.3 \pm 1.7) \times 10^8$ s^{-1} for the entry of ClO_4^- into the first coordination sphere of Gd^{3+} when the water mole fraction = 0.2,[174] whereas ClO_4^- does not enter the first coordination sphere of Gd^{3+} to a detectable extent in aqueous solution.

The application of X-ray methods to nonaqueous Ln^{3+} systems has been less detailed than for aqueous solution. However the stoichiometries of $[Nd(CH_3OH)_6Cl_2]^+$ and $[La(CH_3OH)_8]^{3+}$ have been reported in concentrated solutions of $LnCl_3$ in methanol.[57,175]

B.   *N,N*-Dimethylformamide exchange on [Ln(DMF)$_8$]$^{3+}$

Recent studies of DMF exchange on $[Ln(DMF)_8]^{3+}$ are unique in presenting a systematic examination of a solvent exchange process over a range of Ln^{3+} ions.[155,159,160] The rate of DMF exchange on $[Ln(DMF)_8]^{3+}$, where $Ln^{3+} = Tb^{3+}, Dy^{3+}, Ho^{3+}, Er^{3+}, Tm^{3+}$, and Yb^{3+}, falls within the NMR time scale of the Swift and Connick method[84] and the derived k_{ex}, $\Delta H^{\ddagger}$, $\Delta S^{\ddagger}$, and $\Delta V^{\ddagger}$ values appear in Table 11. The pressure-independent activation volume $\Delta V^{\ddagger}$ was derived through Eq. (31) in which

$$\Delta V^{\ddagger} = \Delta V_0^{\ddagger} - \Delta \beta^{\ddagger} P \tag{31a}$$

$$\ln k_P = \ln k_0 - \Delta V^{\ddagger} P/RT + \Delta \beta^{\ddagger} P^2/RT \tag{31b}$$

$\Delta V_0^{\ddagger}$ is the activation volume in the absence of applied pressure (P), $\Delta \beta^{\ddagger}$ is the compressibility of activation, and k_P and k_0 are k_{ex} in the presence and absence of applied pressure, respectively.

It is apparent from Table 11 that the k_{ex} values for $[Ln(DMF)_8]^{3+}$ are an order of magnitude or so less than those directly determined or estimated for water exchange on $[Ln(OH_2)_9]^{3+}$ (Section III,B and D), which is in accord with the generally observed trend for nonlanthanide ions that nonaqueous solvent exchange tends to be slower than water exchange.[96] The thermodynamic origin of this decreased lability by comparison to that of $[Gd(OH_2)_9]^{3+}$ (Table 4) in the case of $[Tb(DMF)_8]^{3+}$ is an increased $\Delta H^{\ddagger}$ and a more negative $\Delta S^{\ddagger}$, and in the case of $[Yb(DMF)_8]^{3+}$ a much increased $\Delta H^{\ddagger}$. (As will be seen in Section IV,C it appears that a general decrease in Ln^{3+} lability may occur as the coordination number decreases.) Within the $[Ln(DMF)_8]^{3+}$ data set it may be seen (Table 11) that as the series is traversed from Tb^{3+} to Yb^{3+} $\Delta H^{\ddagger}$ increases, $\Delta S^{\ddagger}$ changes from negative to positive, and the positive value of $\Delta V^{\ddagger}$, which indicates the operation of a **d** activation mode, increases in magnitude. Such variations are symptomatic of a mechanistic change, but the DMF exchange rate law required for further mechanistic insight could not be obtained from rates determined in neat solvent.

A study of the variation of k_{ex} with free DMF concentration, [DMF], in noncoordinating CD_3NO_2 revealed differing kinetic characteristics with change in Ln^{3+}, as may be seen for Tb^{3+}, Er^{3+}, and Tm^{3+} in Fig. 11. The

TABLE 11

Kinetic parameters for DMF exchange on $[Ln(DMF)_8]^{3+}$ [a]

	Tb^{3+}	Dy^{3+}	Ho^{3+}	Er^{3+}	Tm^{3+}	Yb^{3+}
k_{ex} (298 K) (s^{-1})	$(1.9 \pm 0.1) \times 10^7$	$(6.3 \pm 0.3) \times 10^6$	$(3.6 \pm 0.6) \times 10^6$	$(1.3 \pm 0.4) \times 10^7$	$(3.1 \pm 0.3) \times 10^7$	$(9.9 \pm 0.9) \times 10^7$
$\Delta H^{\ddagger}$ (kJ mol^{-1})	14.09 ± 0.4	13.76 ± 0.4	15.31 ± 0.8	23.64 ± 1.8	33.18 ± 0.5	39.30 ± 0.6
$\Delta S^{\ddagger}$ (J K^{-1} mol^{-1})	-58.25 ± 2.1	-68.54 ± 1.6	-68.13 ± 4.0	-29.57 ± 8.6	9.85 ± 2.4	39.95 ± 2.7
$\Delta V^{\ddagger}$ (cm^3 mol^{-1})	5.2 ± 0.2	6.1 ± 0.2	5.2 ± 0.5	5.4 ± 0.3	7.4 ± 0.3	11.8 ± 0.4
$10^2 \Delta \beta^{\ddagger}$ (cm^3 mol^{-1} MPa^{-1})	-0.9 ± 0.7	-0.7 ± 0.5	-0.3 ± 1.0	0.2 ± 0.9	0.03 ± 0.9	2.0 ± 0.9

[a] From Refs. [155, 159, and 160].

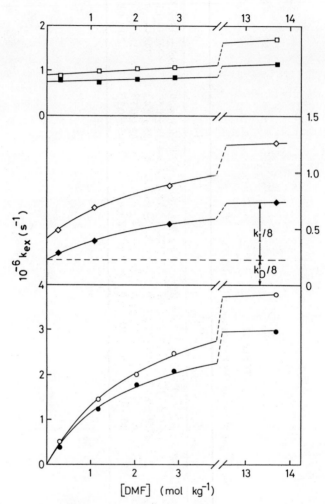

Fig. 11. Plots of k_{ex} for the exchange of DMF on $[M(DMF)_8]^{3+}$ in CD_3NO_2 solution against the concentration of bulk DMF, [DMF]. In the upper plot M = Tm and the upper and lower data sets refer to 248 and 242 K. In the middle plot M = Er and the upper and lower data sets refer to 244 and 234 K. In the lower plot M = Tb and the upper and lower data sets refer to 239 and 231 K. (From Ref. [160] with permission.)

variations of k_{ex} with [DMF] are consistent with a general DMF exchange rate law as shown in Eq. (32).

$$\text{exchange rate}/[\text{Ln(DMF)}_8{}^{3+}] = 8k_{ex} = k_D + k_1 K_0[\text{DMF}]/(1 + K_0[\text{DMF}])$$

$$(32)$$

Exchange through a **D** mechanism in which the rate-determining step is the formation of $[\text{Ln(DMF)}_7]^{3+}$ is characterized by k_D, and k_1 characterizes exchange through the encounter complex $[Ln(DMF)_8]^{3+} \cdots DMF$ in an I_d mechanism (see Section II,A). Thus, DMF exchange on $[Tb(DMF)]^{3+}$ occurs through an I_d mechanism alone and the best fit of the k_{ex} data to Eq. (32) yields K_0 (at 239.2 K) $= 0.42 \pm 0.01$ kg mol^{-1}, which is a reasonable value for the interaction of a 3+ species with a neutral species. In the case of $[Er(DMF)_8]^{3+}$ the variation of k_{ex} is consistent with the simultaneous operation of **D** and I_d mechanisms for DMF exchange. A best fit of k_{ex} to Eq. (30) yields $k_D = (3.4 \pm 0.3) \times 10^6$ s^{-1}, $k_I = (8.4 \pm 0.4) \times 10^5$ s^{-1}, and $K_0 = 0.30 \pm 0.06$ kg mol^{-1} at 244 K. The smaller variation of k_{ex} with [DMF] observed for $[Tm(DMF)_8]^{3+}$ has been ascribed to a general solvent effect occurring as the [DMF]/[CD$_3$NO$_2$] varies, but if this variation does arise from a small k_I contribution, it is clear that this component of the rate law is much diminished by comparison to the cases of $[Er(DMF)_8]^{3+}$ and $[Tb(DMF)_8]^{3+}$ and that the **D** exchange mechanism predominates for $[Tm(DMF)_8]^{3+}$.

The variation in $\Delta V^{\ddagger}$ observed in Table 11 is consistent with the trend toward a **D** mechanism operating alone for the heavier Ln^{3+}. It is interesting to note that the values of $\Delta V^{\ddagger} = 7.4 \pm 0.4$ and 11.8 ± 0.4 cm^3 mol^{-1} characterizing DMF exchange on $[Tm(DMF)_8]^{3+}$ and $[Yb(DMF)_8]^{3+}$, respectively, are similar in magnitude to $\Delta V = -9.8 \pm 1.1$ cm^3 mol^{-1} characterizing the dissociation of DMF from $[Nd(DMF)_9]^{3+}$ (see Table 10), which indicates that these two species are approaching the limiting $\Delta V^{\ddagger}$ for a **D** mechanism. The trends in $\Delta H^{\ddagger}$ and $\Delta S^{\ddagger}$ are also consistent with the dissociative character of the transition state increasing toward the heavier Ln^{3+}. It may be anticipated that the degree of bond forming or associative character in the transition state will increase from Gd^{3+} to La^{3+} and that the exchange mechanism will progress from I_a to **A**, consistent with the observation of $[Ln(DMF)_9]^{3+}$ ground state species in equilibrium with $[Ln(DMF)_8]^{3+}$ for the lighter Ln^{3+} ions (see Section IV,A). Such a variation in mechanism through the sequence **A**, I_a, I_d, **D** from La^{3+} to Lu^{3+} for $[Ln(DMF)_8]^{3+}$ is similar to that observed for solvent exchange on the hexasolvento di- and trivalent first-row transition metal ions.[96] However, in that case the mechanistic change has been largely ascribed to the increasing electronic occupancy of the t_{2g} orbitals causing a change from **a** to **d** activation modes, but in the case of Ln^{3+} the dominant cause of mechanistic change is probably the lanthanide contraction.

C. Solvent exchange on [Sc(solvent)$_6$]$^{3+}$ and related species

There appears to be no reported isolation of a hydrate of scandium(III) in the solid state, and the rate of water exchange on $[Sc(OH_2)_n]^{3+}$ (where n is probably > 6) is evidently very rapid. In contrast, Pisaniello and Lincoln have isolated a number of $[ScS_6]^{3+}$ salts in which S is a nonaqueous solvent, and which remain six coordinate in solution and exhibit moderate solvent exchange rates. For the solvents trimethyl phosphate (TMP),[176,177] dimethyl methylphosphonate (DMMP),[178] *N,N*-dimethylacetamide (DMA),[179,180] *N,N*-diethylacetamide (DEA),[180] *N*-methylacetamide (NMA),[180] and *N,N,N',N'*-tetramethylurea (TMU),[180] crystalline salts of the formula $[ScS_6](ClO_4)_3$ are readily isolated. In the presence of free S in non-coordinating diluents such as CD_3NO_2 and CD_3CN, and under conditions of slow exchange on the NMR time scale, integration of the resonances arising from free and coordinated S show $[ScS_6]^{3+}$ to be the only stable scandium(III) species in solution. Similar observations have been made in the cases of $[M(TMU)_6]^{3+}$ where $M^+ = Y^{3+},$[181] Tm^{3+}, $Yb^{3+},$[182] and $Lu^{3+}.$[183] The rate of S exchange on the diamagnetic $[MS_6]^{3+}$ complexes are readily determined from analysis of the coalescence of the resonances of free and coordinated S, as shown for $[Y(TMU)_6]^{3+}$ in Fig. 12. The general rate law for exchange of S on $[ScS_6]^{3+}$ is given in Eq. (33) and k_1, k_2, and their activation parameters appear in Table 12.

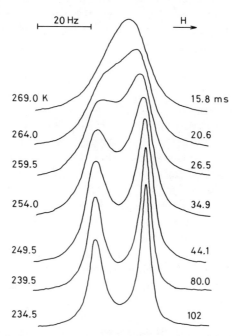

Fig. 12. The temperature variation of the ^{1}H NMR (270 MHz) spectra of a CD_3CN solution of $[Y(TMU)_6]^{3+}$ (0.0602 mol dm^{-3}) and TMU (0.429 mol dm^{-3}). The downfield resonance arises from $[Y(TMU)_6]^{3+}$. The experimental temperature (K) appears to the left, and the lifetimes (ms) of a single TMU in $[Y(TMU)_6]^{3+}$ derived through complete line-shape analysis appear to the right. (From Ref. [181] with permission.)

$$\text{exchange rate}/6[\text{ScS}_6{}^{3+}] = k_{ex} = k_1 + k_2[\text{S}] \tag{33}$$

The magnitudes of k_1 and k_2 depend on the nature of S and the noncoordinating solvent. For a given S there is a tendency for k_1 to increase with respect to k_2 as steric crowding close to the metal center increases, and for k_1 to increase with respect to k_2 in CD_3CN diluent by comparison to the case in CD_3NO_2 diluent. The effect of steric crowding is readily understood if the k_1 and k_2 terms characterize **d**- and **a**-activated exchange mechanisms, respectively, as is discussed below. The variation of k_1 and k_2 with diluent does not seem amenable to simple explanations, but similar effects have been observed for the exchange of S on $[\text{BeS}_4]^{2+[184,185]}$ and $[\text{UO}_2\text{S}_5]^{2+}$.[186] It seems probable that contributions from outside the first coordination sphere contribute to the $[\text{ScS}_6]^{3+}$ activation energetics, and the possible origins of such effects have been discussed elsewhere.[187,188].

The assignment of a **D** mechanism to those $[\text{ScS}_6]^{3+}$ exchange processes which exhibit a k_1 term only, as exemplified by TMP exchange on $[\text{Sc(TMP)}_6]^{3+}$ in CD_3CN,[177] and TMU exchange on $[\text{Sc(TMU)}_6]^{3+}$ in CD_3CN and CD_3NO_2,[179] seems reasonable, because the alternative assignment of an **I** mechanism operating under conditions where $K_0[\text{S}] \gg 1$ would require that $K_0 \geq 300$ dm^3 mol^{-1},[181,183] a value substantially greater than that predicted by the Fuoss equation[117] or the values of 0.30–0.50 kg mol^{-1} determined for $[\text{Ln(DMF)}_8]^{3+}$ under similar conditions[159,160] (see Section IV,B). The single k_2 term and $\Delta V^{\ddagger} = -18.7 \pm 1.1$ cm^3 mol^{-1}[96] which characterize TMP exchange on $[\text{Sc(TMP)}_6]^{3+}$ in CD_3NO_2 confirms the assignment of an **A** mechanism in this case. The similarity of $\Delta V^{\ddagger} = -18.7 \pm 1.1$ cm^3 mol^{-1} to $\Delta V = -23.8 \pm 1.5$ cm^3 mol^{-1} for the reaction $[\text{Nd(TMP)}_6]^{3+} + \text{TMP} \rightarrow [\text{Nd(TMP)}_7]^{3+}$ (Table 10) indicates that the observed $\Delta V^{\ddagger}$ is approaching the limit for an **A** mechanism. The even larger $\Delta V^{\ddagger} = -23.8 \pm 2.7$ cm^3 mol^{-1} observed for $[\text{Sc(TMP)}_6]^{3+}$ in neat TMP[96] appears to be exactly on this limit. (It is interesting to note that $\Delta V^{\ddagger}$ values of 20.7 ± 0.3 and -22.8 ± 1.1 cm^3 mol^{-1} characterizing TMP exchange on $[\text{Ga(TMP)}_6]^{3+}$ and $[\text{In(TMP)}_6]^{3+}$ in CD_3NO_2,[96,189] respectively, are close to the limits for **D** and **A** mechanisms.) The abrupt change of mechanism from **D** to **A** for TMP exchange on $[\text{Sc(TMP)}_6]^{3+}$ on changing from CD_3CN to CD_3NO_2 solution[177] appears to indicate that the activation energetics for the two mechanisms are so similar that perturbations of the ground state and/or the transition state arising from the second coordination sphere can have dramatic consequences.

The mechanistic assignment of a k_2 term when it appears in the exchange rate law with a k_1 term, as exemplified by DMMP exchange on $[\text{Sc(DMMP)}_6]^{3+}$ in CD_3CN,[178] is not without difficulty. The k_1 term may confidently be assigned to a **D** mechanism, but k_2 may be assigned to I_d or I_a

TABLE 12

Solvent and ligand exchange on $[MS_6]^{3+}$ ($k_1 + k_2[S] = k_{ex}$ = exchange rate/$6[MS_6^{3+}]$)

M^{3+}	S	Non-coordinating diluent	k_1 (at 298.2 K) (s^{-1})	k_2 (at 298.2 K) ($dm^3 mol^{-1} s^{-1}$)	$\Delta H^{\ddagger}$ ($kJ\ mol^{-1}$)	$\Delta S^{\ddagger}$ ($J\ K^{-1}\ mol^{-1}$)	$\Delta V^{\ddagger}$ ($cm^3\ mol^{-1}$)	Assigned mechanism	Ref.
Sc^{3+}	TMP	CD_3NO_2	—	45	26.0 ± 0.9	-126 ± 3		A	[177]
	TMP	CD_3NO_2	—	39	21.2	-143.5	-18.7 ± 1.1	A	[96]
	TMP	CD_3CN	59	—	29.8 ± 0.4	-111 ± 2		D	[177]
	TMP	—	736	$(85)^a$	34.1	-75.6	-23.8 ± 2.7	A	[96]
	DMMP	CD_3NO_2	—	13.0	29.7 ± 1.1	-124 ± 3		A	[178]
	DMMP	CD_3CN	286	—	43.5 ± 1.8	-90.3 ± 5.4		D	[178]
	DMMP	CD_3CN	—	14.2	24.4 ± 1.1	-141 ± 3		A	[178]
	DMA	CD_3NO_2	3.88	—	30.8 ± 2.0	-131 ± 6		D	[179, 180]
		CD_3NO_2	—	105	26.0 ± 0.6	-119 ± 2		A	[179, 180]
		CD_3CN	6.1	—	32.2 ± 3.5	-116 ± 12		D	[179, 180]
		CD_3CN	—	190	27.2 ± 1.2	-112 ± 4		A	
	DEA	CD_3NO_2	0.089	—	82 ± 3	10 ± 7		D	[180]
		CD_3NO_2	—	13.5	28.1 ± 0.6	-129 ± 2		A	[180]
		CD_3CN	10.7	—	43.9 ± 1.0	-78 ± 3		D	[180]
		CD_3CN	—	18.1	23.5 ± 0.9	-142 ± 3		A	[180]
	NMA	CD_3CN	79.2	—	27.3 ± 0.9	-117 ± 4		D	[180]
		CD_3CN	—	380	26.1 ± 1.2	-108 ± 5		A	[180]
	TMU	CD_3NO_2	2.06	—	91.2 ± 2.3	47.8 ± 6.7		D	[179]
		CD_3CN	0.90	—	68.6 ± 1.3	-15.7 ± 3.8		D	[179]

Y³⁺	TMU	CD₃CN	253	—	27.1 ± 0.5	−108 ± 2		D	[181]
	OPMe(OMe)Ph	CD₂Cl₂	312 (215 K)	—	31.4 ± 1.4	−48.4 ± 6.6		D	[193]
OPMe(OMe)Ph		CD₂Cl₂	—	455 (215 K)	35.2 ± 2.8	−27.6 ± 12.6		A	
Lu³⁺	TMU	CD₃CN	41.9	—	41.7 ± 0.6	−74 ± 2		D	[183]
Yb³⁺	TMU	CD₃CN	~60	—	—	—		D	[182]
Tm³⁺	TMU	CD₃CN	145	—	29 ± 0.3	−105 ± 1		D	[182]
Al³⁺	TMP	CD₃NO₂	0.36	—	98.3	76.1		D	[190]
	TMP	CD₃NO₂	0.78	—	85.1	38.2	22.5 ± 0.6	D	[191]
	DMMP	CD₃NO₂	5.1	—	79.5	33.0		D	[96]
	DMF	CD₃NO₂	0.05	—	88.3	28.4	13.7 ± 1.2	D	[189]
Ga³⁺	TMP	CD₃NO₂	5.0	—	87.9	63.2		D	[96]
	TMP	CD₃NO₂	6.4	—	76.5	27.0	20.7 ± 0.3	D	[96]
	DMF	CD₃NO₂	1.72	—	85.1	45.1	7.9 ± 1.6	D	[189]
In³⁺	TMP	CD₃NO₂	—	7.2	35.6	−109.2		A	[96]
	TMP	CD₃NO₂	—	7.6	32.8	−118	−22.8 ± 1.1	A	

a Calculated from the first-order rate constant observed in neat TMP by dividing by TMP molarity.

mechanisms where the magnitude of K_0 is insufficient to cause curvature in the plot of k_{ex} against [DMMP], or to an **A** mechanism. (It was seen in Section IV,B that **D** and $\mathbf{I_d}$ mechanisms operated simultaneously for DMF exchange on $[Er(DMF)_8]^{3+}$.) However, the unambiguous identification of the $[Sc(TMP)_6]^{3+}$ k_2 term as characterizing an **A** mechanism as discussed above adds force to the argument that all of the k_2 terms observed for S exchange on $[ScS_6]^{3+}$ typify **A** mechanisms, and they are so assigned in Table 12.

It is seen from Table 12 that in CD_3NO_2 solution TMP exchanges on $[M(TMP)_6]^{3+}$ through a **D** mechanism when $M^{3+} = Al^{3+}$ (53.5 pm) and Ga^{3+} (62.0 pm), and through an **A** mechanism when $M^{3+} = Sc^{3+}$ (74.5 pm) and In^{3+} (80.0 pm).[96,177,189,190,191] It is apparent that the six-coordinate $r(M^{3+})$ values (shown in parentheses) are substantially the determinants of mechanism, and it is of particular interest that the change from a **D** to an **A** mechanism occurs between Ga^{3+} and Sc^{3+}. It seems reasonable to assume that of the four M^{3+} these two will be the most susceptible to mechanistic change with change in environment. This occurs with Sc^{3+} and there may be conditions under which it occurs for Ga^{3+} also.

The labilities of the $[ScS_6]^{3+}$ species whose kinetic parameters appear in Table 12 are somewhat greater than those of the analogous Al^{3+}, Ga^{3+}, and In^{3+} species, but nevertheless the lability of $[ScS_6]^{3+}$ is much decreased by comparison to that of $[Ln(DMF)_8]^{3+}$. Thus, these $[ScS_6]^{3+}$ species exhibit labilities not very different from those expected for light six-coordinate M^{3+} ions. In contrast, the exchange of DMF, *N*-methylformamide (NMF), and *N*,*N*-diethylformamide (DEF) on Sc^{3+} was too rapid to be measured by the NMR methods used to determine the kinetic parameters for exchange of DMA, NMA, and DEA on $[ScS_6]^{3+}$.[180] The crystalline $[Sc(DMF)_6](ClO_4)_3$ complex and its NMA and DEF analogs were isolated, but in solution the rapidity of ligand exchange precluded the determination of the coordination number of Sc^{3+} and kinetic parameters for the exchange process.

There are two possible explanations for the enhanced lability of these amides on Sc^{3+}, and both depend on the lesser steric crowding caused by the amide proton close to the metal center by comparison to that caused by the acetamide methyl group. (The *N*-alkyl groups appear to be less important in steric crowding considerations.) The first explanation is that exchange still occurs on $[ScS_6]^{3+}$ when S = DMF, NMF, and DEF, but now the k_2 term in Eq. (33) becomes dominant and very large as a consequence of decreased steric crowding such that k_{ex} is outside the NMR time scale over viable experimental concentration ranges. The second explanation is that the species in solution is $[ScS_n]^{3+}$ where $n > 6$, and in consequence of increased bond lengths (reflected in $r(M^{3+})$ of six- and eight-coordinate scandium(III): 74.5 and 87.0 pm, respectively), the lability of Sc^{3+} increases dramatically. A comparison of the Gutmann donor numbers[192] of DMF, DMA, DEF, and DEA, which are

26.6, 27.8, 30.9, and 32.2, respectively, indicates that variation in electron-donating ability is unlikely to be the source of the substantial differences in lability observed for the exchange of the amides and acetamides.

An indication of the possible dependence of lability on coordination number is gained from the kinetic data characterizing the exchange of TMU on $[M(TMU)_6]^{3+}$ where $M^{3+} = Sc^{3+}$, Y^{3+}, Tm^{3+}, Yb^{3+}, and Lu^{3+} shown in Table 12. All five species undergo TMU exchange at modest rates through **D** mechanisms, and it is observed that as the six-coordinate $r(M^{3+})$ increases, $Sc^{3+} < Lu^{3+} < Yb^{3+} < Tm^{3+} < Y^{3+}$ (74.5, 86.1, 86.8, 88.0, 90.0 pm), so k_{ex} (at 298.2 K) increases, $\Delta H^{\ddagger}$ decreases, and $\Delta S^{\ddagger}$ becomes more negative. Therefore, in these cases it is $r(M^{3+})$ which predominantly determines lability and produces the juxtaposition of Lu^{3+} and Y^{3+} from that expected if atomic number or electronic configuration was a dominant factor, thereby confirming that bonding is predominantly ion–dipole in these six-coordinate systems. (The dependence of the labilities of Y^{3+} and Ln^{3+} on $r(M^{3+})$ was noted earlier for the $[Ln(OH_2)_9]^{3+}$ system.)

A comparison of data in Tables 11 and 12 shows $[Yb(DMF)_8]^{3+}$ to be $\sim 10^6$ times more labile than $[Yb(TMU)_6]^{3+}$ and a difference in lability of $\sim 10^5$ exists between the analogous Tm^{3+} species. Thus, a trend emerges in which a decrease in coordination number can dramatically decrease lability. The further exploration of this trend will be constrained by the apparently small number of monodentate ligands available which produce $[MS_n]$ species where $n \leq 6$. Only two other such studies seem to have been reported. In one,[193] separate resonances were observed for $[YS_n]^{3+}$ in CD_2Cl_2 solution for species in which $n = 4, 5, 6$ when $S = OP(NMe_2)_3$ or $OPPh_3$, and $n = 5$ and 6 when $S = OPMe(OMe)Ph$. The kinetic parameters for ligand exchange on the latter species are given in Table 12. In a ^{139}La NMR study[194] it has been shown that the exchange of hexamethylphosphoramide (HMPA) is slow on $[La(HMPA)_6]^{3+}$.

D. Lanthanide shift reagents

The use of lanthanide shift reagents (LSR) to simplify the NMR spectra of a wide range of substrate molecules and ions is probably the most widely practiced form of Ln^{3+} complex chemistry, although its prime aim is not the investigation of that chemistry.[7–10] Lanthanide shift reagents are usually of the form $Ln(RCO \cdot CH \cdot COR')_3$ where Ln^{3+} is paramagnetic. In the most often used LSR, $Ln^{3+} = Eu^{3+}$ or Pr^{3+} and $R = R' = tBu$[8] (with dipivalom-ethanate, DPM) or $R = tBu$ and $R' = C_3F_7$[9] (with 6,6,7,7,8,8,8-heptafluoro-2,2-dimethyl-3,5-octanedione, or FOD; when the tBu group is deuterated the abbreviation FOD-d_9 is used). The LSR induces large chemical shifts, which are largely pseudo-contact in origin, in the substrate molecule S as a consequence of the formation of a ternary complex as shown in Eqs. (34a) and (34b), and thereby simplifies the NMR spectrum of S. Such studies are usually

$$[Ln(FOD)_3] + S \rightleftharpoons [Ln(FOD)_3S] \qquad (34a)$$

$$[Ln(FOD)_3S] + S \rightleftharpoons [Ln(FOD)_3S_2] \qquad (34b)$$

carried out in non- or weakly coordinating solvents such as $CDCl_3$ with [S] in large excess over $[Ln(FOD)_3]$. Usually the exchange of S is in the fast exchange limit of the NMR time scale such that the NMR spectrum of S is environmentally averaged. When structural information is sought from the magnitude of the LSR-induced shifts of S, it is clearly essential that the proportions of the mono- and diadducts should be known.[10] Although the use of LSRs has yielded an abundance of valuable information concerning various aspects of substrate molecules this is not the prime interest of this section. Here the major interest is in the kinetic and mechanistic characteristics of the LSR ternary complex itself.

A particularly comprehensive study of the mechanistic aspects of LSRs is that by Evans and Wyatt.[195] They found that separate ^1H NMR resonances were observable for DMSO, HMPA, TMU, and Et_3N coordinated to $[Ln(FOD-d_9)_3S_n]$ and in the free state, and that n could be 1 or 2 depending on S and the solvent. These data and also the parameters for the exchange of S appear in Table 13. It is seen that in most cases $n = 2$ but in the case of $[Er(FOD-d_9)_3(HMPA)_n]$ and its Yb^{3+} analog, n changes from 2 to 1 when the solvent changes from $C_6D_5CD_3$ to the mixed solvent, a result consistent with solvent interactions being important in ground state energetics. In mixed solvent $[Er(FOD-d_9)_3(DMSO)_2]$ and its Yb^{3+} analog are formed, however, thus demonstrating the importance of substrate size in determining stoichiometry. Similarly, the extra bulk of Et_3N close to the metal center causes $n = 1$, whereas the larger HMPA molecule which has its bulk more distant from the metal center achieves $n = 2$. The increase in k_1 with increase in steric crowding as $r(M^{3+})$ decreases is expected for a **D** mechanism, as is the appearance of a k_2 term characterizing an **A** mechanism for DMSO exchange as $r(M^{3+})$ increases. The simultaneous operation of **D** and **A** mechanisms for DMSO implies the formation of transition states or reactive intermediates of coordination number 7 and 9, respectively, and for the Et_3N systems 6 and 8, respectively, all of which are also exhibited by Ln^{3+} ground state species in solution. When the bidentate ligand is changed from FOD-d_9 to the more bulky DPM it is found that there is a decreased tendency for two substrate molecules to coordinate.

A number of other studies of substrate exchange on LSRs have been reported,[196–200] among which is a study[198] of the exchange of tertiary

TABLE 13

Kinetic parameters for S exchange on $[Ln(FOD\text{-}d_9)_3 S_n]^a$

Ln^{3+}	S	n	Solvent[b]	k_1 (s^{-1})	k_2 $(dm^3\,mol^{-1}\,s^{-1})$	T^c (K)	$\Delta H^\ddagger$ $(kJ\,mol^{-1})$	$\Delta S^\ddagger$ $(J\,K^{-1}\,mol^{-1})$	Mechanism
Pr^{3+}	DMSO	2	Mixed	143	—	186.2	37.6	3	D
Pr^{3+}	DMSO	2	Mixed	—	612	186.2	27.6	−39	A
Nd^{3+}	DMSO	2	Mixed	84.9	—	186.2	38.1	0	D
Nd^{3+}	DMSO	2	Mixed	—	328	186.2	24.7	−60	A
Eu^{3+}	DMSO	2	Mixed	29.7	—	186.2	39.7	0	D
Tb^{3+}	DMSO	2	Mixed	131	—	186.2	—	—	D
Ho^{3+}	DMSO	2	Mixed	223	—	186.2	—	—	D
Er^{3+}	DMSO	2	Mixed	245	—	186.2	—	—	D
Yb^{3+}	DMSO	2	Mixed	549	—	186.2	—	—	D
Pr^{3+}	TMU	2	$C_6D_5CD_3$	6.4	—	186.2	41.5	−3	D
Eu^{3+}	TMU	2	$C_6D_5CD_3$	4.1	—	186.2	41.9	−2	D
Er^{3+}	TMU	2	$C_6D_5CD_3$	141	—	186.2	—	—	D
Yb^{3+}	TMU	2	$C_6D_5CD_3$	247	—	186.2	—	—	D
Pr^{3+}	HMPA	2	$C_6D_5CD_3$	38.0	—	273.2	55.1	−12	D
Eu^{3+}	HMPA	2	$C_6D_5CD_3$	65.9	—	273.2	54.6	−10	D
Tb^{3+}	HMPA	2	$C_6D_5CD_3$	167	—	273.2	54.2	−3	D
Ho^{3+}	HMPA	2	$C_6D_5CD_3$	426	—	273.2	50.3	−10	D
Er^{3+}	HMPA	2	$C_6D_5CD_3$	702	—	273.2	52.5	2	D
Yb^{3+}	HMPA	2	$C_6D_5CD_3$	2570	—	273.2	50.2	3	D
Eu^{3+}	HMPA	2	Mixed	91.9	—	273.2	53.0	−13	D
Tb^{3+}	HMPA	2	Mixed	180	—	273.2	—	—	D
Er^{3+}	HMPA	1	Mixed	—	419	273.2	25.8	−80	A
Yb^{3+}	HMPA	1	Mixed	—	349	273.2	24.6	−87	A
Eu^{3+}	Et_3N	1	$C_6D_5CD_3$	~160	—	243	—	—	D
Eu^{3+}	Et_3N	1	$C_6D_5CD_3$	—	~460	243	—	—	A
Pr^{3+}	Et_3N	1	$C_6D_5CD_3$	~330	—	203	—	—	D
Pr^{3+}	Et_3N	1	$C_6D_5CD_3$	—	~320	203	—	—	A

[a] From Ref. [195].

[b] The mixed solvent was CCl_4; $CDCl_3$; $C_6D_5CD_3$, 1.5:1.8:1 (v/v).

[c] Temperature pertaining to quoted k_1 or k_2.

phosphine oxides on $[Gd(DPM)_3S_2]$, which is considered to proceed through an A mechanism. The exchange of methyldiphenylphosphine oxide and dibutylmethoxyoctylphosphine oxide is characterized by pseudo-first-order rate constants of 1.65×10^5 and 3.16×10^3 s^{-1} at 300 K, respectively. The sterically hindered quinuclidine molecule forms a $[Yb(FOD)_3S]$ species on which it has been shown to exchange through an A mechanism.[199] A rather unusual study is that of the formation of $[Eu(FOD-d_9)_3S]$ species in which S is a tris(β-diketonato)cobalt(III) complex.[200] When the β-diketonates are $CH_3 \cdot CO \cdot CCl \cdot CO \cdot CH_3{}^-$ and $C_6H_5 \cdot CO \cdot CH \cdot CO \cdot CH_3{}^-$ exchange rate constants of 34.1 and 10.5 s^{-1} at 300 K, respectively, are reported, and the exchange mechanism is considered to be **D** on the basis of the independence of the magnitude of these rate constants of reactant concentration variation.

By increasing the magnitudes of chemical shifts LSRs offer the opportunity to extend the dynamic NMR time scale for intramolecular processes occurring in the substrate molecules, provided that the intermolecular substrate exchange is substantially more rapid than the intramolecular process of interest. As it is almost inevitable that the energetics of the substrate intermolecular process will be modified by coordination to Ln^{3+}, a number of studies in this area have appeared. It was found through studies of the rotation about the C–N bonds in trimethylcarbamate[201] and *N,N*-dimethylacet-amide[202] over a range of $[Eu(FOD)_3]$ and $[Pr(FOD)_3]$ concentrations that the barrier to rotation was increased by ~ 4 kJ mol^{-1} upon coordination. Where different rotamers coexist an LSR may change the position of equilibrium, as demonstrated by Kessler and Molter[203] who found that $[Eu(FOD)_3]$ increased the proportion of the E rotamer of Boc α-amino acid esters relative to the Z isomer, which predominates in the free state. Intramolecular rearrangements may be observed directly in substrates in the $[Pr(FOD)_3S_n]$ species, as has been demonstrated in the case of bidentate amines.[204] Such modifications of ligand dynamics are well established for nonlanthanide metal ions.

E. Summary

It is clear that the ability of the Ln^{3+} ions to vary their coordination numbers comes very much to the fore in nonaqueous solvents as exemplified by the $[Ln(DMSO)_9]^{3+}$, $[Ln(DMF)_8]^{3+}$, and $[Ln(TMP)_6]^{3+}$ species. This is largely a consequence of the combined effect of solvent electron-donating power and steric interactions, which is observed to extend to the formation of ternary complexes as evidenced by $[Er(FOD-d_9)_3HMPA]$ and $[Er(FOD-d_9)_3(DMSO)_2]$ in which Er^{3+} is respectively seven and eight coordinate in the same noncoordinating medium. The greatly decreased labilities of $[M(TMU)_6]^{3+}$, where $M = Tm^{3+}$, Yb^{3+}, Lu^{3+}, and Y^{3+}, by comparison to $[Ln(DMF)_8]^{3+}$ demonstrate the strong dependence of Ln^{3+} lability on coordination number. Scandium(III) remains somewhat enigmatic. Those six-coordinate Sc^{3+} species kinetically characterized are not remarkably more labile than In^{3+} or Ga^{3+}, but in solvents such as DMF the lability of $[Sc(DMF)_n]^{3+}$ appears to be substantially increased. Thus, Sc^{3+} seems to occupy a niche between the light M^{3+} ions and the Ln^{3+} ions on the basis of the range of labilities in exhibits.

V. CONCLUDING COMMENTS

It is evident from reading this review that the trivalent lanthanides ex-hibit a chameleon-like character in adjusting to their environment in that they vary their coordination number from 6 to 10 and their lability over many orders of magnitude, as exemplified by k_{ex} (at 298.2 K) = 41.9 s^{-1} and 10.6 $\times$ 10^8 s^{-1} for $[Lu(TMU)_6]^{3+}$ and $[Gd(OH_2)_9]^{3+}$, respectively. Yttrium(III) fits this Ln^{3+} pattern of behavior, but the smaller Sc^{3+}, though exhibiting a range of coordination numbers and labilities, really appears to be in a class of its own somewhere between the light trivalent metal ions and Ln^{3+}. This difference in behavior is substantially attributable to the smaller $r(M^{3+})$ of Sc^{3+} (Table 1), and it is appropriate to review briefly the significance of $r(M^{3+})$ at this point. Within the context of this review $r(M^{3+})$ has little meaning without the identification of the coordination number.[14] It is seen from Table 1 that as the coordination number decreases so does $r(M^{3+})$, which is directly derived from Ln^{3+}–donor atom distances and which takes into account unequal bond distances within Ln^{3+} complexes. Hence, the de-crease in lability of the predominantly oxygen donor Ln^{3+} complexes with decrease in $r(M^{3+})$ may be largely attributed to an increase in bond energies in the ground state. (Recently, Marcus[205] has proposed a scale of $r(M^{3+})$ for the "bare" or "unhydrated" Ln^{3+} in aqueous solution which approx-imate to the seven-coordinated $r(M^{3+})$ published by Shannon.[14] The pre-viously discussed strictures concerning coordination number apply equally to this proposed $r(M^{3+})$ scale.)

While quite complicated mechanistic schemes have been proposed for ligand substitution on Ln^{3+}, Y^{3+}, and Sc^{3+} in aqueous solution, particularly for the polyaminocarboxylate complexes, it is only in the study of nonaqueous systems that it has been possible to assign **a** and **d** activation modes with confidence. The nonaqueous systems are less labile than the aqueous systems and their coordination numbers, solvent and ligand exchange rate laws, and $\Delta V^{\ddagger}$ values are determined with relative ease. In contrast, it has proved impracticable to determine the rate law for water exchange on $[Ln(OH_2)_9]^{3+}$ in the presence of a noncoordinating diluent, and the lability of this species has so far proved to be too great for a $\Delta V^{\ddagger}$ determination. Thus, the fundamental requirement for deduction of the mechanism of ligand substitution in water, a knowledge of the mechanism of water exchange, is not yet met. Notwithstanding such frustrations, the Ln^{3+}, Y^3, and Sc^{3+} ions are likely to constitute a fertile field for mechanistic investigations for many years to come, in extension of the studies discussed in this review and also into new areas, as exemplified by recent mechanistic studies of di- and trivalent europium and ytterbium cryptates,[206] lanthanide α-alkyl transfer reactions,[207] and the intramolecular exchange of the alterdentate ninhydrin and alloxan radical ions on Y^{3+} and La^{3+}.[208]

Acknowledgments

This article was written while the author was on leave at the University of Lausanne under the sponsorship of the Convention Intercantonale Romande pour l'Enseignement du 3e Cycle en Chimie. The author expresses much gratitude to both institutions, as he also does to A. E. Merbach, J-C. G. Bünzli, Y. Ducommum (Lausanne), G. Geier (Zurich), and H. B. Silber (Texas) who spent much time in discussions with him.

References

[1] Geier, G. *Ber. Bunsenges. Phys. Chem.* **1965,** *69,* 617.
[2] Moeller, T.; Martin, D. F.; Thompson, L. C.; Ferrus, R.; Feistel, G. R.; Randall, W. J. *Chem. Rev.* **1965,** *65,* 1.
[3] Koppikar, D. K.; Sivapullaiah, P. V.; Ramakrishnan, L.; Soundarajan, S. *Struct. Bond.* **1978,** *34,* 135.
[4] Sinha, S. P. *Struct. Bond.* **1976,** *25,* 67.
[5] Jorgensen, C. K.; Reisfeld, R. *Top. Curr. Chem.* **1982,** *100,* 127.
[6] Bünzli, J-C. G.; Wessner, D. *Coord. Chem. Rev.* **1985,** in press.
[7] Hinckley, C. C. *J. Am. Chem. Soc.* **1969,** *91,* 5160.
[8] Sanders, J. K. M.; Williams, D. H. *J. Chem. Soc. Chem. Commun.* **1970,** 422.
[9] Rondeau, R. E.; Sievers, R. E. *J. Am. Chem. Soc.* **1971,** *93,* 1522.
[10] Reuben J.; Elgavish, G. A. *In* "Handbook on the Physics and Chemistry of Rare Earths", Gschneider, K. A.; Eyring, L., Eds; Elsevier: Amsterdam, 1979, p. 483.
[11] Nieboer, E. *Struct. Bond.* **1975,** *22,* 1.
[12] Horrocks, W. DeW.; Sudnick, D. R. *Acc. Chem. Res.* **1981,** *14,* 384.
[13] Williams, R. J. P. *Struct. Bond. 1982, 50,* 79.
[14] Shannon, R. D. *Acta Crystallogr.* **1976,** *A32,* 751.
[15] Kefelaar, J. A. A. *Physica* 1937, *4,* 619.
[16] Helmholz, L. *J, Am. Chem. Soc.* **1939,** *61,* 1549.
[17] Harrowfield, J. McB.; Kepert, D. L.; Patrick, J. M.; White, A. H. *Aust. J. Chem.* **1983,** *36,* 483.
[18] Albertsson, J.; Elding, I. *Acta Crystallogr.* **1977,** *B33,* 1460.
[19] Hubbard, C. R.; Quicksall, C. O.; Jacobson, R. A. *Acta Crystallogr.* **1974,** *B30,* 2613.

[20] Fitzwater, D. R.; Rundle, R. E. *Z. Kristallogr. Kristallgeom. Kristallchem.* **1959**, *112*, 362.
[21] Broach, R. W.; Williams, J. M.; Felcher, G. P.; Hinks, D. G. *Acta Crystallogr.* **1979**, *B35*, 2317.
[22] Sikka, S. K. *Acta Crystallogr.* **1969**, *A25*, 621.
[23] Glaser, J.; Johansson, G. *Acta Chem. Scand.* **1981**, *A35*, 639.
[24] Keppert, D. L. *In* "Inorganic Stereochemistry"; Vol. 6, Springer-Verlag: Berlin, 1982.
[25] Favas, M. C.; Keppert, D. L. *Prog. Inorg. Chem.* **1981**, *28*, 309.
[26] Bisi, C. C.; Giusta, A. D.; Coda, A.; Tazzoli, V. *Cryst. Struct. Commun.* **1974**, *3*, 381.
[27] Al-Karaghouli, A. R.; Wood, J. S. *Inorg. Chem.* **1979**, *18*, 1177.
[28] Bisi, C. C.; Gorio, M.; Cannillo, E.; Coda, A.; Tazzoli, V. *Acta Crystallogr.* **1975**, *A31*, 8134.
[29] Bisi, C. C.; Coda, A.; Tazzoli, V. *Cryst. Struct. Commun.* **1981**, *10*, 703.
[30] Aslanov, L. A.; Ionov, V. M.; Sotman S. S. *Kristallografiya* **1976**, *21*, 1200.
[31] Aslanov, L. A.; Ionov, V. M.; Stepanov, A. A.; Porai-Koshits, M. A.; Lebedev, V. G.; Kulikovskii, B. N.; Gilyalov, O. N.; Novoderezhkina, T. L. *Izv. Akad. Nauk. SSSR, Neorg. Mater.* **1975**, *11*, 1331.
[32] Anderson, T. J.; Neuman, M. A.; Melson, G. A. *Inorg. Chem.* **1974**, *13*, 1884.
[33] Davis, A. R.; Einstein, F. W. B. *Inorg. Chem.* **1974**, *13*, 1880.
[34] Anderson, T. J.; Neuman, M. A.; Melson, G. A. *Inorg. Chem.* **1974**, *13*, 159.
[35] Anderson, T. J.; Neuman, M. A.; Melson, G. A. *Inorg. Chem.* **1973**, *12*, 927.
[36] Lincoln, S. F. *Coord. Chem. Rev.* **1971**, *6*, 309.
[37] Fratiello, A., *Prog. Inorg. Chem.* **1972**, *17*, 57.
[38] Burgess, J. "Metal Ions in Solution"; Ellis Horwood: Chichester, 1978.
[39] (a) Martell, A. E. "Stability Constants of Metal Ion Complexes"; The Chemical Society, London, 1971. (b) Sillén, L. G.; Martell, A. E. "Stability Constants"; The Chemical Society Special Publication No. 25, London, 1971.
[40] Anderegg, G. "Critical Survey of Stability Constants of EDTA Complexes"; IUPAC Chemical Data Series No. 14, 1976.
[41] Schwarzenbach, G.; Gut, R.; Anderegg, G. *Helv. Chim. Acta* **1954**, *37*, 937.
[42] Powell, J. E. *In* "Handbook on Physics and Chemistry of Rare Earths"; Gscheider, K. A.; Eyring, L., Eds.; North-Holland Publ.: Amsterdam, 1979.
[43] Chmutov, K. V.; Nazarov, P. P.; Chuveleva, E. A.; Kharitonov, O. V. *Sov. Radiochem.* (*Engl. Transl.*) **1977**, *19*, 431.
[44] Powell, J. E.; Potter, M. W.; Burkholder, H. R.; Potter, E. D. H.; Tse, P.-K. *Polyhedron* **1982**, *1*, 277.
[45] Horrocks, W. De W.; Sudnick, D. R. *J. Am. Chem. Soc.* **1979**, *101*, 334.
[46] Habenschuss, A.; Spedding, F. H. *J. Chem. Phys.* **1979**, *70*, 3758.
[47] Geier, G. *Chimica* **1969**, *23*, 148.
[48] Geier, G.; Karlen U.; Zelewsky, A. V. *Helv. Chim. Acta* **1969**, *52*, 1967.
[49] Geier, G.; Karlen, U. *Helv. Chim. Acta* **1971**, *54*, 135.
[50] Geier, G.; Jorgensen, C. K. *Chem. Phys. Lett.* **1971**, *9*, 263.
[51] Habenschuss, A.; Spedding, F. H. *J. Chem. Phys.* **1979**, *70*, 2797.
[52] Habenschuss, A.; Spedding, F. H. *J. Chem. Phys.* **1980**, *73*, 442..
[53] Narten, A. H.; Hahn, R. L. *Science* **1982**, *217*, 1249.
[54] Ohtaki, H.; Yamaguchi, T.; Maeda, M. *Bull. Chem. Soc. Jpn.* **1976**, *49*, 701.
[55] Smith, L. S.; Wertz, D. L. *J. Am. Chem. Soc.* **1975**, *97*, 2365.
[56] Steele, M. L.; Wertz, D. L. *J. Am. Chem. Soc.* **1976**, *98*, 4424.
[57] Steele, M. L.; Wertz, D. L. *Inorg. Chem.* **1977**, *16*, 1225.
[58] Smith, L. S.; Wertz, D. L. *J. Inorg. Nucl. Chem.* **1977**, *39*, 95.
[59] Brady, G. W. *J. Chem. Phys.* **1960**, *33*, 1079.
[60] Breen, P. J.; Horrocks, W. De W. *Inorg. Chem.* **1983**, *22*, 536, and references therein.
[61] Bel'skii, N. K.; Struchkov, Yu. T.; *Crystallography* (*Sov. Phys.*) (*Engl. Transl.*) **1965**, *10*, 15.

[62] Marezio, M.; Plettinger, H. A.; Zachariasen, W. H. *Acta Crystallogr.* **1961,** *14,* 234.

[63] Hewish, N. A.; Neilson, G. W.; Enderby, J. E. *Nature (London)* **1982,** *297,* 138.

[64] Kanno, H.; Hiraishi, J. *J. Phys. Chem.* **1982,** *86,* 1488.

[65] Karraker, D. G. *Inorg. Chem.* **1968,** *7,* 473.

[66] Nakamura, K.; Kawamura, K. *Bull. Chem. Soc. Jpn.* **1971,** *44,* 330.

[67] Bünzli, J.-C. G.; Yersin, J-R. *Inorg. Chem.* **1979,** *18,* 605.

[68] Spedding, F. H., Pikal, M. J.; Ayres, B. O. *J. Phys. Chem.* **1966,** *70,* 2440.

[69] Swaddle, T. W.; Mak, M. K. S. *Can. J. Chem.* **1983,** *61,* 473.

[70] Swaddle, T. W. *Adv. Inorg. Bioinorg. Mech.* **1983,** *2,* 96.

[71] Brücher, E.; Kukri, Cs. E.; Király, R. *Inorg. Chim. Acta* **1984,** *95,* 135.

[72] Alsadi, B. M.; Rossotti, F. J. C.; Williams, R. J. P. *J. Chem. Soc., Dalton Trans.* **1980,** 2147.

[73] Anderegg, G.; Wenk, F. *Helv. Chim. Acta* **1971,** *54,* 216.

[74] Haas, Y.; Stein, G. *J. Phys. Chem.* **1971,** *75,* 3677.

[75] Southwood-Jones, R. V.; Earl, W. L.; Newman, K. E.; Merbach, A. E. *J. Chem. Phys.* **1980,** *73,* 5909.

[76] Lind. M. D.; Lee, B.; Hoard, J. L. *J. Am. Chem. Soc.* **1965,** *87,* 1612.

[77] Swaddle, T. W. *In* "Mechanistic Aspects of Inorganic Reactions"; Rorabacher, D. B.; Endicott, J. F., Eds; *ACS Symp. Ser.* **1982,** *198,* 39.

[78] Margerum, D. W.; Cayley, G. R.; Weatherburn, D. C.; Pagenkopf, G. W. "Coordination Chemistry", Vol. 2, Ch. 1; Martell, A. E., Ed.; American Chemical Society: Washington, DC; 1978.

[79] Langford, C. H.; Gray, H. B. "Ligand Substitution Processes"; Benjamin: New York, 1966.

[80] Basolo, F.; Pearson, R. G. "Mechanisms of Inorganic Reactions"; 2nd ed., Wiley: New York, 1967.

[81] Ducommun, Y.; Newman, K. E.; Merbach, A. E. *Inorg. Chem.* **1980,** *19,* 3696.

[82] Sasaki, Y.; Sykes, A. G. *J. Chem. Soc., Dalton Trans.* **1975,** 1048.

[83] Espenson, J. H. *Inorg. Chem.* **1969,** *8,* 1554.

[84] Swift, T. J.; Connick, R. E. *J. Chem. Phys.* **1962,** *37,* 307; **1964,** *41,* 2553.

[85] Marianelli, R. Ph.D. thesis, University of California; Lawrence Radiation Laboratory Report UCRL-17069, 1966.

[86] Reuben, J.; Fiat, D. *J. Chem. Phys.* **1969,** *51,* 4909, 4918.

[87] Granot, J.; Fiat, D. *J. Magn. Reson.* **1975,** *19,* 372.

[88] Hugi, A. Doctoral thesis, Université de Lausanne, 1984.

[89] a) Hugi-Cleary, D. M.; Helm, L.; Merbach, A. E. *Helv. Chim. Acta.* **1985,** *65,* 545.
b) Hugi-Cleary, D. M. Personal communication.

[90] Glass, G. E.; Schwabacher, W. B.; Tobias, R. S. *Inorg. Chem.* **1968,** *7,* 2471.

[91] Neely, J. W.; Connick, R. E. *J. Am. Chem. Soc.* **1970,** *92,* 3476.

[92] Earl, W. L. Ph. D. thesis, University of California; Lawrence Berkeley Laboratory Report LBL-3712, 1975.

[93] Poupko, R.; Luz, Z. *J. Chem. Phys.* **1972,** *57,* 3311.

[94] West, R. J.; Lincoln, S. F. *J. Chem. Soc., Dalton Trans.* **1974,** 281.

[95] Helm, L.; Lincoln, S. F., Merbach, A. E.; Zbinden, D. Unpublished data.

[96] Merbach, A. E. *Pure Appl. Chem.* **1982,** *61,* 473.

[97] Grant, M. W.; Dodgen, H. W.; Hunt, J. P. *J. Am. Chem. Soc.* **1971,** *93,* 6828.

[98] Zetter, M. S.; Grant, M. W.; Wood, E. J.; Dodgen, H. W.; Hunt, J. P. *Inorg. Chem.* **1972,** *11,* 2701.

[99] Wilkins, R. G. *Adv. Inorg. Bioinorg. Mech.* **1982,** *2,* 139, and references therein.

[100] Caldin, E. F. "Fast Reactions in Solution"; Wiley: New York, 1964.

[101] Purdie, N.; Farrow, M. M. *Coord. Chem. Rev.* **1973,** *11,* 189.

[102] Bonsen, A.; Knoche, W.; Berger, W.; Giese, K.; Petrucci, S. *Ber. Bunsenges. Phys. Chem.* **1978,** *82,* 678.

[103] Eigen, M.; Tamm, K. *Z. Electrochem.* **1962,** *66,* 93, 107.

[104] Jackopin, L. G.; Yeager, E. *J. Phys. Chem.* **1970**, *74*, 3766.

[105] Fay, D. P.; Purdie, N. *J. Phys. Chem.* **1970**, *74*, 1160.

[106] Fay, D. P.; Purdie, N. *J. Phys. Chem.* **1969**, *73*, 544.

[107] Reidler, J.; Silber, H. B. *J. Phys. Chem.* **1973**, *77*, 1275.

[108] Farrow, M. M.; Purdie, N. *J. Solution Chem.* **1973**, *2*, 513.

[109] Farrow, M. M.; Purdie, N. *Inorg. Chem.* **1974**, *13*, 2111.

[110] Garnsey, R.; Ebdon, D. W. *J. Am. Chem. Soc.* **1969**, *91*, 50.

[111] Darbari, G. S.; Fittipaldi, F.; Petrucci, S. *Acustica* **1971**, *25*, 125.

[112] Silber, H. B.; Shienen, N.; Atkinson, G.; Grecsek, J. J. *J. Chem. Soc. Faraday Trans. I* **1972**, *68*, 1200.

[113] Wang, H. C.; Hemmes, P. *J. Phys. Chem.* **1974**, *78*, 261.

[114] Graffeo, A. J.; Bear, J. L. *J. Inorg. Nucl. Chem.* **1968**, *30*, 1577.

[115] Silber, H. B.; Swinehart, J. H. *J. Phys. Chem.* **1967**, *71*, 4344.

[116] Silber, H. B.; Farina, R. D.; Swinehart, J. H. *Inorg. Chem.* **1969**, *8*, 819.

[117] a) Fuoss, R. M. *J. Am. Chem. Soc.* **1958**, *80*, 5059;
   b) Fuoss, R. M. *J. Phys. Chem.* **1978**, *82*, 2427.

[118] Beck, M. T. *Coord. Chem. Rev.* **1968**, *3*, 91.

[119] Garza, V. L.; Purdie, N. *J. Phys. Chem.* **1970**, *74*, 275.

[120] Doyle, M.; Silber, H. B. *J. Chem. Soc. Chem. Commun.* **1972**, 1067.

[121] Nyssen, G. A.; Margerum, D. W. *Inorg. Chem.* **1970**, *9*, 1814.

[122] Laurenczy, G.; Brücher, E. *Inorg. Chim. Acta* **1984**, *95*, 5.

[123] Ryhl, T. *Acta Chem. Scand.* **1973**, *27*, 303.

[124] Brücher, E.; Szarvas, P. *Inorg. Chim. Acta* **1970**, *4*, 632.

[125] Hoard, J. L.; Lee, B.; Lind, M. D. *J. Am. Chem. Soc.* **1965**, *87*, 1612.

[126] Kostromina, N. A.; Ternovaya, T. V. *Teoret. Eksper. Khim.* **1971**, *7*, 115.

[127] Kostromina, N. N.; Tananaeva, N. N. *Russ. J. Inorg. Chem.* (*Engl. Transl.*) **1971**, *16*, 1256.

[128] Sidall, T. H.; Stewart, W. E. *Inorg. Nucl. Chem. Lett.* **1969**, *5*, 421.

[129] Ryhl, T. *Acta Chem. Scand.* **1972**, *26*, 4001.

[130] Alsaadi, B. M.; Rossotti, F. J. C.; Williams, R. J. P. *J. Chem. Soc., Dalton Trans.* **1980**, 2151.

[131] Gennaro, M. C.; Mirti, P.; Casalino, C. *Polyhedron* **1983**, *2*, 13.

[132] Kolat, R. S.; Powell, J. E. *Inorg. Chem.* **1962**, *1*, 485.

[133] Horrocks, W. DeW.; Arkle, V. K.; Liotta, F. J.; Sudnick, D. R. *J. Am. Chem. Soc.* **1983**, *105*, 3455.

[134] Anderegg, G. *Helv. Chim. Acta* **1963**, *46*, 1833.

[135] Leisegang, G. W. *J. Am. Chem. Soc.* **1981**, *103*, 953.

[136] Kauffmann, E.; Dye, J. L.; Lehn, J.-M.; Popov, A. I. *J. Am. Chem. Soc.* **1980**, *102*, 2274.

[137] Mei, E.; Popov, A. I.; Dye, J. L. *J. Am. Chem. Soc.* **1977**, *99*, 6532.

[138] Glentworth, P.; Wiseall, B.; Wright, C. L.; Mahmood, A. J. *J. Inorg. Nucl. Chem.* **1968**, *30*, 967.

[139] Glentworth, P.; Newton, D. A. *J. Inorg. Nucl. Chem.* **1971**, *33*, 1701.

[140] Southwood-Jones, R. V.; Merbach, A. E. *Inorg. Chim. Acta* **1978**, *30*, 77.

[141] Desreux, J. F. *Inorg. Chem.* **1980**, *19*, 1319.

[142] D'Olieslager, W.; Oeyen, A. *J. Inorg. Nucl. Chem.* **1978**, *40*, 1565.

[143] D'Olieslager, W.; Wevers, M.; De Jonghe, M. *J. Inorg. Nucl. Chem.* **1981**, *43*, 423.

[144] Brücher, E.; Laurenczy, G. *Inorg. Chem.* **1983**, *22*, 338.

[145] Margerum, D. W.; Zabin, B. A. *J. Phys. Chem.* **1962**, *66*, 2214.

[146] Choppin, G. R.; Williams, K. R. *J. Inorg. Nucl. Chem.* **1973**, *35*, 4255.

[147] Furrer, M. Doctoral Dissertation, No. 5339, ETH, Zurich, 1974.

[148] Geier, G.; Furrer, M.; Gehrig, R. *Proc. ICCC, 16th,* **1974**, 3.28.

[149] Southwood-Jones, R. V.; Merbach, A. E. *Inorg. Chim. Acta* **1978**, *30*, 135.

[150] Gfeller, Y.; Merbach, A. E. *Inorg. Chim. Acta* **1978**, *29*, 217.

[151] Merbach, A.; Gnaegi, F. *Helv. Chim. Acta* **1971**, *54*, 691.

[152] Cassat, J. C.; Wilkins, R. G. *J. Am. Chem. Soc.* **1968,** *90,* 6045.

[153] Chastellain, F.; Merbach, A. E. *Helv. Chim. Acta* **1975,** *58,* 1.

[154] Krishnamurthy, S. S.; Soundararajan, S. *Can. J. Chem.* **1969,** *47,* 995.

[155] Pisaniello, D. L.; Merbach, A. E. *Helv. Chim. Acta* **1982,** *65,* 573.

[156] Lugina, L. N.; Davidenko, N. K.; Zabotina, L. N.; Yatsimirskii, K. B. *Russ. J. Inorg. Chem.* **1974,** *19,* 1456.

[157] Bünzli, J.-C. G.; Yersin, J. R. *Helv. Chim. Acta* **1982,** *65,* 2498.

[158] Pisaniello, D. L.; Nichols, P. J.; Ducommun, Y.; Merbach, A. E. *Helv. Chim. Acta* **1982,** *65,* 1025.

[159] Pisaniello, D. L.; Helm, L.; Meier, P.; Merbach, A. E. *J. Am. Chem. Soc.* **1983,** *105,* 4528.

[160] Pisaniello, D. L.; Helm, L.; Zbinden, D.; Merbach, A. E. *Helv. Chim. Acta* **1983,** *66,* 1872.

[161] Yatsimirski, K. B.; Davidenko, N. K. *Coord. Chem. Rev.* **1979,** *27,* 223.

[162] Bünzli, J.-C. G.; Vuckovic, M. M. *Inorg. Chim. Acta* **1984,** *95,* 105.

[163] Bünzli, J.-C. G.; Vuckovic, M. M. *Inorg. Chim. Acta* **1983,** *73,* 53.

[164] Bünzli, J.-C. G.; Yersin, J.-R.; Mabillard, C. *Inorg. Chem.* **1982,** *21,* 1471.

[165] Bünzli, J.-C. G.; Yersin, J.-R. *Inorg. Chim. Acta* **1984,** *94,* 301.

[166] Bünzli, J.-C. G.; Mabillard, C.; Yersin, J.-R. *Inorg. Chem.* **1982,** *21,* 4214.

[167] Eriksson, B.; Larsson, L. O.; Niinistö, L.; Valkonen, J. *Inorg. Chem.* **1980,** *19,* 473.

[168] Milinski, N.; Ribar, B.; Sataric, M. *Cryst. Struct. Commun.* **1980,** *9,* 473.

[169] Quarton, M.; Svoronos, D. *J. Solid State Chem.* **1982,** *42,* 325.

[170] Silber, H. B.; Mioduski, T. *Inorg. Chem.* **1984,** *23,* 1577.

[171] Silber, H. B.; Gilbert, D. M.; Riddle, M. R. *J. Less Common Met.* **1983,** *94,* 319.

[172] Silber, H. B.; Bouler, D.; White, T. *J. Phys. Chem.* **1978,** *82,* 775.

[173] Silber, H. B.; Kromer, L. U. *J. Inorg. Nucl. Chem.* **1980,** *42,* 103.

[174] Silber, H. B.; Pezzica, A. *J. Inorg. Nucl. Chem.* **1976,** *38,* 2053.

[175] Wertz, D. L.; Finch, S. T. *Inorg. Chem.* **1979,** *18,* 1590.

[176] Pisaniello, D. L.; Lincoln, S. F.; Williams, E. H. *J. Chem. Soc. Chem. Commun.* **1978,** 1047.

[177] Pisaniello, D. L.; Lincoln, S. F.; Williams, E. H. *J. Chem. Soc., Dalton Trans.* **1979,** 1473.

[178] Pisaniello, D. L.; Lincoln, S. F. *Inorg. Chim. Acta* **1979,** *36,* 85.

[179] Pisaniello, D. L.; Lincoln, S. F. *J. Chem. Soc., Dalton Trans.* **1980,** 699.

[180] Pisaniello, D. L.; Lincoln, S. F. *Inorg. Chem.* **1981,** *20,* 3689.

[181] Pisaniello, D. L.; Lincoln, S. F.; Williams, E. H.; Jones, A. J. *Aust. J. Chem.* **1981,** *34,* 495.

[182] Lincoln, S. F.; White, A. Unpublished material.

[183] Lincoln, S. F.; Hounslow, A. M.; Jones, A. J. *Aust. J. Chem.* **1982,** *35,* 2393.

[184] Tkaczuk, M. N.; Lincoln, S. F. *Ber. Bunsenges. Phys. Chem.* **1982,** *86,* 147.

[185] Tkaczuk, M. N.; Lincoln, S. F. *Ber. Bunsenges. Phys. Chem.* **1982,** *86,* 221.

[186] Honan, G. J.; Lincoln, S. F.; Williams, E. H. *J. Chem. Soc., Dalton Trans.* **1979,** 320.

[187] Caldin, E. F.; Bennetto, H. P. *J. Solution Chem.* **1973,** *2,* 217.

[188] Fischer, P.; Hoffmann, H.; Platz, G. *Ber. Bunsenges. Phys. Chem.* **1972,** *76,* 1060.

[189] Rodehüser, L.; Rubini, P. R.; Delpuech, J.-J. *Inorg. Chem.* **1977,** *16,* 2837.

[190] Frankel, L. S.; Danielson, E. R. *Inorg. Chem.* **1972,** *11,* 1964.

[191] Delpuech, J.-J.; Khaddar, M. R.; Peguy, A. A.; Rubini, P. R. *J. Am. Chem. Soc.* **1975,** *97,* 3373.

[192] Gutmann, V.; Schmid, R. *Coord. Chem. Rev.* **1974,** *12,* 263.

[193] Pisaniello, D. L.; Lincoln, S. F. *Aust. J. Chem.* **1981,** *34,* 1195.

[194] Evans, D. F.; Missen, P. H. *J. Chem. Soc., Dalton Trans.* **1982,** 1929.

[195] Evans, D. F.; Wyatt, M. *J. Chem. Soc., Dalton Trans.* **1974,** 765.

[196] Grotens, A. M.; Backus, J. J. M.; Pijpers, F. W.; deBoer, E. *Tetrahedron Lett.* **1973,** *17,* 1467.

[197] Hart, F. A.; Newbery, J. E.; Shaw, D. *J. Inorg. Nucl. Chem.* **1970,** *32,* 3585.

[198] Quaegebeur, J. P.; Belaid, S.; Chachaty, C.; LeBall, H. *J. Phys. Chem.* **1981,** *85,* 417.

[199] Bovée, W. M. M. J.; Alberts, J. H.; Peters, J. A.; Smidt, J. *J. Am. Chem. Soc.* **1982,** *104,* 1632.

[200] Lindoy, L. F.; Louie, H. W. *J. Am. Chem. Soc.* **1979,** *101,* 841.

[201] Tanny, S. R.; Pickering, M.; Springer, C. S. *J. Am. Chem. Soc.* **1973,** *95,* 6227.

[202] Cheng, H. N.; Gutowsky, H. S. *J. Phys. Chem.* **1980,** *84,* 1039.

[203] Kessler, H.; Molter, M. *J. Am. Chem. Soc.* **1976,** *98,* 5969.

[204] Evans, D. F.; de Villardi, C. G. *J. Chem. Soc., Dalton Trans.* **1978,** 315.

[205] Marcus, Y. *J. Solution Chem.* **1983,** *12,* 271.

[206] Yee, E. L.; Gansow, O. A.; Weaver, M. J. *J. Am. Chem. Soc.* **1980,** *102,* 2278.

[207] Watson, P. L.; Roe, D. C. *J. Am. Chem. Soc.* **1982,** *104,* 6471.

[208] Daul, C.; Deiss, E.; Gex, J.-N.; Perret, D.; Schaller, D.; von Zelewsky, A. *J. Am. Chem. Soc.* **1983,** *105,* 7556.

Index